Synthesis and Techniques in Inorganic Chemistry

Synthesis and Techniques in Inorganic Chemistry

Editor: Kason Smith

NY RESEARCH PRESS
New York

Published by NY Research Press
118-35 Queens Blvd., Suite 400,
Forest Hills, NY 11375, USA
www.nyresearchpress.com

Synthesis and Techniques in Inorganic Chemistry
Edited by Kason Smith

International Standard Book Number: 978-1-64725-453-7 (Hardback)

Cataloging-in-Publication Data

Synthesis and techniques in inorganic chemistry / edited by Kason Smith.
p. cm.
Includes bibliographical references and index.
ISBN 978-1-64725-453-7
1. Chemistry, Inorganic. 2. Inorganic compounds--Synthesis. I. Smith, Kason.
QD151.3 .I56 2023
546--dc23

Contents

Preface

Every book is initially just a concept; it takes months of research and hard work to give it the final shape in which the readers receive it. In its early stages, this book also went through rigorous reviewing. The notable contributions made by experts from across the globe were first molded into patterned chapters and then arranged in a sensibly sequential manner to bring out the best results.

Inorganic chemistry is concerned with the procedures involved in the synthesis and behavior of inorganic compounds such as metals, minerals and organometallic compounds. In practice, there may be more than one way to synthesize an inorganic compound. However, there is no strict criterion for determining the suitability of a synthesis procedure. The only way to determine that a synthesis is suitable is to estimate whether a new or modified experimental procedure is applicable to a variety of related compounds, and ensuring that at least one of those compounds is critically important for industrial use or for research. Research in this field aims at developing innovative synthesis procedures for compounds that are not readily available on a commercial basis at reasonable prices. This book will provide a detailed account of the techniques that are used in inorganic chemistry for synthesizing inorganic compounds. It is compiled in such a manner, that it will provide an in-depth knowledge about inorganic compound synthesis techniques. The book aims to serve as a resource guide for students and experts alike and contribute to the growth of research in inorganic chemistry.

It has been my immense pleasure to be a part of this project and to contribute my years of learning in such a meaningful form. I would like to take this opportunity to thank all the people who have been associated with the completion of this book at any step.

Editor

1

End-Grafted Polymer Chains onto Inorganic Nano-Objects

Demetra S. Achilleos [1,2] **and Maria Vamvakaki** [1,2,*]

[1] Institute of Electronic Structure and Laser, Foundation for Research and Technology–Hellas, P. O. Box 1527, 711 10 Heraklion, Crete, Greece; E-Mail: achill@iesl.forth.gr (D.S.A.)

[2] Department of Materials Science and Technology, University of Crete, P. O. Box 2208, 710 03 Heraklion, Crete, Greece

* Author to whom correspondence should be addressed; E-Mail: vamvakak@iesl.forth.gr

---

Abstract: Organic/inorganic nanohybrid materials have attracted particular scientific and technological interest because they combine the properties of the organic and the inorganic component. Inorganic nanoparticles exhibit interesting electrical, optical, magnetic and/or catalytic properties, which are related with their nano-scale dimensions. However, their high surface-to-volume ratio often induces agglomeration and leads to the loss of their attractive properties. Surface modification of the inorganic nano-objects with physically or chemically end-tethered polymer chains has been employed to overcome this problem. Covalent tethered polymer chains are realized by three different approaches: the "grafting to", the "grafting from" and the "grafting through" method. This article reviews the synthesis of end-grafted polymer chains onto inorganic nanoparticles using "controlled/living" polymerization techniques, which allow control over the polymer characteristics and the grafting density of the end-tethered polymer chains.

Keywords: nanohybrids; core-shell nanoparticles; end-grafted polymers

---

Abbreviations

AGET activators generated by electron transfer
ATRP atom-transfer radical polymerization
BPEA 2-(2-bromopropionyloxy)ethyl acrylate
CMS p-chloromethyl styrene

CRP controlled radical polymerization
CT computed tomography
FRET fluorescence resonance energy transfer
LbL layer-by-layer
LCST lower critical solution temperature
MC merocyanine
MeO-PEGMA methoxy-poly(ethylene glycol) methacrylate
NMP nitroxide-mediated radical polymerization
P(MeO-PEGMA)-*b*-PNIPAM
poly(methoxy-poly(ethylene glycol) methacrylate)-b-poly(*N*-isopropylacrylamide)
P2VP poly(2-vinyl pyridine)
P3HT poly(3-hexylthiophene)
P3VP poly(3-vinyl pyridine)
P4VP poly(4-vinyl pyridine)
PAA poly(acrylic acid)
PAN polyacrylonitrile
PB-PEG polybutadiene-poly(ethylene glycol)
PCEMA poly(2-(carbazol-9-yl)ethyl methacrylate)
PCYPAAM poly(*N*-cyclopropylacrylamide)
PDEAEMA poly(2-diethylamino)ethyl methacrylate)
PDMA poly(*N*,*N*-dimethylacrylamide)
PDMAEMA poly(2-dimethylamino)ethyl methacrylate)
PEO poly(ethylene oxide)
PHEA poly(hydroxyethyl acrylate)
PHEMA poly(2-hydroxyethyl mehacrylate)
PMA poly(methyl acrylate)
PMMA poly(methyl methacrylate)
P*n*BA poly(*n*-butyl acrylate)
P*n*BMA poly(butyl methacrylate)
PNIPAM poly(*N*-isopropylacrylamide)
POEMA poly(oxyethylene methacrylate)
POSS polyhedral oligomeric silsesquioxanes
PS Polystyrene
PS-*b*-P*n*BA polystyrene-*b*-poly(*n*-butyl acrylate)
PSP-*co*-PMMA Poly(spirobenzopyran)-*co*-poly(methyl methacrylate)
PSS poly(styrene sulfonic acid)
P*t*BA poly(*t*-buyl acrylate)
RAFT reversible addition-fragmentation chain transfer
ROMP ring-opening metathesis polymerization
ROP ring-opening polymerization
SAMs self-assembled monolayers
SCVP self-condensing vinyl polymerization

SFRP stable free radical polymerization
si-ATRP surface-initiated atom-transfer radical polymerization
SP spiropyran
TOPO trioctylphosphine oxide
UCST upper critical solution temperature

1. Introduction

The synthesis of core-shell nano-objects comprising of an inorganic core and a polymer shell has recently gathered great scientific interest with a view towards new opportunities for constructing functional nanostructured materials [1–2]. Inorganic-polymer core-shell structures combine the fascinating electronic, optical and magnetic properties of the core with the desired properties of the polymer such as solubility, film formation, processability and compatibility to the environment [3–4]. Several types of nanoparticles have been employed such as silica, metal, magnetic and semiconducting colloids, which depending on their composition and ordering, can find size-depended fascinating applications in a variety of technologies such as diffractive optics, electro-optical devices, information storage and reinforced composite materials. On the other hand, advances in polymer chemistry have allowed control over the chemical composition and function of the polymer shell that governs the interactions with the surrounding media, leading to complex self-organization phenomena in core-shell systems.

The intense interest towards the modification of the inorganic nanomaterials surface with polymer chains has been fueled by the continual tendency of the former to agglomerate due to the strong adhesion forces among the particles originated from their high surface-to-volume ratio. Most of the properties of the nanoparticles are dictated by their nano-sized dimensions and are dramatically altered upon aggregation. Polymer-coated particles are sterically stabilized in interparticle lengths scales in the nanometer range compared to the shorter separation distances accessed by short alkyl chain stabilizers. Moreover, the polymer coating has a lot of free volume, which not only improves the dispersion stability of the nanoparticles in organic and aqueous media, but can also prevent the segregation of the inorganic fillers in polymer matrices due to the high affinity of the particle modifier for the organic medium. Two different approaches have been employed for the preparation of polymer-coated nanoparticles; the first involves the reversible physisorption of the polymer chains onto the nanoparticles surface *via* weak interactions (van den Waals forces, hydrogen bonds, *etc.*) and the second the permanent attachment of the macromolecular chains onto the nanoparticles surface *via* chemisorption. The covalent binding of the polymeric modifiers onto the curved surface can be achieved either by the "grafting to" method, upon which pre-prepared macromolecules are grafted onto the nanomaterials surface using the affinity of the polymer end-group for the functionalities present on the surface of the particles [5], or by the *in situ* growth of the polymer chains from appropriately functionalized nano-objects known as the "grafting from" method.

Several polymerization techniques such as living anionic [3], living cationic [6–10], ring-opening polymerization (ROP) [11–15], ring-opening metathesis polymerization (ROMP) [16–17] and controlled/"living" radical polymerization (CRP) have been employed for the irreversible functionalization of inorganic nano-materials with end-tethered polymer chains. Among them, the development of the CRP techniques has offered lately great potential for the synthesis of well-defined core-shell nanostructures. CRP techniques, namely atom-transfer radical polymerization (ATRP), reversible addition-fragmentation chain transfer polymerization (RAFT) and nitroxide-mediated radical polymerization (NMP), have attracted considerable attention because they provide a facile approach for the development of end-grafted macromolecules and a high degree of synthetic flexibility towards the introduction of a variety of functional polymers onto the surface of different nanoparticles. CPR techniques also allow the covalent attachment of macromolecules with well defined grafting densities and controlled film thicknesses. Densely packed arrays of polymer chains can be attained which are forced to adopt a stretched conformation, due to excluded volume interactions, leading to coatings of high thicknesses [18].

However, there are certain advantages and limitations inherent to each of the above methods. In ATRP the initiators, transition metal catalyst systems and ligands used are commercially available. An important advantage of ATRP is the polymerization of a wide range of monomers including not only the more reactive meth(acrylates), styrenes and acrylonitrile, but also less reactive monomers such as vinyl acetate and vinyl chloride [19]. However, a problem often encountered in ATRP is the complete removal of the transition metal catalyst from the final product which can be critical for certain applications, *i.e.,* in electronics. In contrast, NMP and RAFT do not employ metal catalyst systems however, the commercial availability of alkoxyamines and dithioesters used as radical capping agents is quite limited. Moreover, RAFT is incompatible with basic monomers, while the NMP of methacrylates and other less reactive molecules is still challenging [20].

In this review, the focus is on the surface modification of inorganic nanoparticles using the "grafting to" and "grafting from" techniques to prepare uniform core-shell nanostructures. The synthesis, properties and potential applications of the hybrid structures are reviewed for six different types of nanoparticles including, silica, silsequioxane, metal-oxide semiconducting, magnetic, metal and quantum dots. Particular emphasis is given on responsive polymer coatings which endow the inorganic nanoparticles with new exceptional physical and chemical properties.

2. Results and Discussion

2.1. Polymer-grafted silica nanoparticles

2.1.1. "Grafting to" method

Silica particles are widely used as reinforcing fillers to improve the mechanical and thermal properties of a polymer matrix. They can be easily prepared by the Stöber method [21], which allows control over the nanoparticle size and size distribution. The grafting of polymer chains onto the surface of silica nanoparticles has been achieved by the synthesis of linear macromolecules using a controlled/living polymerization technique followed by organic reactions such as Huisgen cycloaddition and other coupling reactions to immobilize the polymers onto the nanoparticle surface (Figure 1). Click chemistry has been used successfully employed to immobilize alkyne-terminated homopolymers and diblock copolymers synthesized by RAFT and/or ATRP onto the surface of azide-modified silica particles [22]. The same strategy was successfully applied to load mesoporous silica with polymers without blocking the pores of the inorganic material [23]. A more direct method which does not require the pre-functionalization of the inorganic nanoparticles employed a combination of RAFT polymerization and siloxane coupling reactions to functionalize the surface of silica nanoparticles with pre-synthesized polymer chains. In this work, trimethoxysilane-functionalized homopolymers and block copolymers were first synthesized by RAFT polymerization mediated by S-methoxycarbonylphenylmethyl S'-trimethoxysilylpropyltrithiocarbonate and were linked onto the surface of the nanoparticles *via* the coupling reaction of the hydroxyl groups of the solid surface with the trimethoxysilane end-groups of the macromolecules [24].

Figure 1. Schematic representation of the synthetic procedure followed for the grafting of polymer chains onto the surface of silica particles *via* the "grafting to" approach using click chemistry (a) and silane coupling (b).

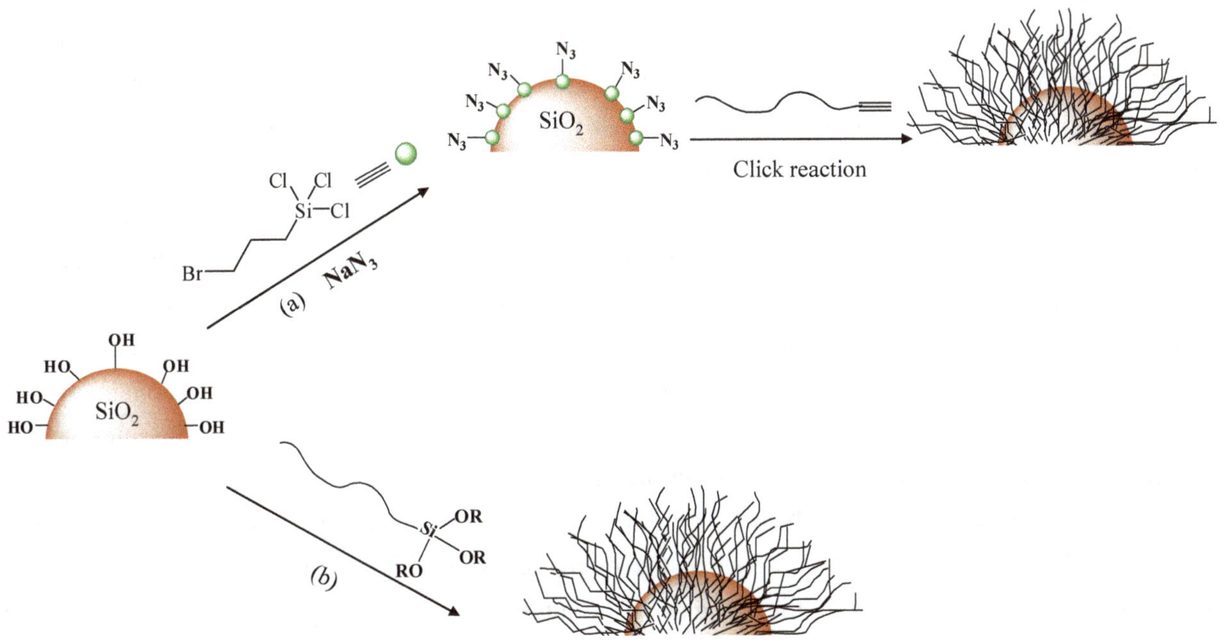

An interesting approach used the covalent layer-by-layer (LbL) assembly of acetylene- and azido-functionalized poly(N-isopropylacrylamide) PNIPAM random copolymers synthesized by ATRP onto silica nanoparticles to prepare ultrathin thermo-responsive microcapsules *via* click chemistry. PNIPAM is a thermo-responsive polymer, which responds to changes of the solution temperature and exhibit the so-called lower critical solution temperature (LCST). The LCST corresponds to the region in the phase diagram of a polymer at which the enthalpic contribution of hydrogen-bonded water to the polymer chain becomes lower than the entropic gain of the system as a whole. At this point the polymer chains undergo a coil-to-globule transition due to hydrophobic interactions and phase separate [25–26]. The LbL assembly was based on the 1,3-dipolar cycloaddition of the azide groups located at the surface of the silica particles with the acetylene-functionalities of a PNIPAM copolymer in the solution. The remaining free acetylene groups located at the outermost surface polymer layer were reacted with an azido-functionalized PNIPAM random copolymer in the next assembly step leading to the LbL formation. Upon etching the silica core of the assembly with HF hollow PNIPAM microcapsules that exhibited reversible thermo-induced swelling/deswelling behavior were obtained, which are particularly attractive for use on controlled drug delivery applications [27].

2.1.2. "Grafting from" approach

2.1.2.1. "Hairy" homopolymer and block copolymer silica particles

The surface hydroxyl groups exposed on the periphery of silica permit their facile derivatization to appropriate initiating moieties (Table 1) for the controlled growth of polymer chains from the surface of the nanoparticles (Figure 2). The si-ATRP from silica nanoparticles was first reported by Patten *et al.* to polymerize polystyrene (PS) and poly(methyl methacrylate) PMMA end-grafted chains [28–29]. Since then, a variety of polymeric stabilizers including homopolymers and copolymers of styrenes, meth(acrylates) and acrylamides have been grown from the surface of spherical silica colloids [30–39]. Recently, the controlled polymerization of dense PMMA chains, with a very narrow molecular weight distribution from a silica core was achieved by ATRP in bulk. A high grafting-density hydrophobic brush was successfully prepared using an initiator carrying a triethoxysilane anchor group [40]. The control of the grafting density resulted in the formation of a colloidal crystal when the hybrids were suspended in a good solvent for the polymer in a certain concentration range which decreased with increasing the graft polymer chain length, L_c. Long polymer chains led to a nearest neighbor interparticle distance D_{dis} within the crystal in the micrometer length scale. The colloidal crystal included both hexagonal close-packed (hcp) and face-centered cubic (fcc) lattice arrangements with the fcc arrangement likely to increase when increasing the polymer chain length. This transition of the crystalline structure from a nearly random stacking to an fcc arrangement was ascribed to a qualitative change of the conformation of the grafted chains and hence the interparticle potential curve at the crossover of the concentrated polymer brush to the semidilute polymer brush in the effective graft density of the polymer layer [41–42]. Moreover, the grafting density of the polymer had a significant effect on the "softness" of the hybrid colloids. Significant interpenetration of the grafted polymer chains took place for particles in the intermediate brush regime and decreased as the polymer grafting density increased. As a result, the threshold concentration for graft polymer interpenetration of dense particle brushes was found to increase by about an order of magnitude as compared to the intermediate brush analogue [43–44]. These finding offer unique possibilities for fundamental and applied research of colloidal systems for application in photonic materials, sensors and elsewhere.

Figure 2. Schematic representation of the synthetic procedure followed for the preparation of silica-polymer core-shell hybrids using the "grafting from" method.

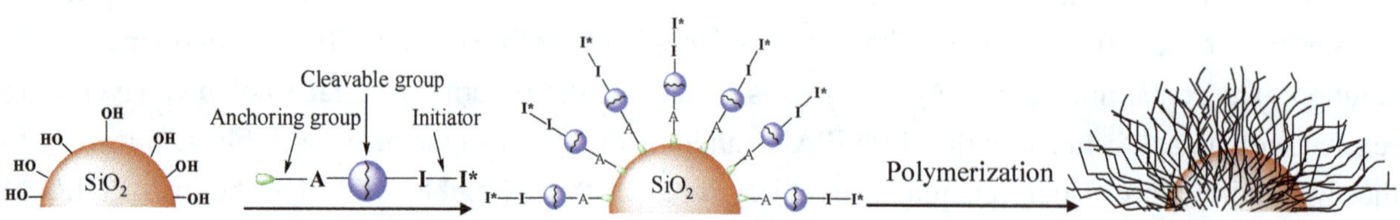

Table 1. Initiator molecules grafted on the surface of silica particles.

No	Initiator	Ref.
1	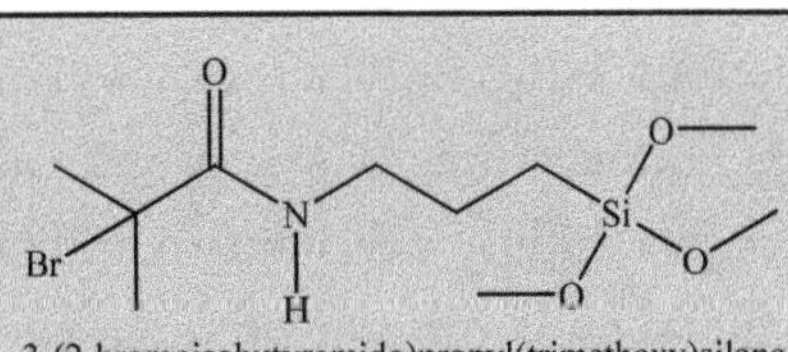 3-(2-bromoisobutyramido)propyl(trimethoxy)silane	[49,53–60]
2	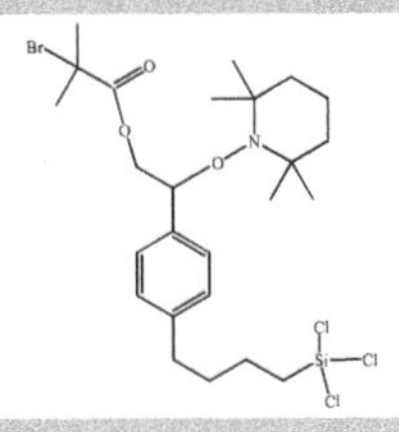	[61–62]
3	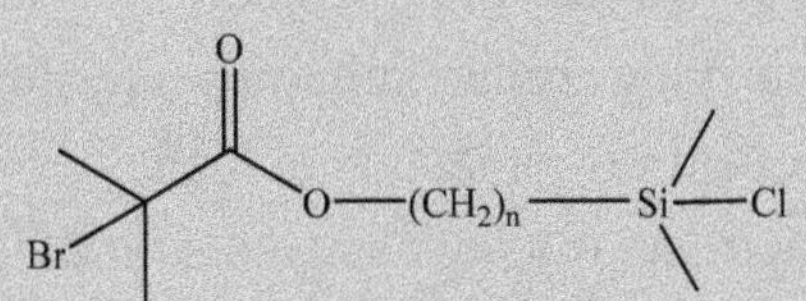	[63–66]
4	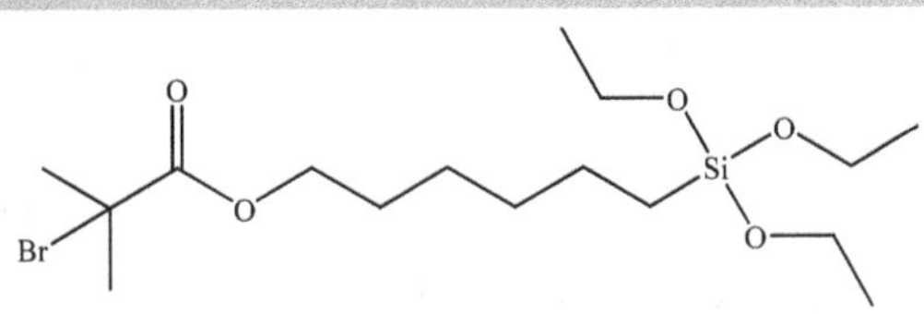 (2-Bromo-2-methyl)propionyloxyhexyltriethoxysilane	[40,67]
5	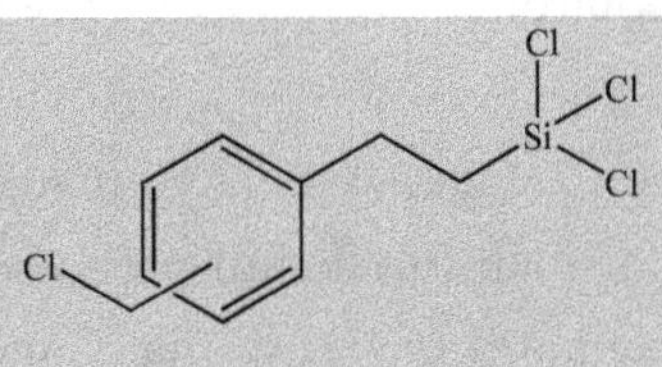 2-(m,p-chloromethylphenyl)ethyltrichlorosilane	[68–71]
6	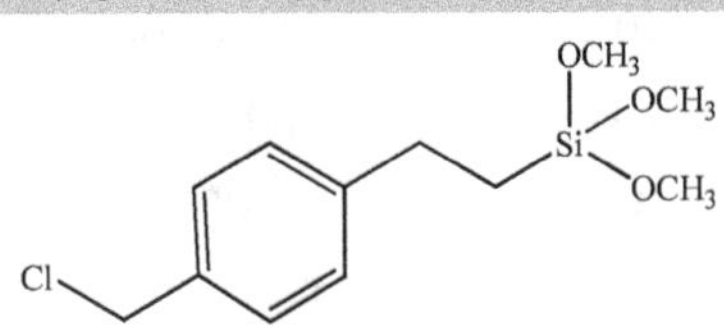 (4-(chloromethyl)phenethyl)trimethoxysilane	[72]
7	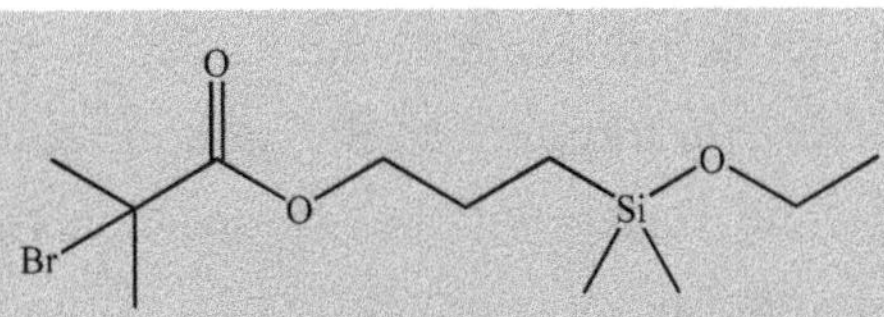 3-(dimethylethoxysilyl)propyl-2-bromoisobutyrate	[45–46]

The synthesis of hydrophilic methacrylate polymers from the surface of silica particles *via* ambient temperature ATRP in protic media, utilizing a siloxane-grafted initiator [45], have been shown to suffer from the hydrolytic cleavage of the grafted polymer chains from the nanoparticle surface [46]. To overcome this problem and circumvent the need for surface pre-treatment of the silica sol with siloxane initiators prior to the polymerization, a cationic ATRP initiator was electrostatically absorbed onto ultrafine anionic silica for subsequent polymer grafting of hydrophilic methacrylates in protic media at ambient temperature. Nonionic and cationic monomers have been polymerized directly in one-pot syntheses with reasonably high conversions, however the growth of anionic polymers was hindered due to the electrostatic interactions between the negatively charged monomer with the cationic macroinitiator-particles [47–48]. Furthermore, this approach also allowed the synthesis of zwitterionic polymer coatings [49] which are highly resistant to nonspecific protein adsorption in complex media and improve the bio-interfacial properties of the silica particles protecting them against aggregation in complex physiological conditions [50].

Another hybrid system was developed by the si-ATRP of acrylonitrile from the surface of silica nanopartricles. The "hairy" silica-polyacrylonitrile (PAN) hybrids were used as precursors for the synthesis of nanoporous carbon thin films with appreciable adsorption capacities after the pyrolysis of the PAN stabilizers, followed by the removal of silica by HF etching [51]. Using a similar protocol, mesoporous carbons were obtained by the surface polymerization of acrylonitrile from ordered and disordered mesoporous silica templates carrying surface initiating sites followed by the stabilization of the polymer layer under air and its subsequent carbonization at higher temperatures [52].

Silica-polymer core-shell hybrids are particularly advantageous for the facile preparation of uniform hollow polymer microspheres by the removal of the silica core of the hybrids using HF etching. Homopolymer and block copolymer coated particles have been used as precursors to obtain polymeric hollow spheres of uniform size and good dispersibility in aqueous or organic media. The structural integrity of the hollow shell or capsule is preserved by either cross-linking the polymer chains [67,73], or by dispersing the hollow structure in a bad solvent for the polymer [72]. Such hollow capsules are very attractive for the encapsulation and controlled delivery of active compounds in drug delivery and other biomedical applications.

NMP is based on a similar concept to that of ATRP involving the activation–deactivation equilibrium between dormant species and a small fraction of propagating macroradicals. Although this technique has not been extensively employed as a surface-initiated method there are a few examples of alkoxyamine systems in the literature used to polymerize homopolymers and diblock copolymers of styrenes and meth(acrylates) from the surface of silica nanoparticles [74–84]. In order to overcome the complicated synthesis of specific surface grafted NMP initiators an alternative method, that of "trapping of carbon radicals" has been used [85–86] which provides the feasible trapping of the growing carbon radicals by the NMP initiator. However, this method suffers from secondary reactions, such as cross-linking or formation of inactive molecules, which reduce the efficiency of the grafting procedure. The initiating process by the *in situ* trapping of the carbon radicals is also hindered by the presence of unreacted double bonds on the surface of the nanoparticles upon grafting the initiator. Thus, an alternative approach, the "*in situ* thermo-dependant trapping of the carbon radicals," has been developed. This method provides good control over the molecular weight and the grafting density of the bound macromolecules leading to hybrid suspensions with long-term colloidal stability [76,87].

Well-defined hybrid silica-polymer core-shell structures have been synthesized as well *via* the RAFT technique. The application of this versatile method to prepare polymer brushes on the surface of silica nanoparticles is unproportionally limited compared with the large number of studies on the surface initiated-ATRP (si-ATRP) from silica particles. Although the RAFT technique is compatible with the polymerization of almost all the conventional monomers, the covalent attachment of the

RAFT agents onto the nanoparticles surface is still challenging thus restricting its utilization. Most studies in the literature investigate the anchoring of the RAFT agents onto the nanoparticles surface. Silane chemistry [88–89] and the condensation of mercaptothiazoline-activated RAFT agents with amino groups present of the nanoparticles surface [90] have been the most successful methods at present. For the RAFT polymerization of styrenics and (meth)acrylates on solid supports two different approaches have been utilized; the R-group approach [88–92] and the Z-group approach [93–94]. In the R-group approach the chain transfer agent (Table 2) is attached to the backbone *via* the leaving and reinitiating R-group, while in the Z-group approach the chain transfer agent is attached to the backbone *via* the stabilizing Z-group. Both methods exhibit advantages and limitations. For instance, the R-group approach can provide grafted polymers of higher molecular weight and grafting density however, the molecular weight distribution of the grafts may be broadened by possible side reactions such as chain coupling. On the other hand, the Z-group approach can yield well-defined grafted polymers with monomodal molecular weight distributions but of lower grafting densities due to shielding effects.

Table 2. RAFT agents bound on silica nanoparticles (R-group approach).

No	RAFT agent	Polymer	Ref.
1	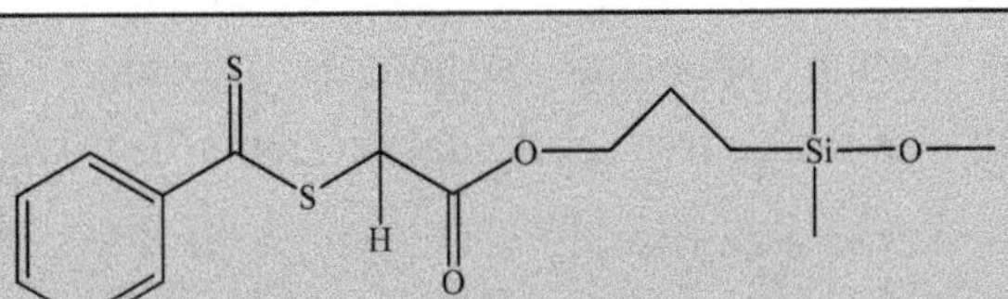	P*n*BA, PS-b-P*n*BA	[88]
2	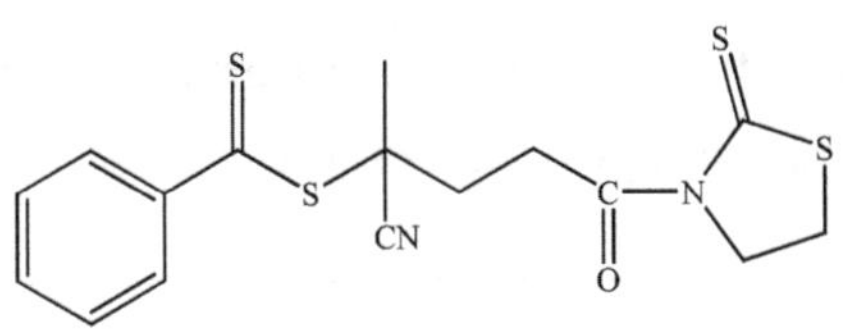	PMMA	[90]
3	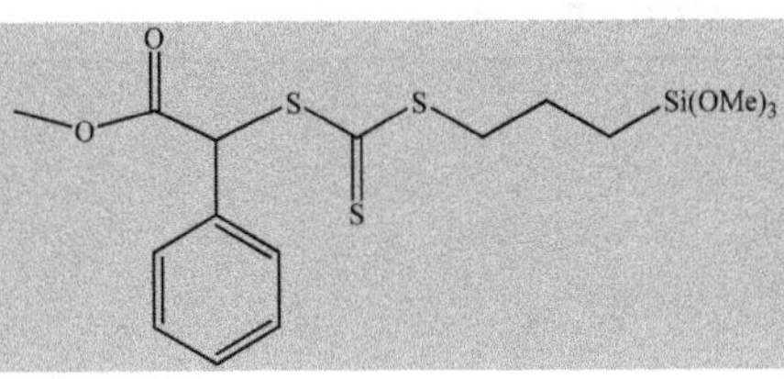	PMA, P*n*BA, PMMA, PS, PNIPAM, PDMA	[89]
4	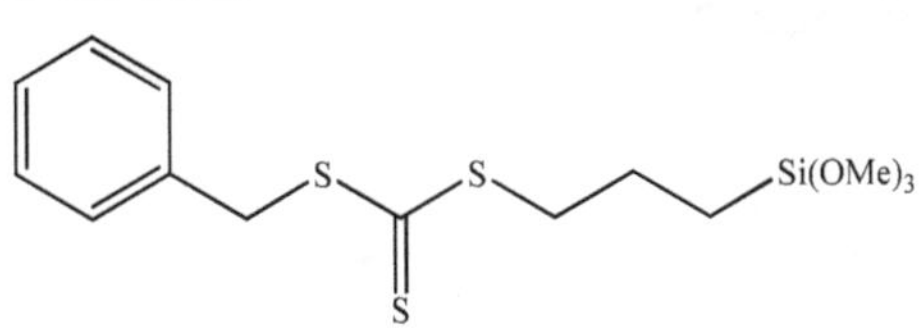	PMA, P*n*BA, PMMA, PS, PNIPAM, PDMA	[89]

Finally, living-anionic surface-initiated polymerization has been also successfully employed for the growth of PS chains from the surface of silica nanoparticles pre-functionalized with a 1,1-diphenylethylene derivative bearing chlorosilane end-groups. Although a living anionic polymerization mechanism was reported, attributed to the linear increase of the molecular weight of the bound polymer upon increasing the amount of monomer, the excess initiator present in solution and the observed nanoparticle aggregation limit the applicability of this method for the preparation well-defined core-shell structures [95].

2.1.2.2. Grafted comb and branched polymer architectures

Polymer brushes of more complex architectures can be grown effectively from the surface of silica nanoparticles. One approach for the synthesis of comb-coil polymer brushes from the nanoparticle surface employed a three step reaction; first poly(2-hydroxyethyl methacrylate) (PHEMA) brushes were grown from the surface of initiator-functionalized silica particles by ATRP, followed by the ring-opening polymerization of DL-lactide initiated by the hydroxyl side-groups of PHEMA in the second step and the activators generated by electron transfer (AGET) ATRP of an acrylic monomer initiated by the bromine end-groups of the PHEMA chains in the final step [55]. Alternatively, a more convenient method involving a one-step reaction for the preparation of hyperbranched polymer-coated silica particles is the surface-initiated self-condensing vinyl polymerization (SCVP) of a vinyl AB* inimer from an initiator functionalized silica surface. 2-(2-bromopropionyloxy)ethyl acrylate (BPEA) [66] and p-chloromethyl styrene (CMS) [56] (Table 3) have been employed as inimers to grow dendritic structures from the nanoparticle surface. These core-shell particles were used for the polymerization of a second monomer, initiated by the end-groups at the outermost surface of the hyperbranched polymer shell, to obtain superstar structures bearing linear hydrophobic or polyelectrolyte outer chains which are attractive colloidal systems for use in surface coatings and paints.

Table 3. Inimer molecules employed for the synthesis of grafted hyperbranched polymers on silica particles by self-condensing vinyl polymerization.

No	Inimer	Ref.
1	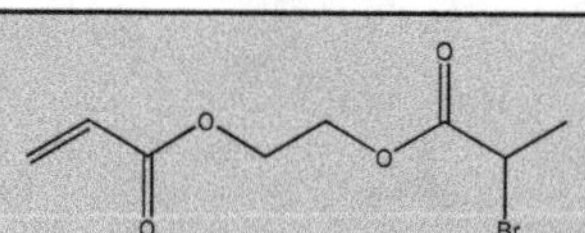 2-(2-bromopropionyloxy)ethyl acrylate (BPEA)	[66]
2	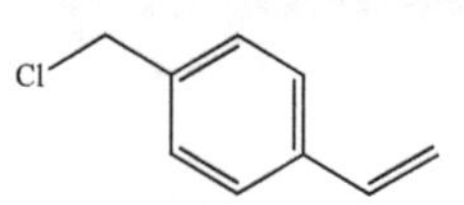 p-chloromethyl styrene (CMS)	[56]

Lately, a new versatile method for the synthesis of industrially relevant branched polyolefin brushes, of a narrow molecular weight distribution, onto silica particles has been developed. The synthesis employed the surface functionalization of silica with acrylate moieties, which are used as anchoring sites for the covalent attachment of Pd-diimine complexes. The silica-supported Pd-diimine complexes catalyze the surface-initiated "living" polymerization of ethylene at 5 °C and an ethylene pressure of 400 psi to yield well-defined grafted polyethylene [96].

2.1.2.3. Responsive nanohybrids

pH- and temperature-sensitive core-shell systems

Recent advances in the field of responsive materials focus on the coating of flat or curved surfaces with polymers exhibiting a stimuli-sensitive behavior and changing their properties in response to an applied stimulus. External stimuli, *i.e.,* temperature, pH and solvent quality, have been investigated leading to changes in the grafted polymer chain conformation and the physicochemical properties of the surface. Silica nanoparticles coated with responsive polymers are attractive building blocks for the development of "smart", environmentally responsive nanostructured materials. The most extensively studied system is PNIPAM-grafted polymer chains onto silica nanoparticles which were shown to exhibit a non-conventional temperature-induced double phase transition behavior [60]. The phase transition at lower temperature is ascribed to the cluster-induced collapse of the inner region of the PNIPAM brushes, close to the silica surface, which posses a high chain density, while the transition at higher temperature is attributed to the outer region of the PNIPAM shell where the chain density is lower. The thermo-induced phase transition of the grafted polymer chains is shifted to lower temperatures and occurs over a broader temperature range compared to that of the free polymer in solution. This is attributed to interchain interactions in the brush layer [63]. The characteristic dehydration of temperature-sensitive polymers in densely-packed brushes can be crucial for the application of such systems as a thermo-responsive stationary phase in chromatography for the separation of active molecules. It was shown that the retention time of steroids increased over a broad range of temperatures for highly grafted PNIPAM-silica hybrids, while at lower grafting densities the retention time increased only in a narrow range around the LCST of the polymer [68].

The surface-initiated polymerization of responsive monomers has also been used for the modification of the pore properties of mesoporous materials. PNIPAM chains were grown in the pore walls of mesoporous silica leading to materials with temperature responsive pore properties. The long-range ordered hexagonal mesostructure and the thermo-responsive behavior of PNIPAM were retained after the synthesis of the interpenetrating silica-PNIPAM network rendering these materials attractive for use in bioseparations or molecular transport switches [97].

Another interesting effect induced by the stimuli responsive behavior of the polymer coating is the reversible transportation of stimuli-sensitive core-shell hybrids between two immiscible phases due to the reversible hydration-dehydration of the polymer shell. The reversible migration of temperature-sensitive polymer-grafted silica nanoparticles from an aqueous to an immiscible organic phase upon heating and cooling the system, respectively, was shown [64,98], which is consistent with the de Gennes concept [99].

Recently, it was proposed that pH-sensitive core-shell structures can be used as transducers of signals in logic operation systems. More specifically, "smart" signal-responsive hybrid systems with built-in Boolean logic were developed by grafting poly(2-vinyl pyridine) (P2VP) chains onto the surface of silica particles. The hybrids were coupled with an enzyme-based system and were used to detect signals induced by the enzyme function. The biochemical input signals, that is pH changes originated from the AND/OR logic operations of the enzyme systems, were detected by the self-assembly changes of the pH-sensitive hybrids in solution [100] rendering these materials very promising for use in the development of biochips and other biomedical devices.

Another responsive system of particular interest are “hairy” silica nanoparticles comprising mixed homopolymer brushes. These hybrids possess interesting responsive properties induced by changes of the solvent quality. To ensure the uniformity of the mixed polymer layer on the surface of the nanoparticles Y-shaped multifunctional colloidal initiators carrying dual initiating functionalities have been employed. A combination of ATRP followed by NMP chemistry resulted in the formation of the mixed brush by sequential polymerization. The phase behavior of the mixed poly(*t*-buyl acrylate) (P*t*BA)/PS brushes on the silica nanoparticles showed a lateral phase separation of the two polymers under equilibrium melt conditions which resulted in worm-like patterns with a feature size of around 10 nm [62]. Furthermore, hydrolysis of the *tert*-butyl groups produced amphiphilic mixed poly(acrylic acid) (PAA)/PS brushes which rendered the particles dispersible in both chloroform which is a selective solvent for the PS chains and in methanol which is selective for the PAA grafts *via* the chain reorganization on the surface of the “hairy” particles [61].

Photo-responsive hybrids

Core-shell hybrids carrying photo-responsive groups, whose properties can be reversibly altered by applying a remote stimulus such as light of specific wavelength, are highly attractive nowadays. Poly(spirobenzopyran)-*co*-poly(methyl methacrylate) (PSP-*co*-PMMA) copolymer brushes have been synthesized onto the surface of silica particles by ATRP. A progressive decrease of both the coloration and decoloration rates of the SP moieties incorporated in the polymer chains was observed, accompanied by the H-stacking of the chromophore moieties in dense polymer brushes upon irradiation [65]. This effect is very attractive and has been used for the formation of two-dimensional and three-dimensional structures by a light induced assembly and phase separation of the hybrid nanocolloids. The latter is based on the UV induced isomerization of the spiropyran (SP) moieties to merocyanine (MC) groups which exhibit a high dipolar moment and form SP-MC interchain complexes in less polar media leading to the aggregation of the particles. Both the SP content of the polymer and the solvent polarity were shown to affect the photo-induced aggregation of the polymer shell [101]. Two dimensional-patterned colloid assemblies were obtained *via* the appropriate modification of a flat substrate with a PSP-*co*-PMMA polymer film which was next immersed in a stable colloidal dispersion of the photo-responsive hybrids. Upon UV irradiation, the MC moieties formed on the flat surface due to the SP photo-isomerization, aggregate with the SP moieties of the polymer brushes of the colloidal hybrids and hold the two surfaces together [102]. Analogous three-dimensional porous structures have been fabricated by direct laser writing of a dense colloidal suspension of the photo-responsive spheres. A two photon absorption process at a focal point of a near-IR pulsed laser, focused tightly in the colloidal dispersion, induce gelation of the hybrids and the isomerization of the SP photo-responsive moieties of the polymer brush to MC groups. The MC-SP and MC-MC *H*-aggregates formed, induce strong intermolecular and interparticle bonding in the system and lead to the 3D structure fabrication [103]. Recently, a similar photo-responsive core-shell hybrid system was proposed for use as a sensitive ratiometric fluorescent thermometer. The hybrid comprised silica particles coated with a densely grafted PNIPAM shell, labeled with a fluorescence resonance energy transfer (FRET) donor in its inner layer and a photo-responsive SP acceptor in the outer layer. Upon irradiation with UV the non-fluorescent SP form isomerizes to the fluorescent MC form and facilitates the FRET process between the donor and the acceptor moieties. The thermo-modulated collapse/swelling of the PNIPAM shell can be used as a temperature probe which tunes the relative distance of the donor and the acceptor in the polymer shell and influences the FRET efficiency [59].

Finally, photolabile polymer shells have been used to tune the surface wettability of nanocolloids upon irradiation with light. Photolabile hydrophobic coatings were efficiently transformed to hydrophilic shells thus switching the surface from hydrophobic to hydrophilic. The switch can be augment by increasing the surface roughness leading to tunable superhydrophilic/superhydrophobic surfaces [104].

Hybrids with gating properties

An ongoing challenge in materials science relies on the development of hybrid assemblies displaying gating properties. It was recently shown, that transport through colloidal nanoporous assemblies can be effectively modulated by modifying the surface of the colloids with "smart", environmentally-responsive polymer brushes in such a way to mimic the gating function in biological nanoporous systems. The first example in this field was reported by Schepelina and Zharov. They used poly(acrylamide) brushes grown by ATRP to modify the surface of the nanopores in films of closed-packed face-centered cubic (fcc) lattices of submicrometer silica spheres. Flux measurements of redox species across these opal films verified that the polymer film thickness increased with the polymerization time leading to a decrease of the pore size [70]. The use of temperature-responsive PNIPAM brushes allowed tuning the flow as a function of temperature, while the film thickness had also a significant effect on the diffusion through the nanopores. Nanopores bearing a thin PNIPAM layer possessed a positive gating behavior and the diffusion rates increased with increasing temperature, while the nanopores with the thicker PNIPAM layers exhibited the opposite effect (negative gating behavior) and diminished the flow through the nanopores at higher temperatures due to the dehydration of the densely packed polymer inside the nanopores [71]. An alternative approach to develop "smart" hybrid materials displaying gating properties involves the simple modification of ordered mesoporous networks or porous nanocolloids with stimuli responsive end-grafted polymer chains. The temperature-dependent uptake and release of small molecules by these porous materials was modulated by the polymer phase transition and the porosity of the materials. The presence of the hydrated and extended polymer within the porous structure at low temperature inhibited the transport of solutes, while at higher temperatures the collapse of the hydrophobic polymer in the globule conformation within the pores allowed solute diffusion [105]. The inherent disadvantage of the above system was the leakage of the active molecules even at low temperatures. To overcome this problem a different mode of action has been targeted which relies on the uptake and release of the actives below rather than above the LCST. This was achieved by "grafting to" temperature-responsive polymer chains onto the surface of micro-to-mesoporous silica nanoparticles. Below the LCST, the grafted polymer chains adopted a random coil configuration and facilitated the uptake and release of the molecules through the pores. However, upon increasing the temperature above the LCST the polymer chains collapsed, blocked the pores and retarded the diffusion of the molecules [106].

Finally, pH-responsive polymer coatings exhibit a rather complex pH- and ion-responsive flux [107] behavior which is dependent on the pH, the charge of the active species and the ionic strength of the medium have been developed as gating materials. The stretching of the polymer chains due to the ionization of the monomer repeat units block the nanopores and decrease the diffusion rate. The flow can be restored by increasing the salt concentration which screens the electrostatic repulsion forces [57]. pH-responsive polymer shells were proposed for the development of organic-inorganic hybrid membranes exhibiting high proton conductivity. To this direction, sintered silica and self-assembled nanoporous silica colloidal crystals modified with surface-bound polymer brushes carrying acidic groups were shown to exhibit high water uptake and proton conductivity. However, despite the superior proton conductivity of the self-assembled membranes their poor mechanical properties limit their use for practical purposes [58].

Pickering emulsions and Janus nanohybrids

Janus particles, which are characterized by an asymmetric nature in structure, have attracted particular attention, because they can self-assemble into fascinating hierarchical structures and serve as building blocks for complex superstructures. The preparation of such synthetic asymmetric particles is effectively assisted by Pickering emulsions, which are emulsions stabilized by particles rather than surfactant molecules. Pickering emulsions can be effectively stabilized by hybrid core-shell particles based on strong polyelectrolyte brushes grown from the surface of silica colloids. Anionic poly(styrenesulfonate) chains have been used because they are "quenched" [108] and highly charged and can thus form long-term stabilized oil-in-water emulsions which are stable against coalescence even at low particle concentrations. The emulsifying properties of the particles are based on the hydrophilic character of the polymer charged side-groups and the hydrophobic character of the polymer backbone which drive their absorption at the oil/water interface [69]. Amphiphilic Janus particles were prepared in a one-step synthesis by the adsorption of azide-modified silica spheres around the monomer droplets of a styrene/water emulsion with one hemisphere being immerse in the aqueous phase and the other in the organic phase. The emulsion-assisted synthesis of amphiphilic Janus particles comprising PS- and poly(sodium methacrylate)-coated hemispheres used initiator-modified particles to simultaneously grow the polymer chains from the surface of the particles in the two hemispheres. These hybrids can form supramolecular assemblies upon dispersion in solvents being selective for only one hemisphere [109]. In another approach, azide-modified silica particles have been used for the preparation of Janus particles using *in situ* click chemistry. Adsorption of these particles at a polymer/solvent interface allowed first the binding of one type of alkylated molecules on the solvent exposed hemisphere whereas next, the removal of the polymer template exposed the second hemisphere for the immobilization of a second alkylated polymer *via* click reaction with the residual azide groups. Using this general protocol both small molecules and long polymer chains can be introduced on the two particle hemispheres. Zhang *et al.* conjugated biotin molecules and poly(ethylene oxide) (PEO) chains onto azide functionalized silica particles. The aggregation of the biotin-conjugated hybrids upon the addition of avidin in solution, verified the bioavailability of the biotin-moieties to avidin which disposes four biotin-binding sites and can act as a cross-linker between the particles [53].

A challenge in the development of Janus particles is the ability to affect their self-organization properties by the application of an external stimulus. Following the Pickering emulsion approach of Granick and co-workers [110], stimuli-responsive oppositely charged Janus nanoparticles were synthesized using a sequential "grafting from"/"grafting to" approach to decorate the particles with temperature and/or pH responsive polymer chains, respectively. These switchable Janus particles form hierarchically structured aggregates upon altering the solution pH and are proposed for use in the controlled stabilization of emulsions and the regulation of molecular transport across the interface of two immiscible phases [54].

2.2. Polymer-grafted Silsequioxane nanoparticles

A lot of synthetic effort has focused on the preparation and properties of polyhedral oligomeric silsesquioxanes (POSS)-polymer hybrid materials due to the observed enhancement of the properties of the organic matrix (increased thermal stability, reduced flammability and dielectric constant) upon incorporation of POSS molecules [111–116]. POSS molecules have a cage-shaped three-dimensional structure with a general chemical formula $(RSiO_{1.5})_n$. Among them, octasilsesquioxanes $(R_8Si_8O_{12})$ (n=8) have been widely investigated owing to their precise molecular structure; they consist of a rigid, cubic silica core with a 0.53 nm side length, and an organic group attached to each of the eight corners. These organic groups can be either inert or reactive, and control the reactivity and the solubility of the POSS molecules [117–118]. Intense research effort has recently led to the preparation of POSS-containing hybrid polymers of different topological structures such as star-shaped, block and tadpole-shaped hybrid polymers. Telechelic and hemitelechelic POSS-containing hybrids have been also synthesized.

2.2.1. Star-shaped hybrids

Star-like hybrids with inorganic POSS cores can be prepared by the "core-first" method in which the polymerization is initiated by appropriately functionalized POSS molecules. Well-defined end-tethered homopolymers and block copolymers were grown from the polysilsesquioxane $(SiO_{1.5})$ nanoparticles by ATRP leading to layered core-shell structures [119–120]. A similar method employed POSS macroinitiators with about 58 functions to grow water soluble glycopolymer functionalized silsequioxane particles of approximately 25 arms. Such well-defined branched structures are attractive for a wide range of applications in biosciences as well as hybrid thickeners in coatings and paints.

An interesting approach was followed for the synthesis of a unique hybrid architecture comprising a POSS core surrounded by four macrocyclic polymer arms to form quatrefoil-shaped star-cyclic PS. The synthesis involved a combination of ATRP chemistry to form an 8-arm star with a POSS core and eight end-functionalized polymer arms carrying azide functionalities. Click reaction between the azide functionalities and a difunctional linker yielded a polymer loop with both ends attached on the POSS molecule. The star-cyclic PS processed a higher glass transition temperature compared to the star-linear PS due to the absence of chain ends in the former molecule [121]. These molecules are also expected to exhibit interesting rheological properties and dynamics related to the ring polymer structure.

2.2.2. Tadpole-shaped hybrids

An interesting POSS-based hybrid architecture combines a particle with one covalently bound polymer chain in a single entity. Initial work focused on the growth of methacrylate polymers from a functionalized POSS however, low degrees of polymerization [122] were obtained attributed to the steric hindrance by the bulky POSS molecule [123]. Recently, tadpole-shaped polymeric hybrids with an inorganic POSS "head" and a well-defined polymer "tail" were synthesized using an incompletely condensed POSS molecule bearing a highly reactive trisodium silanolate group as the efficient and versatile initiator for copper-mediated ATRP. This allowed the synthesis of POSS-polymer hybrids with good control over the polymer molecular weight and molecular weight distribution and gave hybrid materials with enhanced degradation and glass transition temperature compared to those of the model polymers without the POSS moiety [124]. POSS-PMMA block copolymer hybrids have been employed for the preparation of low-dielectric-constant, nanoporous thin films by the sequential secondary condensation of the silanol groups and the decomposition of PMMA which led to pore formation within the film [125]. It was also shown that PMMA-POSS hybrids exhibit specific interactions with phenolic resins to produce totally miscible amorphous blends in which at least three competing association mechanisms take place: self-association of the phenolic groups (hydroxyl-hydroxyl interactions), hydroxyl-siloxane inter-association between the phenolic moieties and POSS, and hydroxyl-carbonyl inter-association between the phenolic groups and PMMA [126]. POSS-diblock copolymer hybrids can form well-defined self-assembled long-range lamellae structures [127]. The formation of these hierarchical nanostructures is driven by the low polydispersity and the high molecular weight of the polymers synthesized by living anionic polymerization. The morphology of POSS-based block-copolymer thin films showed vertically oriented lamellae and cylinders [128] which could be tuned to spheres, cylinders, and lamellae by varying the grafted polymer chain length. A helix-like structure was also formed to relieve the steric crowding of the POSS units which upon thermal annealing packed into an orthorhombic lattice structure driven by POSS aggregation [129].

Tadpole-shaped fluorinated POSS molecules bearing well-defined grafted PMMA chains were blended in a PMMA matrix and were shown to migrate at the air/polymer interface with the outermost layer of the film covered almost completely by the POSS heads. Owing to this unique structure, the films exhibited strong resistance against Ar+ ion etching, despite their low overall POSS content [130] and are attractive for various applications.

Recently, responsive polymer-POSS hybrids comprising temperature-responsive or polyelectrolyte polymer chains have been prepared leading to amphiphilic tadpole-shaped structures. These molecules self-assemble into well-defined temperature-responsive POSS-core micelles or pH-responsive particulate aggregates in aqueous media [131–132] and are proposed for use in controlled drug delivery.

2.2.3. Telechelic and hemitelechelic POSS-containing hybrids

Telechelic polymers are linear chains containing associating 'sticker' groups only at the chain-ends and are analogous to triblock copolymers which self assemble in selective media. Telechelic and hemitelechelic POSS-containing hybrid polymers have been prepared by direct urethane linkage between the hydroxyl end-groups of the polymer and the monoisocyanate groups of the POSS molecules [133]. The use of a hydrophilic polymer resulted in the formation of amphiphilic telechelics the hydrophilic/hydrophobic balance of which was controlled by the molecular weight of the polymer. In bulk, the bulky nature of POSS influences the crystallization of the polymer with respect to the crystalline lamellae dimensions [134] while POSS moieties, attached to both polymer ends aggregate and form nanocrystals [135]. In dilute solution the POSS-polymer telechelics associate to form nanostructures which are influenced by the polymer architecture and the solvent polarity [136].

2.3. Polymer-grafted metal-oxide semiconducting nanoparticles

The functionalization of low-dimensional semiconductor materials such as metal oxide (TiO_2, ZnO, SnO_2) nanoparticles, nanotubes and nanowires with grafted polymer chains has attracted particular attention owing to their intriguing physical properties and their potential use in a wide range of applications. For example hybrid TiO_2 particles can be incorporated into capacitors and thin film transistors leading to the enhancement of the dielectric constant (K) and to mobilities approaching 0.2 $cm^2/V\times s$ [137]. However, the surface chemistry of these materials is rather complex and thus the immobilization of small organic molecules or long polymer chains, is not necessarily straightforward.

2.3.1. Polymer-grafted TiO_2 nanoparticles

The surface of TiO_2 nanoparticles can be modified by the "grafting through" the "grafting from" or the "grafting to" method. The latter requires pre-prepared polymer chains bearing an appropriate anchoring group. Dopamine moieties have been the most efficient anchoring groups used so far allowing the attachment of long stabilizing polymer chains which lead to the effective dispersion of the TiO_2 nanoparticles in organic media [138–140].

The use of initiator functionalized particles has been also investigated for the growth of polymer coatings directly from the nanoparticles surface (Table 4). However, the surface chemistry of the TiO_2 particles is different to that of silica and thus the development of new initiator anchoring groups is required. The attachment of organic moieties to the surface of these particles by strong covalent bonding requires the use of highly reactive organic groups such as carboxylic acid [141–144], phosphoric acid [145–147] and catechol derivatives [148]. Recently, an approach involving the reaction of an acid chloride-terminated initiator with the nanoparticle surface was also applied for the immobilization of an ATRP initiator onto the nanoparticles surface [149]. Three possible coordination mechanisms have been proposed for the binding of the carboxylic acid groups onto the TiO_2 surface; the monodentate, the chelating and the bridging coordination mode (Scheme 1). The preferred coordination mechanism is highly dependent on the nanoparticle size which affects the coordinate states of the surface Ti atoms of the nanoparticle. Biomimetic catechol binding sites are advantageous because they facilitate the intramolecular ligand-to-particle charge transfer within the surface Ti-catechol complex [150]. This approach has been used for the growth of metal-binding polymer chains on the surface of the titania nanoparticles followed by the templated growth of metal nanoparticles within the polymer shell to prepare a ternary nanostructure with electrocatalytic properties [151].

Scheme 1. Coordinating modes of the carboxylate groups on the titania surface.

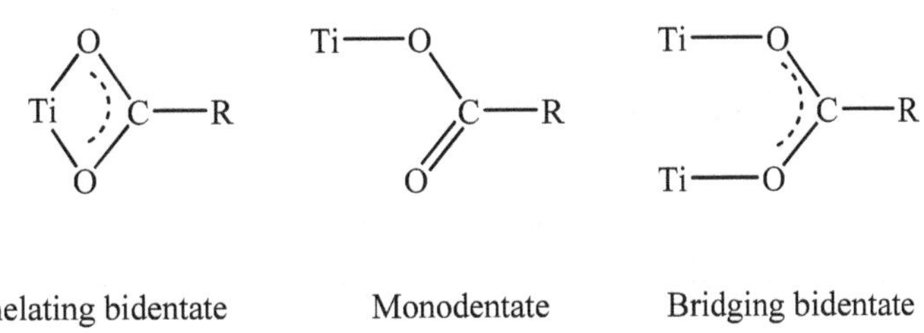

Another route for the preparation of initiator-functionalized metal oxide nanoparticles involves the use of metal alkoxides appropriately derivatized to carry polymerization initiating sites. Upon hydrolysis and condensation of these precursor molecules, amorphous titanium, zirconium and vanadium oxo clusters bearing surface initiating sites are obtained which can serve as multifunctional initiators for si-ATRP [152–153].

2.3.2. Polymer-grafted ZnO nanoparticles

Si-ATRP has been employed for the functionalization of ZnO nanoparticles with grafted polymer chains. The nanoparticles were pre-functionalized with an appropriate organic initiator using two different anchoring chemistries (Table 4); silane condensation or carboxylic acid coordination.

Both hydrophobic [156] and hydrophilic [155,157] polymers have been grown from the surface of the ZnO nanoparticles by ATRP. Moreover, hyperbranched polymer-functionalized ZnO particles were obtained by surface-initiated self-condensing vinyl polymerization (si-SCVP) from initiator-functionalized nanoparticles [154].

However, the surface functionalization of metal-oxide nanoparticles is far from being resolved. There are undoubtedly still many challenges to be meet in the synthesis, functionalization, characterization and potential applications of these materials.

Table 4. Initiator molecules grafted on the surface of TiO_2 and ZnO nanoparticles.

No	Initiators	Polymer	Ref.
1	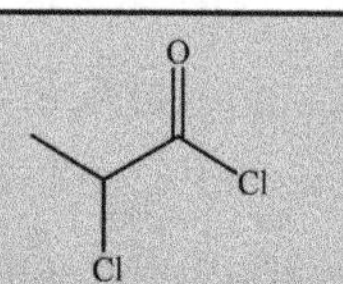 2-Chloropropionyl chloride	POEMA, PSS	[149]
2	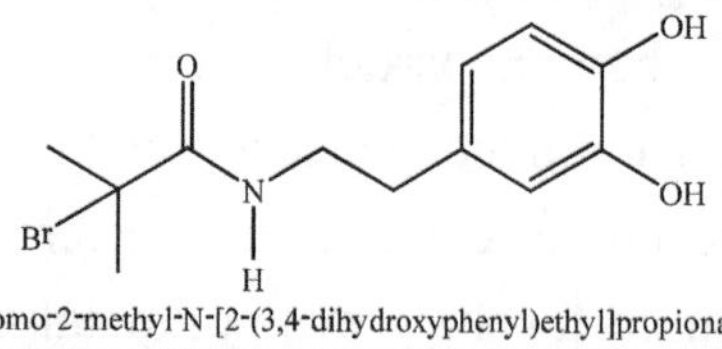 2-Bromo-2-methyl-N-[2-(3,4-dihydroxyphenyl)ethyl]propionamide	PMMA, PDMAEMA	[150–151]
3	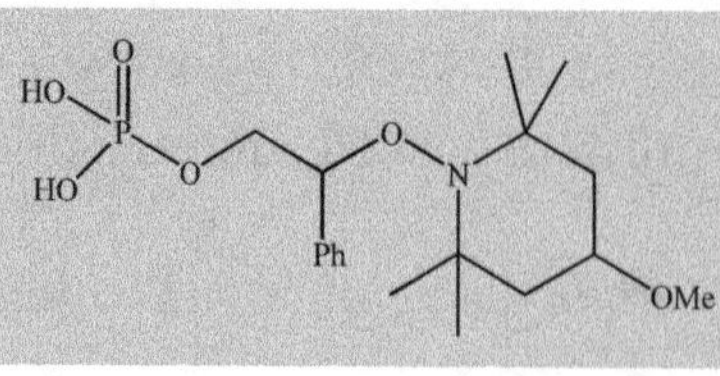	PS, P3VP	[146–147]
4	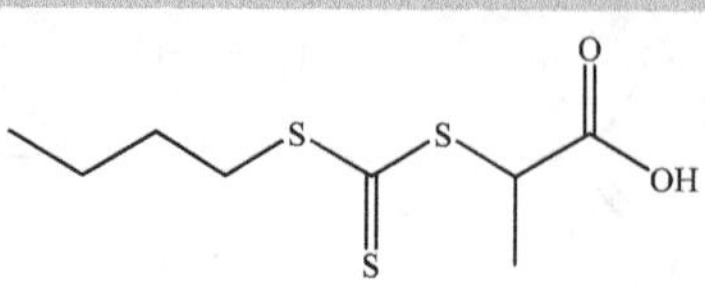 2-{[(butylsulfanyl)carbonothioyl]sulfanyl}propanoic acid	PAA	[142]

Table 4. *Cont.*

No	Initiators	Polymer	Ref.
5	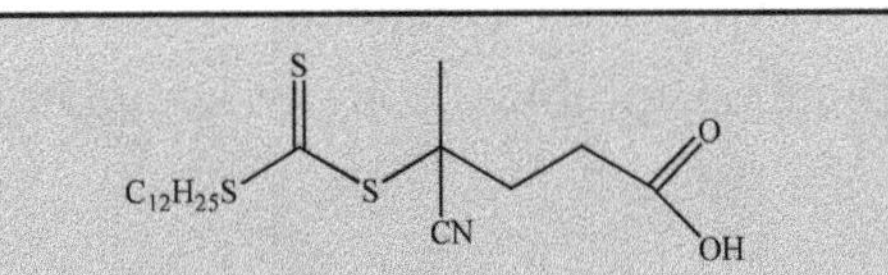 4-cyano-4-(dodecylsulfanylthiocarbonyl) sulfanyl pentanoic acid	PMMA	[143]
6*	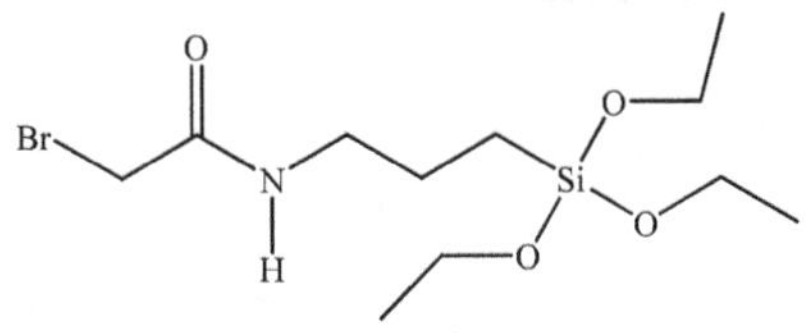	PHEA	[154–155]
7*	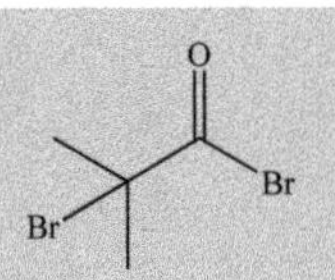 2-bromoisobutyryl bromide	MeO-PEGMA, PMMA, PCEMA	[156–157]

* Initiators used for ZnO functionalization.

2.4. Polymer-grafted magnetic nanoparticles

Magnetic nano-objects have been deemed to hold immense promise for use in *in vivo* biomedical applications including magnetic resonance imaging (MRI) contrast enhancement [158–161], targeted drug delivery [162–163], hyperthermia [164–165], magnetic field assisted radionuclide and cancer therapy [166], *in vitro* automated separations and isolation of biomolecules and cells [167–169] and magnetically separable nanocatalytic systems [170–171]. Recently, magnetic nanoparticles have been used to selectively separate "living" from dead polymer chains prepared by a RAFT process using Z-functional RAFT agents with high affinity for the surface of the nanoparticles [172]. However, due to van der Waals attraction forces and their anisotropic dipolar attraction, the magnetic nanoparticles tend to agglomerate into larger clusters and the properties associated with their nanometer dimensions are eliminated. Their functionalization with organic stabilizers [173–174] can prevent aggregation and improve their dispersion and the confinement of the inorganic material in a polymer matrix [175–176].

Several reports have described the surface functionalization of magnetic nanoparticles with initiating sites (Table 5) for the growth of polymeric stabilizers utilizing CRP techniques and showed that the magnetic properties were not significantly affected [177–178].

Lattuada and Hatton proposed a flexible methodology for the preparation of magnetic nanoparticles coated with poly(lactic acid) end-tethered chains grown by surface-initiated ROP and poly(methacrylate) end-grafted chains obtained by ATRP [179]. Poly(2-dimethylamino)ethyl methacrylate) (PDMAEMA) end-grafted chains were also synthesized from the surface of γ-F_2O_3 nanoparticles by ATRP affording an "amphibious" character to the hybrids. These core-shell nano-objects being soluble in both aqueous and common organic solvents remain in the bulk phase in water/oil biphasic systems [180]. PS- and PAA-functionalized F_3O_4 nanoparticles were also obtained by a RAFT-mediated process utilizing peroxide and hydroperoxide functionalities generated on the surface of the nanoparticles upon ozone pretreatment [181]. F_2O_3 [182] and $MnFe_2O_4$ [183] core-PS shell hybrids were synthesized by surface-initiated ATRP utilizing an initiator bearing a carboxyl anchoring group. However, the non-covalent linkage of the macromolecules onto the surface facilitated the exchange of the polymer chains with other competing molecules from the solution possessing –COOH groups. One way to overcome this problem is by cross-linking the PS chains onto the nanoparticle surface [184].

Table 5. ATRP initiators anchored on the surface of magnetic nanoparticles.

No	Initiators	Ref.
1	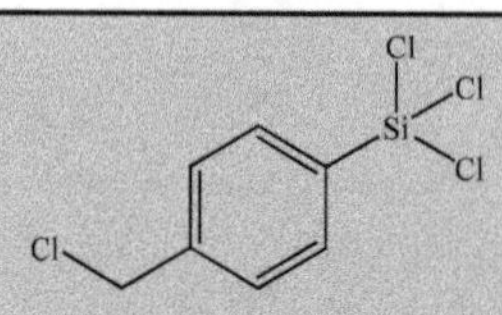 [4-(chloromethyl)phenyl]trichlorosilane	[185]
2	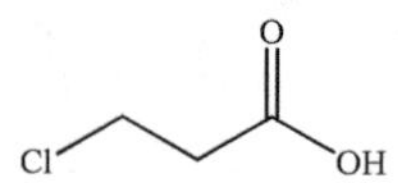 3-chloropropionic acid	[183]
3	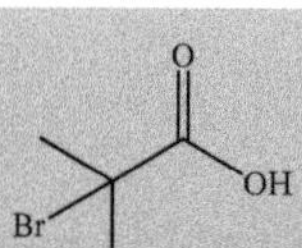 2-bromo-2-methylpropionic acid (BMPA)	[180,182,186–189]
4	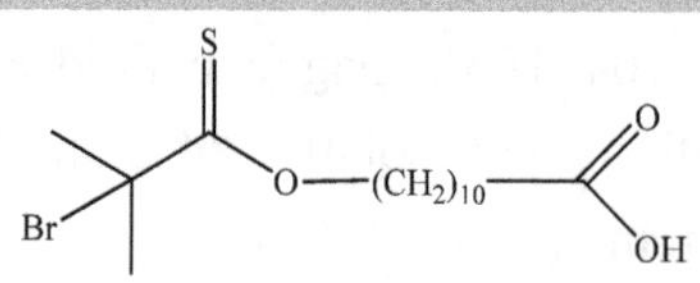 10-carboxydecanyl 2-bromo-2-methyl-thiopropanoate	[184]
5	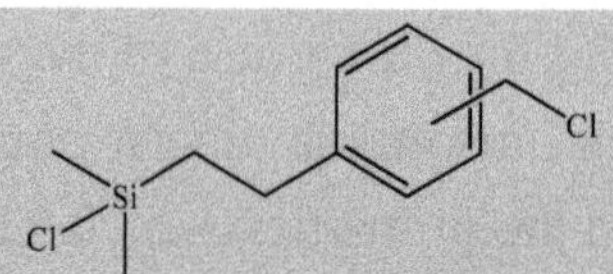 ((chloromethyl)phenylethyl)-dimethylchlorosilane (CDS)	[177]

Recently, Xu *et al.* studied the self-assembly process of nanocomposite films consisting of a lamellar-forming PS-*b*-PMMA copolymer and PMMA-coated magnetite nanoparticles as a function of the concentration of the nanoparticles and the molecular weight of the polymer chains [190]. The particles coated with short PMMA chains were dispersed at low particle concentration and aggregated only upon increasing the concentration of the nanoparticles. However, the increase of the molecular weight of the end-tethered PMMA chains induced nanoparticle aggregation even at low particle concentration. The size of these aggregates was larger than the copolymer domain size and forced the block copolymer to form an onion-like morphology around the aggregates. PMMA-coated magnetic nanoparticles synthesized by ATRP were also introduced in a thin polymer film of a symmetric P2VP-*b*-PMMA copolymer. The structural polymorphism of the film was studied in the presence of carbon tetrachloride vapors, leading to the formation of hexagonal and lamellar microphase-separated morphologies at different exposure times due to the selective solvation of the PMMA block in the solvent. The colloidal hybrids segregated in the PMMA domains of the diblock copolymer due to the affinity of the PMMA-shell for the PMMA block. However, the self-assembled morphology of the film disappeared upon application of a magnetic field generated by the interaction of a magnetized atomic force microscopy tip with the magnetic nanoparticles [191].

Tuning the composition and functionality of the macromolecules introduced onto the surface of the nanocolloids allows the tailoring of the novel physical, chemical and biological functions of the nanoparticles. In particular, the functionalization of Fe_3O_4 particles with poly(ethylene glycol methacrylate) (PEGMA) chains does not only improve the dispersability of the nanoparticles in an aqueous solution but also increases the in-vivo circulation time of the magnetic nanoparticles. PEGMA renders the surface neutral and hydrophilic and increases its resistance to the adsorption of proteins or macromolecules bearing receptors recognized by the macrophages, thus preventing the endocytosis/phagocytosis of the particles by the macrophase cells [185]. Fe_3O_4/SiO_2-g-poly(styrene sulfonate sodium salt) hybrids were also proposed for use as solid supports for the immobilization of enzymes such as pectinase. Biochemical studies proved that the enzyme retained its activity for 30 days, at a wider pH and temperature range compared to the free enzyme [192].

A challenging issue that has emerged lately is the formation of clusters, comprising polymer-coated magnetic nanoparticles, in response to an external trigger. These clusters are advantageous for certain potential applications as for example in high gradient magnetic separation processes in which clustering enables the facile scavenging of the particles after the adsorption process [193]. The development of such "smart" hybrids which undergo relatively large and abrupt changes in their physical and/or chemical and colloidal properties in response to external environmental conditions requires the functionalization of magnetic nano-objects with a polymer shell being sensitive to an external stimulus such as the temperature, pH, ionic strength, magnetic field, *etc.* The thermo-responsive dispersion of F_3O_4@PHEMA hybrids in methanol [186], F_3O_4@PS particles in cyclohexane [188–189] and poly(ε-caprolactone) (PCL)-coated colloids in dimethyl sulfoxide [187] has been studied. The solubility of the polymer as a function of the solvent quality is determined by the enthalpy and the entropy of the system and is described by the Florry and Huggins theory [194]. Below a critical temperature known as the theta temperature (T_θ) the polymer chains are in a collapsed state due to the predominant attractive forces between the polymer segments. At T_θ the entropy determines the polymer chain conformation which behaves as an ideal chain. Above T_θ the dominant forces are the repulsive interchain and the favorable chain-solvent interactions causing the polymer chains to expand. This thermo-responsive behavior of PS-coated superparamagnetic nanoparticles allowed the development of magneto-responsive Pichering emulsions in a cyclohexane/water mixture above the T_θ of PS [188], which offer new opportunities for the encapsulation and controlled release of lipophilic compounds using the magnetic heating of the emulsion in an ac magnetic field to break the emulsion on demand.

Of particular interest is the ability to control the aggregation of core-shell systems employing the property of several polymers to collapse in aqueous media upon heating above the LCST. This approach has been used to enhance the contrast in magnetic resonance imaging by introducing temperature sensitive Fe_3O_4 nanoparticles into blood cells and causing their reversible agglomeration in response to temperature changes [195]. The hybrid-loaded cells exhibited better magnetic response and enhanced magnetic contrast above the LCST due to aggregation. Besides the direct temperature change of the hybrid dispersion, the application of an AC magnetic field to generate heat, which is transferred to the temperature-responsive polymer-shell, can be also used to indirectly cause the aggregation of the nanocolloids [196]. A great challenge in this field is the ability to control the cluster size upon aggregation. Janus particles possessing stimuli responsive polymer coatings were shown to meet these requirements. Due to their asymmetric surface functionalization these nano-objects become amphiphilic and self-assemble into well-defined clusters in response to an external stimulus. Janus magnetite nanoparticles coated at one hemisphere with a pH-responsive polymer and on the second hemisphere with a pH-independent or a temperature-responsive polymer can self-assemble into micelle-like structures of finite-size at low pH or high temperatures driven by the individual response of each of the polymer coatings [197]. Moreover, polyampholyte Janus nanoparticles comprising acidic and basic hemispheres have shown an interesting pH-responsive behavior forming clusters at low pH-values due to the excess charge on the one hemisphere of the nanoparticles followed by aggregation and precipitation of the nanocolloids at intermediate pH-values due to charge neutralization on the particle surface [193].

2.5. Polymer-grafted metal nanoparticles

2.5.1. Hybrid gold nanocolloids

Functionalized metal nanoparticles such as Au, Ag, Pt, and Cu have been the subject of intense research activity over the last decade. The unique electrical properties of these particles [198–202] as well as their optical and photophysical features (size-controlled plasmon absorbance, photonic electron–hole pair generation, and fluorescence) have attracted great attention [203–204]. Gold nanoparticles are probably the most attractive metal colloids for potential applications, in electronic and optical materials [205–211] and in biomedicine due to their biocompatibility and non-toxicity [212–213]. However, gold nanoparticles have a high tendency for aggregation which is usually accompanied by a red-shift and a broadening of their plasmon band thus limiting their use in the above applications [214]. The modification of the surface properties of gold nanoparticles was thus urged by the need to stabilize them in both organic and/or aqueous media. Polymeric stabilizers have been employed using either the "grafting to" or the "grafting from" approach and were shown to effectively prevent nanoparticle aggregation. Although the "grafting to" method is more facile and allows better control of the molecular weight and molecular weight distribution of the grafted polymer chains, the "grafting from" technique has lately emerged as the preferred method to prepare densely coated nanoparticles with enhanced colloidal stability.

2.5.1.1. "Grafting to" approach

The "grafting to" approach almost exclusively employs the relatively strong sulfur-gold interactions to bind sulfur-terminated surfactant or polymers onto the gold surface. This method is particularly useful because it allows control over the molecular characteristics of the polymer ligand which is usually synthesized by a controlled/living polymerization technique prior to its grafting onto the nanoparticle surface. Among others, RAFT has emerged as a very promising controlled radical polymerization technique for the preparation of gold nanoparticle stabilizers not only because of its versatility and simplicity, but also because polymers prepared by the RAFT technique bear dithioester end-groups which are easily reduced to thiol end-groups in the presence of a reducing agent. Thiol-modified gold colloids are very stable and behave as robust large molecules being amenable to characterization by a variety of analytical methods.

There are several reports in the literature on the modification of gold nanoparticles with neutral, anionic, cationic and switterionic pre-synthesized polymers [215–216]. Wuelfing at al. first reported the grafting of thiol terminated PEO onto the surface of gold nanoparticles [217]. These nanohybrids can be useful as computed tomography (CT) contrast agents in blood pool and hepatoma imaging. Combination of the antibiofouling properties of PEG with the high X-ray absorption properties of the gold colloid provides an efficient CT contrast agent with a long circulation time that can avoid the shortcomings of current iodine-based CT contrast agents [218]. Thiol terminated PS synthesized by anionic polymerization have been also grafted onto the surface of gold nanoparticles and their homogeneous dispersion in a PS matrix was confirmed [219]. The use of faceted gold nanoparticles for the grafting of thiol-terminated hydrophobic or hydrophilic homopolymers allowed the grafting of very dense polymer brushes with grafting densities 1.2– to 23.5–fold greater than those accessible in a self-assembled monolayer onto two-dimensional gold surfaces [220]. An interesting approach for the synthesis of hydrophobic and hydrophilic arms uniformly attached in an alternating fashion on the surface of gold nanoparticles used a V-shaped polybutadiene-poly(ethylene glycol) (PB-PEG) amphiphile containing a carboxyl group at its junction point to graft the molecules onto mercaptophenol-functionalized gold nanoparticles *via* an esterification reaction. These V-shaped amphiphilc arms render the hybrid gold particles soluble in water as well as in organic media, and provide an exceptional thermal and solvent stability to the Au clusters, which remain in solution without precipitation, agglomeration, or decomposition for more than two years [221].

Temperature sensitive polymeric stabilizers have been extensively used for the modification of the surface of gold nanoparticles. PNIPAM homopolymers prepared by the RAFT technique using *S*-benzyl dithiobenzoate as the RAFT agent were immobilized onto gold particles by the "grafting to" method following reduction of the dithioester end-group with sodium borohydride to provide the required thiol group. These thermo-sensitive hybrids retained the LCST of PNIPAM and exhibited a sharp, reversible phase transition in solution between 25 and 30 °C [222]. The addition of salt was found to promote interparticle associations leading to a red shift in the plasmon band and an increase in the hydrodynamic size, attributed to aggregate formation [223]. This feature can be useful for the development of sensing devices due to the enhancement of the optical response obtained from the functionalized gold nanoparticles upon their interaction with specific substrates [224]. Responsive three-layer core-shell-corona microstructures are accessible by this method by grafting thiol-terminated responsive diblock copolymers onto the surface of gold nanoparticles [225]. These nanostructures can be shell cross-linked at high particle concentrations, due to the protection afforded by the hydrophilic corona, resulting in the formation of locked layered nanoparticles with improved colloidal stability. Particles exhibiting a complex pH- and temperature-responsive behavior and reversible swelling properties both before and after the cross-linking of the shell were obtained.

A facile approach to stabilize gold nanoparticles within a polymeric matrix employs the use of star block copolymers [226] or diblock copolymer micellar structures into which the nanoparticles are incorporated by simple polymer physisorption on the nanoparticle surface or hydrophobic interactions between hydrophobized particles and the micelle core [227]. Even though these core/shell structures can be permanently fixed by cross-linking the polymer shell a disadvantage of this method is found in the difficulty to control the shell thickness and the number of nanoparticles within each polymer nanostructure.

2.5.1.2. "Grafting from" method

The "grafting to" method described above often requires a large concentration of thiol-terminated stabilizers in solution which can interfere with the surface functionalization. Moreover, polymer adsorption onto nanometer-sized metal particles cannot usually provide compact core-shell systems because of steric hindrance among the adsorbed and the approaching polymer chains. Gold nanoparticle core nanostructures with a dense polymer layer of well-defined homopolymers and copolymers are easily prepared using a controlled/living polymerization technique to grow the polymer chains from initiator-functionalized gold nanoparticles (Table 6).

Table 6. ATRP Initiators immobilized on the surface of gold nanoparticles.

No	Initiator	Polymer	Ref.
1	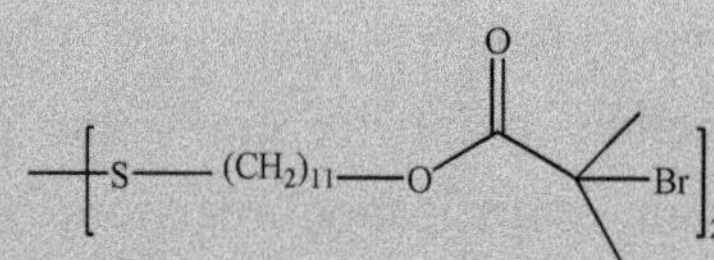 11,11'-dithiobis[1-(2-bromo-2-methylpropionyloxy)undecane] (DTBU)	P4VP, PMMA, P(MeO-PEGMA)-*b*-PNIPAM, PNIPAM, PDMAEMA-*co*-PDEAEMA-*co*-P(PEGMA)	[228–234,236]
2	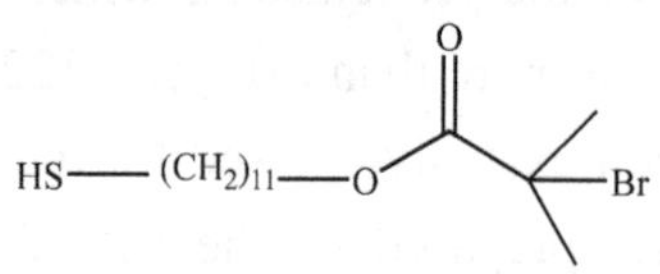	PMMA, P*n*BA, P*t*BA	[237-239]
3	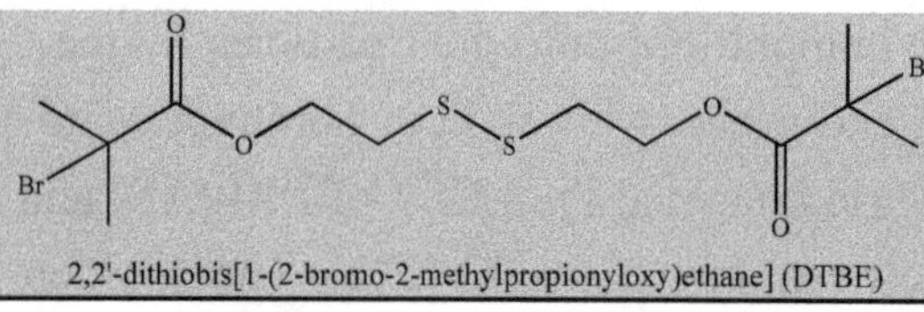 2,2'-dithiobis[1-(2-bromo-2-methylpropionyloxy)ethane] (DTBE)	PDMAEMA	[234]

Watson *et al.* reported the transition-metal catalyzed ring-opening metathesis polymerization of norbornenyl-containing monomers carrying pendent electrochemical sensitive ferrocene groups from the surface of gold nanoparticles [240]. Dense polymer brushes were synthesized from the surface of gold particles by surface-initiated living cationic polymerization of 2-oxazolines in "one-pot multistep" reaction. For this, end-functionalized three-dimensional self-assembled monolayers (SAMs) on gold nanoparticles were used to initiate the living cationic ring-opening polymerization reaction directly on the gold nanoparticle surface. The resulting gold/polymer nanocomposite consisted of "brush-type" shells of linear macromolecules which upon the introduction of a terminal mesogen by means of a quantitative termination reaction resulted in an amphiphilic core-shell structures of well-defined hydrophilic/lipophilic balance [241]. Among the syntheses used, si-ATRP from self-assembled monolayers of bromofunctionalized thiolates on gold nanoparticles appears as the most versatile and facile method for the development of structurally controlled hybrids with a gold core of several nanometers in diameter and a shell of variable thickness composed of well-defined, dense polymer brushes [237–238]. These highly grafted hybrids exhibit an intense surface plasmon absorption band which is blue shifted when increasing the grafted polymer chain length suggesting the improved stabilization of the gold colloids as the size of the polymer stabiliser increases [230]. Moreover, they can form a two-dimensional array with a high degree of structural order due to their long range interparticle interactions which is comparable to the full length of the grafted polymer chains [231].

Among the different polymer shells, stimuli-responsive polymers are particularly attractive because they allow the manipulation of the optical signal obtained from the gold nanoparticles by external triggers. Cationic, pH-responsive polymer brushes grown from gold nanoparticles *via* si-ATRP, were shown to exhibit a reversible shrinkage of the polymer chains which wrap around the gold particle surface. A careful investigation of the pH-response of the grafted polymer chains revealed a two-step transition ascribed to the initial polymer collapse which induced nanoparticle agglomeration in a second stage [229]. These hybrids can serve as pH-stimuli bimetallic catalysts because they can entrap transition metal ions by their efficient coordinating segments followed by *in situ* metal reduction [228]. Polymer-coated gold nanoparticles have been also used for the development of pH-sensitive, colloidally stable, hydrophilic polymer capsules which are attractive for the controlled encapsulation and release of drug molecules. The capsules were obtained by first cross-linking the polymer shells followed by the cleavage of the Au-S bonds to remove the gold cores [234]. Temperature-responsive polymers, PNIPAM and poly(*N*-cyclopropylacrylamide) (PCYPAAM), exhibiting an LCST behavior have been also grown from gold nanoparticles using the ATRP [216] or RAFT polymerization technique. The optical properties of these gold particles can be varied by temperature, based on the thermal-responsive behavior of the polymer monolayer. During the collapse of the PNIPAM grafts the surface plasmon not only decreases in intensity but also blue shifts due to the decrease of the hydrophilicity of the nanoparticle surface as it becomes covered by the collapsed polymer [242].

Besides, the growth of linear polymers the synthesis of cross-linked temperature responsive gels onto the surface of gold nanoparticles has been explored. Gold nanoparticles coated with a cross-linked PNIPAM shell have been proposed for use in nanoparticle trapping using the thermo-modulated "breathing process" [235] of the polymer shell [236].

Moreover, Janus gold-nanoparticles with two types of different polymer chains decorated on the opposite sides of the nanoparticles are accessible. A novel approach was employed which combined a "solid-state grafting-to" method for the immobilization of crystalline polymer chains on one hemisphere and a "grafting-from" technique for the growth of a second type of polymer from the "free" surface of the nanoparticles. The symmetric decoration of the Janus hybrids was confirmed by the selective *in situ* formation of platinum nanoparticles on one of the hemispheres [239]. Although this solid-state approach has a higher throughput compared with self-assembled monolayers it is limited to the grafting of crystalline polymers in one of the two hemispheres.

Finally, binary thermo-sensitive nanocomposites can be obtained by the growth of block copolymers comprising two thermo-responsive blocks on the surface of the gold nanoparticles. These core-shell nanostructures possess two transition temperatures corresponding to the thermally induced conformational transition of the inner and the outer polymer, respectively. Upon cross-linking the polymer shells the nanostructures can be used as temperature-sensitive catalytic systems with tunable catalytic activity of the incorporated metal nanoparticles as a function of the solution temperature [232].

2.5.2. Polymer-coated platinum nanoparticles

Similar to gold nanoparticles, platinum nanoparticles have a high surface-to-volume ratio and consequently a large fraction of metal atoms are exposed at the surface and are accessible to reactant molecules. Few reports in the literature have described the synthesis of polymer chains from the surface of platinum nanoparticles using the "grafting from" method in combination with a CRP technique. Recently, Carrot *et al.* have grown poly(butyl methacrylate) (P*n*BMA) chains from initiator-derivatized platinum nanoparticles using ATRP. Small-angle neutron scattering studies in solution showed that the hybrid structure is closer to a star architecture rather than a dense polymer corona surrounding the inorganic core. The formation of 2D arrays (Langmuir film) of these polymer-grafted-platinum hybrids at the air-water interface was shown, in which the interparticle distance was controlled in the range of tens of nanometers to a few nanometers, by the polymer chain length and the surface pressure [243].

2.6. Polymer-grafted quantum dots

Nanometer-sized semiconductor nanoparticles or quantum dots such as CdSe, CdTe, CdS, and ZnS have been extensively studied and still remain an active research area because of their unique size- and surface-dependent nonlinear optical, physicochemical, and electronic properties [244]. Due to their size in the nanometer range, quantum dots show interesting physical behavior which is totally different to that found in bulk materials. This effect is known as the quantum confinement effect.

Because of these unique electronic and optical properties, including their broad absorption spectra, narrow absorption profiles and their discrete energy bands, semiconductor nanoparticles are candidates for many applications in sensors, solar cells, biological imaging, diagnostics, data storage, light-emitting diodes, single-molecule transistors, and substrates for biological tags and detection devices [245-253]. The modification of the surface of the nanoparticles with polymers is highly advantageous since the polymer endows the quantum dots with new physical and chemical properties including solubility, processability, chemical and biological stability, biocompatibility, and further functionalization possibility.

2.6.1. "Grafting to" method

The grafting of polymer chains onto the surface of quantum dots requires the appropriate end-functionalization of the polymer chains and/or the pre-functionalization of the surface of the nanoparticles. Initial studies have shown the grafting of preformed thiol-terminated PCL of controlled molecular weight and narrow molecular weight distribution onto the surface of CdS nanoparticles. The hybrids were highly stable both in solution, in films and in polymer matrices [254].

Responsive PNIPAM-stabilized CdS particles have been prepared *via* the grafting of pre-synthesized thiolated-PNIPAM chains to the surface of CdS quantum dots. Association and dissociation studies of these hybrids induced by the increase and the decrease of the solution temperature, respectively (Figure 3), showed a hysteresis in the heating-cooling cycle attributed to the formation of large aggregates which initially dissociate into smaller aggregates as the temperature decreases, however, their complete dissociation is retarded by a balanced effect of inter- and intrachain hydrogen bonding as well as hydrophobic interactions between the PNIPAM chains and the CdS particles [255].

Figure 3. Reversible thermo-induced aggregation of temperature-sensitive hybrids.

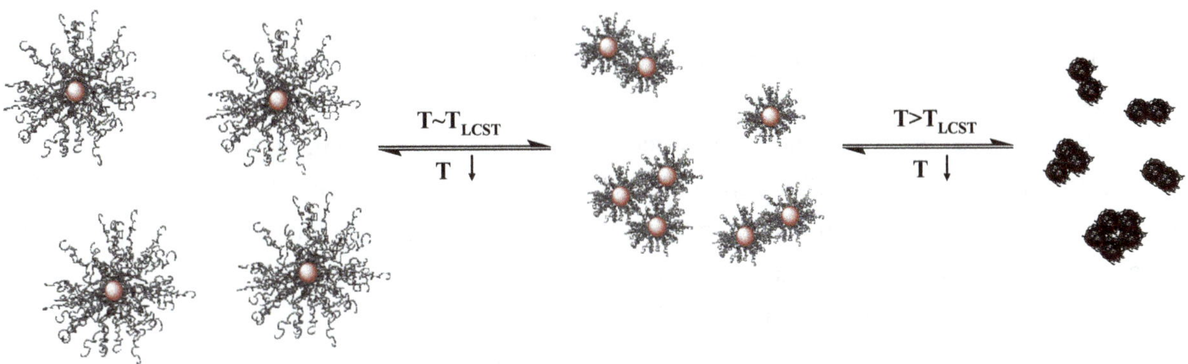

Recently, CdSe nanoparticles pre-functionalized with [(4-bromophenyl)methyl]dioctylphosphine oxide were used to covalently bind long vinyl-terminated poly(3-hexylthiophene) (P3HT) chains *via* a mild palladium-catalyzed Heck coupling. The solid-state emission spectra of the nanocomposites suggested an efficient charge transfer mechanism from the polymer to the quantum dots, which is three times faster compared to the energy transfer found for P3HT/CdSe composites prepared by physically mixing the pristine quantum dots in the polymer matrix [256].

Poly(*para*-methyl triphenylamine-*b*-cysteamine acrylamide) coated CdSe-ZnS quantum dots prepared by the "grafting to" method using thiol chemistry showed improved performance in light-emitting diodes attributed to the uniform film deposition of the hybrid layer as well as the facilitated hole injection into the quantum dots by the directly linked polymer [257].

2.6.2. "Grafting from" approach

Conventional polymerization methods used to modify the surface of nanoparticles with polymers are not compatible with semiconductor nanoparticles. For example, anionic conditions are too severe for use in the presence of the quantum dots, and radicals generated by conventional radical polymerization (e.g., azo-bisisobutyronitrile initiated) degrade the nanoparticles and quench their photoluminescence. One of the first attempts for the surface modification of quantum dots came from Patten and Farmer who encapsulated CdS quantum dots in a silica shell and polymerized PMMA chains from the silica surface by ATRP [258]. These hybrid nanoparticles were casted into films which retained the photoluminescence of the precursor CdS nanoparticles and showed good dispersion of the polymer-coated CdS/SiO_2 hybrids throughout the PMMA matrix. However, the latter advances in CRP techniques, which are accompanied with low radical concentrations, have allowed the growth of functional polymers from the surface of quantum dots.

"Unprotected" quantum dots without the intermediate silica layer have been used lately as substrates for the development of core-shell hybrids providing an intimate connection of the polymer to the nanoparticle core. Sill and Emrick reported the NMP synthesis of PS homopolymer and PS-*co*-

PMMA copolymer chains from the surface of CdSe nanoparticles pre-functionalized with initiating molecules bearing phosphine oxide derivatives as the anchor groups [259]. The PS-coated CdSe nanoparticles not only showed a uniform dispersion in a polystyrene matrix but also exhibited enhanced fluorescence compared to the initiator functionalized particles due to their thermal treatment during the polymerization. At the same time a range of well-defined styrenes and (meth)acrylate homopolymers, random and block copolymers were grown by the RAFT technique from the surface of CdSe nanoparticles modified with trithiocarbonate moieties [260]. Upon surface modification with the polymeric stabilizers the particles maintained their optical properties without the requirement for a protective inorganic shell.

Another approach used activators generated by electron transfer (AGET) ATRP in a miniemulsion system to synthesize well-defined polymer chains from the surface of pre-functionalized CdS nanoparticles [261]. The hybrids exhibited a high homogeneity, while quantum confinement effects were revealed by a blue shift at the onset of the trioctylphosphine oxide (TOPO) quantum dot absorption compared to the absorption of bulk CdS.

Polyolefin-coated CdSe nanoparticles have been prepared by ruthenium-catalyzed ROMP from particles pre-modified with *p*-vinylbenzyl-di-*n*-octylphosphine oxide [262]. This approach allowed the incorporation of nanoparticles of potential technological importance within simple and commercially available polymers and could potentially lead to industrially relevant applications.

Moreover, an interesting approach was used to synthesize PCL *via* coordination-insertion ROP from the surface of CdS particles bearing hydroxyl-groups. In the first step, the reaction of mercaptoethanol and thioglycerol with cadmium acetate and thiourea afforded the hydroxy-coated CdS nanoparticles. The surface hydroxyl groups were next activated into aluminum alkoxide species, by reaction with $AlEt_3$, which selectively initiated the ROP of (ε-caprolactone) from the nanoparticles surface. The covalent grafting of the hydrophobic polyester chains onto the semiconducting nanoparticles modified the surface properties of the nanoparticles and facilitated their dispersion in organic solvents such as toluene [11].

Hydroxyl-terminated germanium nanoclusters have been also appropriately functionalized to yield polymer-coated nanoclusters *via* ATRP. The original nanoclusters retain their photophysics upon altering the termination group which is consistent with a stable nanocluster surface [263].

Hyperbranched polyglycerol-modified quantum dots which are attractive for use in biomedical applications due to the reactivity of the hydroxyls towards functional biomolecules have been prepared by *in situ* anionic ROP from the surface of CdTe nanoparticles. Their good dispersion in water along with the strong fluorescence signal supported their use in cytotoxicity studies, which revealed the low toxicity of the hybrids in human lung cancer cells [264].

Finally, thermo-responsive quantum dots with PDMAEMA-grafted polymer chains, grown by surface-initiated oxyanionic vinyl polymerization, were shown to self-assemble into large spheres above the LCST of the grafted polymer and are attractive for use as "smart" materials in biological applications [265].

3. Conclusions and Perspectives

This review has given an overview of the range of hybrid core-shell nanostructures that can be prepared by end-grafting polymer chains onto the surface of inorganic nanocolloids using the direct covalent "grafting to"/"grafting from" strategies. CRP techniques have allowed the synthesis of well-defined advanced functional materials. It was shown that different mechanisms are possible and have been successfully applied for a variety of nanoparticles. However, although the surface functionalization of some types of nanoparticles such as silica and gold, have been extensively studied and, to a great extend, understood, more research is required for the successful dispersion and functionalization of other nanoparticulate objects, *i.e.,* semiconducting metal oxides, magnetic and quantum dots, which are still prone to aggregation and colloidal or chemical instabilities at various stages of their synthesis. Moreover, the grafting of inert polymer chains has been adequately described in most cases, whereas stimuli-responsive polymer brushes and their potential to control the designed morphology of the nanocomposite have not been fully explored. The controlled synthesis of end-grafted macromolecules with pre-defined molecular weight, composition, functionality and grafting density onto the surface of active nanocolloids will open up new avenues for their use in many high-value applications. Furthermore, much can be gained by understanding the structure-properties relationship of the nanohybrids for the retro-design of materials with the desired macroscopic properties.

Acknowledgements

The general Secretariat for Research and Technology – Hellas is acknowledged for funding D. S. Achilleos' work (PENED 2003, programme 03EΔ623).

References and Notes

1. Caruso, F. Nanoengineering of particle surfaces. *Adv. Mater.* **2001**, *13*, 11–22.
2. Golden, J.H.; Deng, H.; DiSalvo, F.J.; Fréchet, J.M.J.; Thompson, P.M. Monodisperse metal clusters 10 angstroms in diameter in a polymeric host: The "monomer as solvent" approach. *Science* **1995**, *268*, 1463–1466.
3. Zhou, Q.; Wang, S.; Fan, X.; Advincula, R.; Mays, J. Living anionic surface-initiated polymerization (lasip) of a polymer on silica nanoparticles. *Langmuir* **2002**, *18*, 3324–3331.
4. Pyun, J.; Kowalewski, T.; Matyjaszewski, K. Synthesis of polymer brushes using atom transfer radical polymerization. *Macromol. Rapid Commun.* **2003**, *24*, 1043–1059.
5. Mansky, P.; Liu, Y.; Huang, E.; Russell, T.P.; Hawker, C. Controlling polymer-surface interactions with random copolymer brushes. *Science* **1997**, *275*, 1458–1460.
6. Spange, S. Silica surface modification by cationic polymerization and carbenium intermediates. *Prog. Polym. Sci.* **2000**, *25*, 781–849.
7. Eismann, U.; Spange, S. Cationic initiation of vinyl ether polymerization induced by (4-$R_1C_6H_4$)(4-$R_2C_6H_4$)R_3CX in conjunction with silica: Producing highly head group functionalized polymers. *Macromolecules* **1997**, *30*, 3439–3446.
8. Spange, S.; Höhne, S.; Francke, V.; Günther, H. The cationic polymerization of 2-vinylfuran on the surface of silica particles–An NMR spectroscopic study. *Macromol. Chem. Phys.* **1999**, *200*, 1054–1061.
9. Hohne, S.; Spange, S. Surface functionalization of silica with 2-vinylfuran by cationic polymerization. *Macromol. Chem. Phys.* **2003**, *204*, 841–849.

10. Zirbs, R.; Binder, W.; Gahleitner, M.; Machl, D. "Grafting from"-living cationic polymerization of poly(isobutylene) from silica-nanoparticle surfaces. *Macromol. Symp.* **2007**, *254*, 93–96.
11. Carrot, G.; Rutot-Houze, D.; Pottier, A.; Degee, P.; Hilborn, J.; Dubois, P. Surface-initiated ring-opening polymerization: A versatile method for nanoparticle ordering. *Macromolecules* **2002**, *35*, 8400–8404.
12. Yoon, K.R.; Lee, Y.W.; Lee, J.K.; Choi, I.S. Silica/poly(1,5-dioxepan-2-one) hybrid nanoparticles by "direct" surface-initiated polymerization. *Macromol. Rapid Commun.* **2004**, *25*, 1510–1513.
13. Joubert, M.; Delaite, C.; Bourgeat-Lami, E.; Dumas, P. Hairy PEO-silica nanoparticles through surface-initiated polymerization of ethylene oxide. *Macromol. Rapid Commun.* **2005**, *26*, 602–607.
14. Joubert, M.; Delaite, C.; Lami, E.B.; Dumas, P. Synthesis of poly(ε-caprolactone)-silica nanocomposites: From hairy colloids to core-shell nanoparticles. *New J. Chem.* **2005**, *29*, 1601–1609.
15. Joubert, M.; Delaite, C.; Bourgeat-Lami, E.; Dumas, P. Ring-Opening polymerization of ε-caprolactone and L-lactide from silica nanoparticles surface. *J. Polym. Sci. A* **2004**, *42*, 1976–1984.
16. Mingotaud, A.F.; Reculusa, S.; Mingotaud, C.; Keller, P.; Sykes, C.; Duguet, E.; Ravaine, S. Ring-opening metathesis polymerization on well defined silica nanoparticles leading to hybrid core-shell particles. *J. Mater. Chem.* **2003**, *13*, 1920–1925.
17. Jordi, M.A.; Seery, T.A.P. Quantitative determination of the chemical composition of silica-poly(norbornene) nanocomposites. *J. Am. Chem. Soc.* **2005**, *127*, 4416–4422.
18. Milner, S.T. Polymer brushes. *Science* **1991**, *251*, 905–914.
19. Percec, V.; Popov, A.V.; Ramirez-Castillo, E.; Monteiro, M.; Barboiu, B.; Weichold, O.; Asandei, A.D.; Mitchell, C.M. Aqueous room temperature metal-catalyzed living radical polymerization of vinyl chloride. *J. Am. Chem. Soc.* **2002**, *124*, 4940–4941.
20. Matyjaszewski, K.; Tsarevsky, N.V. Nanostructured functional materials prepared by atom transfer radical polymerization. *Nat. Chem.* **2009**, *1*, 276–288.
21. Stober, W.; Fink, A.; Bohn, E. Controlled growth of monodisperse silica spheres in the micron size range. *J. Colloid Interface Sci.* **1968**, *26*, 62–69.
22. Ranjan, R.; Brittain, W.J. Combination of living radical polymerization and click chemistry for surface modification. *Macromolecules* **2007**, *40*, 6217–6223.
23. Huang, L.; Dolai, S.; Raja, K.; Kruk, M. "Click" grafting of high loading of polymers and monosaccharides on surface of ordered mesoporous silica. *Langmuir* **2009**, doi: 10.1021/la9026943.
24. Huang, Y.; Liu, Q.; Zhou, X.; Perrier, S.B.; Zhao, Y. Synthesis of silica particles grafted with well-defined living polymeric chains by combination of RAFT polymerization and coupling reaction. *Macromolecules* **2009**, *42*, 5509–5517.
25. De Las Heras Alarcón, C.; Pennadam, S.; Alexander, C. Stimuli responsive polymers for biomedical applications. *Chem. Soc. Rev.* **2005**, *34*, 276–285.
26. Schild, H.G. Poly(N-isopropylacrylamide): Experiment, theory and application. *Prog. Polym. Sci.* **1992**, *17*, 163–249.

27. Huang, C.-J.; Chang, F.-C. Using click chemistry to fabricate ultrathin thermoresponsive microcapsules through direct covalent layer-by-layer assembly. *Macromolecules* **2009**, *42*, 5155–5166.
28. von Werne, T.; Patten, T.E. Preparation of structurally well-defined polymer-nanoparticle hybrids with controlled/living radical polymerizations. *J. Am. Chem. Soc.* **1999**, *121*, 7409–7410.
29. von Werne, T.; Patten, T.E. Atom transfer radical polymerization from nanoparticles: A tool for the preparation of well-defined hybrid nanostructures and for understanding the chemistry of controlled/"living" radical polymerizations from surfaces. *J. Am. Chem. Soc.* **2001**, *123*, 7497–7505.
30. Pyun, J.; Jia, S.; Kowalewski, T.; Patterson, G.D.; Matyjaszewski, K. Synthesis and characterization of organic/inorganic hybrid nanoparticles: kinetics of surface-initiated atom transfer radical polymerization and morphology of hybrid nanoparticle ultrathin films. *Macromolecules* **2003**, *36*, 5094–5104.
31. Carrot, G.; Diamanti, S.; Manuszak, M.; Charleux, B.; Vairon, J.P. Atom transfer radical polymerization of n-butyl acrylate from silica nanoparticles. *J. Polym. Sci. A* **2001**, *39*, 4294–4301.
32. Böttcher, H.; Hallensleben, M.L.; Nuss, S.; Wurm, H. ATRP grafting from silica surface to create first and second generation of grafts. *Polym. Bull.* **2000**, *44*, 223–229.
33. El Harrak, A.; Carrot, G.; Oberdisse, J.; Jestin, J.; Boué, F. Atom transfer radical polymerization from silica nanoparticles using the "grafting from" method and structural study *via* small-angle neutron scattering. *Polymer* **2005**, *46*, 1095–1104.
34. Pyun, J.; Matyjaszewski, K. Synthesis of nanocomposite organic/inorganic hybrid materials using controlled/"living" radical polymerization. *Chem. Mater.* **2001**, *13*, 3436–3448.
35. Carrot, G.; Harrak, A.E.; Oberdisse, J.; Jestin, J.; Boué, F. Polymer grafting from 10-nm individual particles: Proving control by neutron scattering. *Soft Matter* **2006**, *2*, 1043–1047.
36. Lei, Z.; Bi, S. Preparation of polymer nanocomposites of core-shell structure *via* surface-initiated atom transfer radical polymerizations. *Mater. Lett.* **2007**, *61*, 3531–3534.
37. Savin, D.A.; Pyun, J.; Patterson, G.D.; Kowalewski, T.; Matyjaszewski, K. Synthesis and characterization of silica-graft-polystyrene hybrid nanoparticles: Effect of constraint on the glass-transition temperature of spherical polymer brushes. *J. Polym. Sci. B Polym. Phys.* **2002**, *40*, 2667–2676.
38. Zhang, K.; Ma, J.; Zhang, B.; Zhao, S.; Li, Y.; Xu, Y.; Yu, W.; Wang, J. Synthesis of thermoresponsive silica nanoparticle/PNIPAM hybrids by aqueous surface-initiated atom transfer radical polymerization. *Mater. Lett.* **2007**, *61*, 949–952.
39. Perruchot, C.; Khan, M.A.; Kamitsi, A.; Armes, S.P.; Watts, J.F.; Von Werne, T.; Patten, T.E. XPS characterisation of core-shell silica-polymer composite particles synthesised by atom transfer radical polymerisation in aqueous media. *Eur. Polym. J.* **2004**, *40*, 2129–2141.
40. Ohno, K.; Morinaga, T.; Koh, K.; Tsujii, Y.; Fukuda, T. Synthesis of monodisperse silica particles coated with well-defined, high-density polymer brushes by surface-initiated atom transfer radical polymerization. *Macromolecules* **2005**, *38*, 2137–2142.

41. Ohno, K.; Morinaga, T.; Takeno, S.; Tsujii, Y.; Fukuda, T. Suspensions of silica particles grafted with concentrated polymer brush: Effects of graft chain length on brush layer thickness and colloidal crystallization. *Macromolecules* **2007**, *40*, 9143–9150.
42. Morinaga, T.; Ohno, K.; Tsujii, Y.; Fukuda, T. Structural analysis of "semisoft" colloidal crystals by confocal laser scanning microscopy. *Macromolecules* **2008**, *41*, 3620–3626.
43. Bombalski, L.; Dong, H.; Listak, J.; Matyjaszewski, K.; Bockstaller, M.R. Null-scattering hybrid particles using controlled radical polymerization. *Adv. Mater.* **2007**, *19*, 4486–4490.
44. Voudouris, P.; Choi, J.; Dong, H.; Bockstaller, M.R.; Matyjaszewski, K.; Fytas, G. Effect of shell architecture on the static and dynamic properties of polymer-coated particles in solution. *Macromolecules* **2009**, *42*, 2721–2728.
45. Perruchot, C.; Khan, M.A.; Kamitsi, A.; Armes, S.P.; von Werne, T.; Patten, T.E. Synthesis of well-defined, polymer-grafted silica particles by aqueous ATRP. *Langmuir* **2001**, *17*, 4479–4481.
46. Chen, X.; Randall, D.P.; Perruchot, C.; Watts, J.F.; Patten, T.E.; Von Werne, T.; Armes, S.P. Synthesis and aqueous solution properties of polyelectrolyte-grafted silica particles prepared by surface-initiated atom transfer radical polymerization. *J. Colloid Interface Sci.* **2003**, *257*, 56–64.
47. Chen, X.; Armes, S.P. Surface polymerization of hydrophilic methacrylates from ultrafine silica sols in protic media at ambient temperature: A novel approach to surface functionalization using a polyelectrolytic macroinitiator. *Adv. Mater.* **2003**, *15*, 1558–1562.
48. Chen, X.Y.; Armes, S.P.; Greaves, S.J.; Watts, J.F. Synthesis of hydrophilic polymer-grafted ultrafine inorganic oxide particles in protic media at ambient temperature *via* atom transfer radical polymerization: Use of an electrostatically adsorbed polyelectrolytic macroinitiator. *Langmuir* **2004**, *20*, 587–595.
49. Jia, G.; Cao, Z.; Xue, H.; Xu, Y.; Jiang, S. Novel zwitterionic-polymer-coated silica nanoparticles. *Langmuir* **2009**, *25*, 3196–3199.
50. Vaisocherov, H.; Yang, W.; Zhang, Z.; Cao, Z.; Cheng, G.; Piliarik, M.; Homola, J.I.; Jiang, S. Ultralow fouling and functionalizable surface chemistry based on a zwitterionic polymer enabling sensitive and specific protein detection in undiluted blood plasma. *Anal. Chem.* **2008**, *80*, 7894–7901.
51. Tang, C.; Bombalski, L.; Kruk, M.; Jaroniec, M.; Matyjaszewski, K.; Kowalewski, T. Nanoporous carbon films from "hairy" polyacrylonitrile-grafted colloidal silica nanoparticles. *Adv. Mater.* **2008**, *20*, 1516–1522.
52. Kruk, M.; Dufour, B.; Celer, E.B.; Kowalewski, T.; Jaroniec, M.; Matyjaszewski, K. Synthesis of mesoporous carbons using ordered and disordered mesoporous silica templates and polyacrylonitrile as carbon precursor. *J. Phys. Chem. B* **2005**, *109*, 9216–9225.
53. Zhang, J.; Wang, X.; Wu, D.; Liu, L.; Zhao, H. Bioconjugated Janus particles prepared by *in situ* click chemistry. *Chem. Mater.* **2009**, *21*, 4012–4018.
54. Berger, S.; Synytska, A.; Ionov, L.; Eichhorn, K.-J.; Stamm, M. Stimuli-Responsive bicomponent polymer Janus particles by "grafting from"/"grafting to" approaches. *Macromolecules* **2008**, *41*, 9669–9676.
55. Zhao, H.; Kang, X.; Liu, L. Comb-coil polymer brushes on the surface of silica nanoparticles. *Macromolecules* **2005**, *38*, 10619–10622.

56. Mu, B.; Wang, T.; Liu, P. Well-defined dendritic-graft copolymer grafted silica nanoparticle by consecutive surface-initiated atom transfer radical polymerizations. *Ind. Eng. Chem. Res.* **2007**, *46*, 3069–3072.
57. Schepelina, O.; Zharov, I. Poly(2-(dimethylamino)ethyl methacrylate)-modified nanoporous colloidal films with pH and ion response. *Langmuir* **2008**, *24*, 14188–14194.
58. Smith, J.J.; Zharov, I. Preparation and proton conductivity of sulfonated polymer-modified sintered and self-assembled silica colloidal crystals. *Chem. Mater.* **2009**, *21*, 2013–2019.
59. Wu, T.; Zou, G.; Hu, J.; Liu, S. Fabrication of photoswitchable and thermotunable multicolor fluorescent hybrid silica nanoparticles coated with Dye-labeled Poly(N-isopropylacrylamide) brushes. *Chem. Mater.* **2009**, *21*, 3788–3798.
60. Wu, T.; Zhang, Y.; Wang, X.; Liu, S. Fabrication of hybrid silica nanoparticles densely grafted with thermoresponsive Poly(N-isopropylacrylamide) brushes of controlled thickness *via* surface-initiated atom transfer radical polymerization. *Chem. Mater.* **2007**, *20*, 101–109.
61. Li, D.; Sheng, X.; Zhao, B. Environmentally responsive "hairy" nanoparticles: Mixed homopolymer brushes on silica nanoparticles synthesized by living radical polymerization techniques. *J. Am. Chem. Soc.* **2005**, *127*, 6248–6256.
62. Zhao, B.; Zhu, L. Nanoscale phase separation in mixed poly(tert-butyl acrylate)/polystyrene brushes on silica nanoparticles under equilibrium melt conditions. *J. Am. Chem. Soc.* **2006**, *128*, 4574–4575.
63. Li, D.; Jones, G.L.; Dunlap, J.R.; Hua, F.; Zhao, B. Thermosensitive hairy hybrid nanoparticles synthesized by surface-initiated atom transfer radical polymerization. *Langmuir* **2006**, *22*, 3344–3351.
64. Li, D.; Zhao, B. Temperature-induced transport of thermosensitive hairy hybrid nanoparticles between aqueous and organic phases. *Langmuir* **2007**, *23*, 2208–2217.
65. Piech, M.; Bell, N.S. Controlled synthesis of photochromic polymer brushes by atom transfer radical polymerization. *Macromolecules* **2006**, *39*, 915–922.
66. Mori, H.; Seng, D.C.; Zhang, M.; Müller, A.H.E. Hybrid nanoparticles with hyperbranched polymer shells *via* self-condensing atom transfer radical polymerization from silica surfaces. *Langmuir* **2002**, *18*, 3682–3693.
67. Morinaga, T.; Ohkura, M.; Ohno, K.; Tsujii, Y.; Fukuda, T. Monodisperse silica particles grafted with concentrated oxetane-carrying polymer brushes: Their synthesis by surface-initiated atom transfer radical polymerization and use for fabrication of hollow spheres. *Macromolecules* **2007**, *40*, 1159–1164.
68. Nagase, K.; Kobayashi, J.; Kikuchi, A.; Akiyama, Y.; Kanazawa, H.; Okano, T. Effects of graft densities and chain lengths on separation of bioactive compounds by nanolayered thermoresponsive polymer brush surfaces. *Langmuir* **2007**, *24*, 511–517.
69. Saleh, N.; Sarbu, T.; Sirk, K.; Lowry, G.V.; Matyjaszewski, K.; Tilton, R.D. Oil-in-Water emulsions stabilized by highly charged polyelectrolyte-grafted silica nanoparticles. *Langmuir* **2005**, *21*, 9873–9878.
70. Schepelina, O.; Zharov, I. Polymer-Modified opal nanopores. *Langmuir* **2006**, *22*, 10523–10527.
71. Schepelina, O.; Zharov, I. PNIPAAM-Modified nanoporous colloidal films with positive and negative temperature gating. *Langmuir* **2007**, *23*, 12704–12709.

72. Mandal, T.K.; Fleming, M.S.; Walt, D.R. Production of hollow polymeric microspheres by surface-confined living radical polymerization on silica templates. *Chem. Mater.* **2000**, *12*, 3481–3487.
73. Fu, G.D.; Zhao, J.P.; Sun, Y.M.; Kang, E.T.; Neoh, K.G. Conductive hollow nanospheres of polyaniline *via* surface-initiated atom transfer radical polymerization of 4-vinylaniline and oxidative graft copolymerization of aniline. *Macromolecules* **2007**, *40*, 2271–2275.
74. Laruelle, G.; Parvole, J.; Francois, J.; Billon, L. Block copolymer grafted-silica particles: A core/double shell hybrid inorganic/organic material. *Polymer* **2004**, *45*, 5013–5020.
75. Parvole, J.; Montfort, J.P.; Billon, L. Formation of inorganic/organic nanocomposites by nitroxide-mediated polymerization in bulk using a bimolecular system. *Macromol. Chem. Phys.* **2004**, *205*, 1369–1378.
76. Inoubli, R.; Dagréou, S.; Delville, M.H.; Lapp, A.; Peyrelasse, J.; Billon, L. *In situ* thermo-dependant trapping of carbon radicals: A versatile route to well-defined polymer-grafted silica nanoparticles. *Soft Matter* **2007**, *3*, 1014–1024.
77. Sonnenberg, L.; Parvole, J.; Borisov, O.; Billon, L.; Gaub, H.E.; Seitz, M. AFM-Based single molecule force spectroscopy of end-grafted poly(acrylic acid) monolayers. *Macromolecules* **2006**, *39*, 281–288.
78. Bartholome, C.; Beyou, E.; Bourgeat-Lami, E.; Chaumont, P.; Zydowicz, N. Nitroxide-mediated polymerization of styrene initiated from the surface of fumed silica. Comparison of two synthetic routes. *Polymer* **2005**, *46*, 8502–8510.
79. Parvole, J.; Lamelle, G.; Khoukh, A.; Billon, L. Surface initiated polymerization of poly (butyl acrylate) by nitroxide mediated polymerization: First comparative polymerization of a bimolecular and a unimolecular initiator-grafted silica particles. *Macromol. Chem. Phys.* **2005**, *206*, 372–382.
80. Bartholome, C.; Beyou, E.; Bourgeat-Lami, E.; Cassagnau, P.; Chaumont, P.; David, L.; Zydowicz, N. Viscoelastic properties and morphological characterization of silica/polystyrene nanocomposites synthesized by nitroxide-mediated polymerization. *Polymer* **2005**, *46*, 9965–9973.
81. Bartholome, C.; Beyou, E.; Bourgeat-Lami, E.; Chaumont, P.; Lefebvre, F.; Zydowicz, N. Nitroxide-Mediated polymerization of styrene initiated from the surface of silica nanoparticles. *In situ* generation and grafting of alkoxyamine initiators. *Macromolecules* **2005**, *38*, 1099–1106.
82. Bartholome, C.; Beyou, E.; Bourgeat-Lami, E.; Chaumont, P.; Zydowicz, N. Nitroxide-Mediated polymerizations from silica nanoparticle surfaces: "Graft from" polymerization of styrene using a triethoxysilyl-terminated alkoxyamine initiator. *Macromolecules* **2003**, *36*, 7946–7952.
83. Parvole, J.; Billon, L.; Montfort, J.P. Formation of polyacrylate brushes on silica surfaces. *Polym. Int.* **2002**, *51*, 1111–1116.
84. Parvole, J.; Laruelle, G.; Guimon, C.; Francois, J.; Billon, L. Initiator-Grafted silica particles for controlled free radical polymerization: Influence of the initiator structure on the grafting density. *Macromol. Rapid Commun.* **2003**, *24*, 1074–1078.
85. Inoubli, R.; Dagréou, S.; Khoukh, A.; Roby, F.; Peyrelasse, J.; Billon, L. "Graft from" polymerization on colloidal silica particles: Elaboration of alkoxyamine grafted surface by *in situ* trapping of carbon radicals. *Polymer* **2005**, *46*, 2486–2496.

86. Bartholome, C.; Beyou, E.; Bourgeat-Lami, E.; Chaumont, P.; Zydowicz, N. Nitroxide-mediated polymerization of styrene initiated from the surface of fumed silica. Comparison of two synthetic routes. *Polymer* **2005**, *46*, 8502–8510.
87. Deleuze, C.; Delville, M.H.; Pellerin, V.; Derail, C.; Billon, L. Hybrid core-soft shell particles as adhesive elementary building blocks for colloidal crystals. *Macromolecules* **2009**, *42*, 5303–5309.
88. Li, C.; Benicewicz, B.C. Synthesis of well-defined polymer brushes grafted onto silica nanoparticles *via* surface reversible addition-fragmentation chain transfer polymerization. *Macromolecules* **2005**, *38*, 5929–5936.
89. Zhao, Y.; Perrier, S. Reversible addition-fragmentation chain transfer graft polymerization mediated by fumed silica supported chain transfer agents. *Macromolecules* **2007**, *40*, 9116–9124.
90. Li, C.; Han, J.; Ryu, C.Y.; Benicewicz, B.C. A versatile method to prepare RAFT agent anchored substrates and the preparation of PMMA grafted nanoparticles. *Macromolecules* **2006**, *39*, 3175–3183.
91. Hua, D.; Tang, J.; Jiang, J.; Gu, Z.; Dai, L.; Zhu, X. Controlled grafting modification of silica gel *via* RAFT polymerization under ultrasonic irradiation. *Mater. Chem. Phys.* **2009**, *114*, 402–406.
92. Yang, Y.; Yang, Z.; Zhao, Q.; Cheng, X.; Tjong, S.C.; Li, R.K.Y.; Wang, X.; Xie, X. Immobilization of RAFT agents on silica nanoparticles utilizing an alternative functional group and subsequent surface-initiated RAFT polymerization. *J. Polym. Sci. A* **2009**, *47*, 467–484.
93. Zhao, Y.; Perrier, S. Synthesis of well-defined homopolymer and diblock copolymer grafted onto silica particles by Z-supported RAFT polymerization. *Macromolecules* **2006**, *39*, 8603–8608.
94. Zhao, Y.; Perrier, S. Reversible addition−fragmentation chain transfer graft polymerization mediated by fumed silica supported chain transfer agents. *Macromolecules* **2007**, *40*, 9116–9124.
95. Matyjaszewski, K.; Xia, J. Atom transfer radical polymerization. *Chem. Rev.* **2001**, *101*, 2921–2990.
96. Zhang, Y.; Ye, Z. Covalent surface grafting of branched polyethylenes on silica nanoparticles by surface-initiated ethylene "living" polymerization with immobilized Pd-diimine catalysts. *Macromolecules* **2008**, *41*, 6331–6338.
97. Tian, B.-S.; Yang, C. Temperature-Responsive nanocomposites based on mesoporous SBA-15 silica and PNIPAAm: Synthesis and characterization. *J. Phys. Chem. C* **2009**, *113*, 4925–4931.
98. Chung, P.W.; Kumar, R.; Pruski, M.; Lin, V.S.Y. Temperature responsive solution partition of organic-inorganic hybrid poly(N-isopropylacrylamide)-coated mesoporous silica nanospheres. *Adv. Funct. Mater.* **2008**, *18*, 1390–1398.
99. Wagner, M.; Brochard-Wyart, F.; Hervet, H.; de Gennes, P.G. Colloid & Polymer Science - Collapse of polymer brushes induced by n-clusters. *Colloid Polym. Sci.* **1993**, *271*, 621–628.
100. Motornov, M.; Zhou, J.; Pita, M.; Gopishetty, V.; Tokarev, I.; Katz, E.; Minko, S. "Chemical transformers" from nanoparticle ensembles operated with logic. *Nano Lett.* **2008**, *8*, 2993–2997.
101. Bell, N.S.; Piech, M. Photophysical effects between spirobenzopyran-methyl methacrylate-functionalized colloidal particles. *Langmuir* **2006**, *22*, 1420–1427.
102. Piech, M.; George, M.C.; Bell, N.S.; Braun, P.V. Patterned colloid assembly by grafted photochromic polymer layers. *Langmuir* **2006**, *22*, 1379–1382.

103. George, M.C.; Mohraz, A.; Piech, M.; Bell, N.S.; Lewis, J.A.; Braun, P.V. Direct laser writing of photoresponsive colloids for microscale patterning of 3D porous structures. *Adv. Mater.* **2009**, *21*, 66–70.
104. Brown, A.A.; Azzaroni, O.; Huck, W.T.S. Photoresponsive polymer brushes for hydrophilic patterning. *Langmuir* **2009**, *25*, 1744–1749.
105. Fu, Q.; Rao, G.V.R.; Ista, L.K.; Wu, Y.; Andrzejewski, B.P.; Sklar, L.A.; Ward, T.L.; López, G.P. Control of molecular transport through stimuli-responsive ordered mesoporous materials. *Adv. Mater.* **2003**, *15*, 1262–1266.
106. You, Y.-Z.; Kalebaila, K.K.; Brock, S.L.; Oupicky, D. Temperature-Controlled uptake and release in PNIPAM-modified porous silica nanoparticles. *Chem. Mater.* **2008**, *20*, 3354–3359.
107. Zhou, L.; Yuan, W.; Yuan, J.; Hong, X. Preparation of double-responsive SiO_2-*g*-PDMAEMA nanoparticles *via* ATRP. *Mater. Lett.* **2008**, *62*, 1372–1375.
108. Guo, X.; Ballauff, M. Spherical polyelectrolyte brushes: Comparison between annealed and quenched brushes. *Phys. Rev. E Stat. Nonlin. Soft Matter Phys.* **2001**, *64*, 0514061.
109. Zhang, J.; Jin, J.; Zhao, H. Surface-initiated free radical polymerization at the liquid-liquid interface: A one-step approach for the synthesis of amphiphilic Janus silica particles. *Langmuir* **2009**, *25*, 6431–6437.
110. Hong, L.; Jiang, S.; Granick, S. Simple method to produce Janus colloidal particles in large quantity. *Langmuir* **2006**, *22*, 9495–9499.
111. Kopesky, E.T.; Haddad, T.S.; Cohen, R.E.; McKinley, G.H. Thermomechanical properties of poly(methyl methacrylate)s containing tethered and untethered polyhedral oligomeric silsesquioxanes. *Macromolecules* **2004**, *37*, 8992–9004.
112. Li, G.Z.; Wang, L.; Toghiani, H.; Daulton, T.L.; Koyama, K.; Pittman, C.U. Viscoelastic and mechanical properties of epoxy/multifunctional polyhedral oligomeric silsesquioxane nanocomposites and epoxy/ladderlike polyphenylsilsesquioxane blends. *Macromolecules* **2001**, *34*, 8686–8693.
113. Strachota, A.; Kroutilova, I.; Kovarova, J.; Matejka, L. Epoxy networks reinforced with polyhedral oligomeric silsesquioxanes (POSS). Thermomechanical properties. *Macromolecules* **2004**, *37*, 9457–9464.
114. Matejka, L.; Strachota, A.; Plestil, J.; Whelan, P.; Steinhart, M.; Slouf, M. Epoxy networks reinforced with polyhedral oligomeric silsesquioxanes (POSS). Structure and morphology. *Macromolecules* **2004**, *37*, 9449–9456.
115. Waddon, A.J.; Zheng, L.; Farris, R.J.; Coughlin, E.B. Nanostructured polyethylene-POSS copolymers: Control of crystallization and aggregation. *Nano Lett.* **2002**, *2*, 1149–1155.
116. Tamaki, R.; Choi, J.; Laine, R.M. A polyimide nanocomposite from octa(aminophenyl)-silsesquioxane. *Chem. Mater.* **2003**, *15*, 793–797.
117. Brown, J.F. The polycondensation of phenylsilanetriol. *J. Am. Chem. Soc.* **1965**, *87*, 4317–4324.
118. Sprung, M.M.; Guenther, F.O. The partial hydrolysis of ethyltrietboxysilane. *J. Am. Chem. Soc.* **1955**, *77*, 3996–4002.
119. Pyun, J.; Matyjaszewski, K.; Kowalewski, T.; Savin, D.; Patterson, G.; Kickelbick, G.; Huesing, N. Synthesis of well-defined block copolymers tethered to polysilsesquioxane nanoparticles and their nanoscale morphology on surfaces. *J. Am. Chem. Soc.* **2001**, *123*, 9445–9446.

120. Costa, R.O.R.; Vasconcelos, W.L.; Tamaki, R.; Laine, R.M. Organic/inorganic nanocomposite star polymers *via* atom transfer radical polymerization of methyl methacrylate using octafunctional silsesquioxane cores. *Macromolecules* **2001**, *34*, 5398–5407.
121. Ge, Z.; Wang, D.; Zhou, Y.; Liu, H.; Liu, S. Synthesis of organic/inorganic hybrid quatrefoil-shaped star-cyclic polymer containing a polyhedral oligomeric silsesquioxane core. *Macromolecules* **2009**, *42*, 2903–2910.
122. Pyun, J.; Matyjaszewski, K. The synthesis of hybrid polymers using atom transfer radical polymerization: homopolymers and block copolymers from polyhedral oligomeric silsesquioxane monomers. *Macromolecules* **2000**, *33*, 217–220.
123. Pyun, J.; Matyjaszewski, K.; Wu, J.; Kim, G.M.; Chun, S.B.; Mather, P.T. ABA triblock copolymers containing polyhedral oligomeric silsesquioxane pendant groups: Synthesis and unique properties. *Polymer* **2003**, *44*, 2739–2750.
124. Ohno, K.; Sugiyama, S.; Koh, K.; Tsujii, Y.; Fukuda, T.; Yamahiro, M.; Oikawa, H.; Yamamoto, Y.; Ootake, N.; Watanabe, K. Living radical polymerization by polyhedral oligomeric silsesquioxane-holding initiators: precision synthesis of tadpole-shaped organic/inorganic hybrid polymers. *Macromolecules* **2004**, *37*, 8517–8522.
125. Ro, H.W.; Kim, K.J.; Theato, P.; Gidley, D.W.; Yoon, D.Y. Novel inorganic-organic hybrid block copolymers as pore generators for nanoporous ultralow-dielectric-constant films. *Macromolecules* **2005**, *38*, 1031–1034.
126. Huang, C.-F.; Kuo, S.-W.; Lin, F.-J.; Huang, W.-J.; Wang, C.-F.; Chen, W.-Y.; Chang, F.-C. Influence of PMMA-chain-end tethered polyhedral oligomeric silsesquioxanes on the miscibility and specific interaction with phenolic blends. *Macromolecules* **2006**, *39*, 300–308.
127. Hirai, T.; Leolukman, M.; Hayakawa, T.; Kakimoto, M.A.; Gopalan, P. Hierarchical nanostructures of organosilicate nanosheets within self-organized block copolymer films. *Macromolecules* **2008**, *41*, 4558–4560.
128. Hirai, T.; Leolukman, M.; Liu, C.C.; Han, E.; Kim, Y.J.; Ishida, Y.; Hayakawa, T.; Kakimoto, M.-a.; Nealey, P.F.; Gopalan, P. One-Step direct-patterning template utilizing self-assembly of POSS-containing block copolymers. *Adv. Mater.* **2009**, *21*, 4334–4338.
129. Hirai, T.; Leolukman, M.; Jin, S.; Goseki, R.; Ishida, Y.; Kakimoto, M.-a.; Hayakawa, T.; Ree, M.; Gopalan, P. Hierarchical self-assembled structures from POSS-containing block copolymers synthesized by living anionic polymerization. *Macromolecules* **2009**, *42*, 8835–8843.
130. Koh, K.; Sugiyama, S.; Morinaga, T.; Ohno, K.; Tsujii, Y.; Fukuda, T.; Yamahiro, M.; Iijima, T.; Oikawa, H.; Watanabe, K.; Miyashita, T. Precision synthesis of a fluorinated polyhedral oligomeric silsesquioxane-terminated polymer and surface characterization of its blend film with poly(methyl methacrylate). *Macromolecules* **2005**, *38*, 1264–1270.
131. Zhang, W.; Fang, B.; Walther, A.; Muller, A.H.E. Synthesis *via* RAFT polymerization of tadpole-shaped organic/inorganic hybrid poly(acrylic acid) containing polyhedral oligomeric silsesquioxane (POSS) and their self-assembly in water. *Macromolecules* **2009**, *42*, 2563–2569.
132. Zhang, W.; Liu, L.; Zhuang, X.; Li, X.; Bai, J.; Chen, Y. Synthesis and self-assembly of tadpole-shaped organic/inorganic hybrid poly(N-isopropylacrylamide) containing polyhedral oligomeric silsesquioxane *via* RAFT polymerization. *J. Polym. Sci. A* **2008**, *46*, 7049–7061.

133. Cardoen, G.; Coughlin, E.B. Hemi-Telechelic polystyrene-POSS copolymers as model systems for the study of well-defined inorganic/organic hybrid materials. *Macromolecules* **2004**, *37*, 5123–5126.
134. Kim, B.-S.; Mather, P.T. Amphiphilic telechelics incorporating polyhedral oligosilsesquioxane: 1. synthesis and characterization. *Macromolecules* **2002**, *35*, 8378–8384.
135. Kim, B.-S.; Mather, P.T. Morphology, microstructure, and rheology of amphiphilic telechelics incorporating polyhedral oligosilsesquioxane. *Macromolecules* **2006**, *39*, 9253–9260.
136. Kim, B.S.; Mather, P.T. Amphiphilic telechelics with polyhedral oligosilsesquioxane (POSS) end-groups: Dilute solution viscometry. *Polymer* **2006**, *47*, 6202–6207.
137. Maliakal, A.; Katz, H.; Cotts, P.M.; Subramoney, S.; Mirau, P. Inorganic oxide core, polymer shell nanocomposite as a high K gate dielectric for flexible electronics applications. *J. Am. Chem. Soc.* **2005**, *127*, 14655–14662.
138. Tahir, M.N.; Eberhardt, M.; Theato, P.; Faiβ, S.; Janshoff, A.; Gorelik, T.; Kolb, U.; Tremel, W. Reactive polymers: A versatile toolbox for the immobilization of functional molecules on TiO_2 nanoparticles. *Angew. Chem. Int. Ed.* **2006**, *45*, 908–912.
139. Zorn, M.; Meuer, S.; Tahir, M.N.; Khalavka, Y.; Sönnichsen, C.; Tremel, W.; Zentel, R. Liquid crystalline phases from polymer functionalised semiconducting nanorods. *J. Mater. Chem.* **2008**, *18*, 3050–3058.
140. Meuer, S.; Oberle, P.; Theato, P.; Tremel, W.; Zentel, R. Liquid crystalline phases from polymer-functionalized TiO_2 nanorods. *Adv. Mater.* **2007**, *19*, 2073–2078.
141. Zhang, Q.-L.; Du, L.-C.; Weng, Y.-X.; Wang, L.; Chen, H.-Y.; Li, J.-Q. Particle-Size-Dependent distribution of carboxylate adsorption sites on TiO_2 nanoparticle surfaces: Insights into the surface modification of nanostructured TiO_2 electrodes. *J. Phys. Chem. B* **2004**, *108*, 15077–15083.
142. Hojjati, B.; Sui, R.; Charpentier, P.A. Synthesis of TiO_2/PAA nanocomposite by RAFT polymerization. *Polymer* **2007**, *48*, 5850–5858.
143. Hojjati, B.; Charpentier, P.A. Synthesis and kinetics of graft polymerization of methyl methacrylate from the RAFT coordinated surface of nano-TiO_2. *J. Polym. Sci. A* **2008**, *46*, 3926–3937.
144. Lowes, B.J.; Bohrer, A.G.; Tran, T.; Shipp, D.A. Grafting of polystyrene "from" and "through" surface modified titania nanoparticles. *Polym. Bull.* **2009**, *62*, 281–289.
145. Clearfield, A. Metal phosphonate chemistry. In *Progress in Inorganic Chemistry*; John Wiley & Sons: Hoboken, NJ, USA, 1998; pp. 371–510.
146. Matsuno, R.; Otsuka, H.; Takahara, A. Polystyrene-grafted titanium oxide nanoparticles prepared through surface-initiated nitroxide-mediated radical polymerization and their application to polymer hybrid thin films. *Soft Matter* **2006**, *2*, 415–421.
147. Kobayashi, M.; Matsuno, R.; Otsuka, H.; Takahara, A. Precise surface structure control of inorganic solid and metal oxide nanoparticles through surface-initiated radical polymerization. *Sci. Technol. Adv. Mater.* **2006**, *7*, 617–628.
148. Rajh, T.; Saponjic, Z.; Liu, J.; Dimitrijevic, N.M.; Scherer, N.F.; Vega-Arroyo, M.; Zapol, P.; Curtiss, L.A.; Thurnauer, M.C. Charge transfer across the nanocrystalline-DNA interface: Probing DNA recognition. *Nano Lett.* **2004**, *4*, 1017–1023.

149. Park, J.T.; Koh, J.H.; Koh, J.K.; Kim, J.H. Surface-initiated atom transfer radical polymerization from TiO_2 nanoparticles. *Appl. Surf. Sci.* **2009**, *255*, 3739–3744.
150. Fan, X.; Lin, L.; Messersmith, P.B. Surface-initiated polymerization from TiO_2 nanoparticle surfaces through a biomimetic initiator: A new route toward polymer-matrix nanocomposites. *Compos. Sci. Technol.* **2006**, *66*, 1195–1201.
151. Ye, Q.; Wang, X.; Hu, H.; Wang, D.; Li, S.; Zhou, F. Polyelectrolyte brush templated multiple loading of Pd nanoparticles onto TiO_2 nanowires *via* regenerative counterion exchange-reduction. *J. Phys. Chem. C* **2009**, *113*, 7677–7683.
152. Kickelbick, G.; Holzinger, D.; Brick, C.; Trimmel, G.; Moons, E. Hybrid inorganic-organic core-shell nanoparticles from surface-functionalized titanium, zirconium, and vanadium oxo clusters. *Chem. Mater.* **2002**, *14*, 4382–4389.
153. Holzinger, D.; Kickelbick, G. Preparation of amorphous metal-oxide-core polymer-shell nanoparticles *via* a microemulsion-based sol−gel approach. *Chem. Mater.* **2003**, *15*, 4944–4948.
154. Liu, P.; Wang, T. Surface-graft hyperbranched polymer *via* self-condensing atom transfer radical polymerization from zinc oxide nanoparticles. *Polym. Eng. Sci.* **2007**, *47*, 1296–1301.
155. Liu, P.; Wang, T. Poly(hydroethyl acrylate) grafted from ZnO nanoparticles *via* surface-initiated atom transfer radical polymerization. *Curr. Appl Phys.* **2008**, *8*, 66–70.
156. Sato, M.; Kawata, A.; Morito, S.; Sato, Y.; Yamaguchi, I. Preparation and properties of polymer/zinc oxide nanocomposites using functionalized zinc oxide quantum dots. *Eur. Polym. J.* **2008**, *44*, 3430–3438.
157. Peng, X.; Chen, Y.; Li, F.; Zhou, W.; Hu, Y. Preparation and optical properties of ZnO@PPEGMA nanoparticles. *Appl. Surf. Sci.* **2009**, *255*, 7158–7163.
158. Yan, F.; Xu, H.; Anker, J.; Kopelman, R.; Ross, B.; Rehemtulla, A.; Reddy, R. Synthesis and characterization of silica-embedded iron oxide nanoparticles for magnetic resonance imaging. *J. Nanosci. Nanotechnol.* **2004**, *4*, 72–76.
159. Perez, J.M.; Simeone, F.J.; Tsourkas, A.; Josephson, L.; Weissleder, R. Peroxidase substrate nanosensors for MR imaging. *Nano Lett.* **2003**, *4*, 119–122.
160. Bulte, J.W.M. Magnetic nanoparticles as markers for cellular MR imaging. *J. Magn. Magn. Mater.* **2005**, *289*, 423–427.
161. Lacava, L.M.; Lacava, Z.G.M.; Da Silva, M.F.; Silva, O.; Chaves, S.B.; Azevedo, R.B.; Pelegrini, F.; Gansau, C.; Buske, N.; Sabolovic, D.; Morais, P.C. Magnetic resonance of a dextran-coated magnetic fluid intravenously administered in mice. *Biophys. J.* **2001**, *80*, 2483–2486.
162. Häfeli, U.O. Magnetically modulated therapeutic systems. *Int. J. Pharm.* **2004**, *277*, 19–24.
163. Lübbe, A.S.; Bergemann, C.; Riess, H.; Schriever, F.; Reichardt, P.; Possinger, K.; Matthias, M.; Dörken, B.; Herrmann, F.; Gürtler, R.; Hohenberger, P.; Haas, N.; Sohr, R.; Sander, B.; Lemke, A.J.; Ohlendorf, D.; Huhnt, W.; Huhn, D. Clinical experiences with magnetic drug targeting: A phase I study with 4'-epidoxorubicin in 14 patients with advanced solid tumors. *Cancer Res.* **1996**, *56*, 4686–4693.
164. Hilger, I.; Katrin, F.; Andrä, W.; Hiergeist, R.; Hergt, R.; Kaiser, W.A. Heating potential of iron oxides for therapeutic purposes in interventional radiology. *Acad. Radiol.* **2002**, *9*, 198–202.

165. Moroz, P.; Jones, S.K.; Gray, B.N. Magnetically mediated hyperthermia: Current status and future directions. *Int. J. Hyperth.* **2002**, *18*, 267–284.

166. Jordan, A.; Scholz, R.; Wust, P.; Fähling, H.; Felix, R. Magnetic fluid hyperthermia (MFH): Cancer treatment with AC magnetic field induced excitation of biocompatible superparamagnetic nanoparticles. *J. Magn. Magn. Mater.* **1999**, *201*, 413–419.

167. Nam, J.M.; Thaxton, C.S.; Mirkin, C.A. Nanoparticle-based bio-bar codes for the ultrasensitive detection of proteins. *Science* **2003**, *301*, 1884–1886.

168. Patolsky, F.; Weizmann, Y.; Katz, E.; Willner, I. Magnetically amplified DNA assays (MADA): Sensing of viral DNA and single-base mis-matches by using nucleic acid modified magnetic particles. *Angew. Chem. Int. Ed.* **2003**, *42*, 2372–2376.

169. Gu, H.; Ho, P.-L.; Tsang, K.W.T.; Wang, L.; Xu, B. Using biofunctional magnetic nanoparticles to capture vancomycin-resistant enterococci and other gram-positive bacteria at ultralow concentration. *J. Am. Chem. Soc.* **2003**, *125*, 15702–15703.

170. Teunissen, W.; De Groot, F.M.F.; Geus, J.; Stephan, O.; Tence, M.; Colliex, C. The structure of carbon encapsulated NiFe nanoparticles. *J. Catal.* **2001**, *204*, 169–174.

171. Lu, A.H.; Schmidt, W.; Matoussevitch, N.; Bönnemann, H.; Spliethoff, B.; Tesche, B.; Bill, E.; Kiefer, W.; Schuth, F. Nanoengineering of a magnetically separable hydrogenation catalyst. *Angew. Chem. Int. Ed.* **2004**, *43*, 4303–4306.

172. Saoud, F.M.; Tonge, M.P.; Weber, W.G.; Sanderson, R.D. Magnetite nanoparticles for the preparation of ultrapure RAFT polymers. *Macromolecules* **2008**, *41*, 1598–1600.

173. García, I.; Zafeiropoulos, N.E.; Janke, A.; Tercjak, A.; Eceiza, A.; Stamm, M.; Mondracon, I. Functionalization of iron oxide magnetic nanoparticles with poly(methyl methacrylate) brushes *via* grafting-from atom transfer radical polymerization. *J. Polym. Sci. A* **2007**, *45*, 925–932.

174. Marutani, E.; Yamamoto, S.; Ninjbadgar, T.; Tsujii, Y.; Fukuda, T.; Takano, M. Surface-initiated atom transfer radical polymerization of methyl methacrylate on magnetite nanoparticles. *Polymer* **2004**, *45*, 2231–2235.

175. García, I.; Tercjak, A.; Zafeiropoulos, N.E.; Stamm, M.; Mondragon, I. Self-assembling nanomaterials using magnetic nanoparticles modified with polystyrene brushes. *Macromol. Rapid Commun.* **2007**, *28*, 2361–2365.

176. García, I.; Tercjak, A.; Rueda, L.; Mondragon, I. Self-assembled nanomaterials using magnetic nanoparticles modified with polystyrene brushes and poly(styrene-b-butadiene-b-styrene). *Macromolecules* **2008**, *41*, 9295–9298.

177. Ninjbadgar, T.; Yamamoto, S.; Fukuda, T. Synthesis and magnetic properties of the γ-Fe_2O_3/poly-(methyl methacrylate)-core/shell nanoparticles. *Solid State Sci.* **2004**, *6*, 879–885.

178. Gravano, S.M.; Dumas, R.; Liu, K.; Patten, T.E. Methods for the surface functionalization of γ-Fe_2O_3 nanoparticles with initiators for atom transfer radical polymerization and the formation of core-shell inorganic-polymer structures. *J. Polym. Sci. A* **2005**, *43*, 3675–3688.

179. Lattuada, M.; Hatton, T.A. Functionalization of monodisperse magnetic nanoparticles. *Langmuir* **2007**, *23*, 2158–2168.

180. Duan, H.; Kuang, M.; Wang, D.; Kurth, D.G.; Möhwald, H. Colloidally stable amphibious nanocrystals derived from poly[(2-(dimethylamino)ethyl) methacrylate] capping. *Angew. Chem. Int. Ed.* **2005**, *44*, 1717–1720.

181. Wang, W.C.; Neoh, K.G.; Kong, E.T. Surface functionalization of Fe_3O_4 magnetic nanoparticles *via* RAFT-mediated graft polymerization. *Macromol. Rapid Commun.* **2006**, *27*, 1665–1669.
182. Wang, Y.; Teng, X.; Wang, J.-S.; Yang, H. Solvent-free atom transfer radical polymerization in the synthesis of Fe_2O_3@polystyrene core-shell nanoparticles. *Nano Lett.* **2003**, *3*, 789–793.
183. Vestal, C.R.; Zhang, Z.J. Atom transfer radical polymerization synthesis and magnetic characterization of $MnFe_2O_4$/polystyrene core/shell nanoparticles. *J. Am. Chem. Soc.* **2002**, *124*, 14312–14313.
184. Li, G.; Fan, J.; Jiang, R.; Gao, Y. Cross-linking the linear polymeric chains in the ATRP synthesis of iron oxide/polystyrene core/shell nanoparticles. *Chem. Mater.* **2004**, *16*, 1835–1837.
185. Hu; Neoh, K.G.; Cen, L.; Kang, E.-T. Cellular response to magnetic nanoparticles "PEGylated" *via* surface-initiated atom transfer radical polymerization. *Biomacromolecules* **2006**, *7*, 809–816.
186. Gelbrich, T.; Feyen, M.; Schmidt, A.M. Magnetic thermoresponsive core-shell nanoparticles. *Macromolecules* **2006**, *39*, 3469–3472.
187. Kaiser, A.; Gelbrich, T.; Schmidt, A.M. Thermosensitive magnetic fluids. *J. Phys. Condens. Matter* **2006**, *18*, S2563-S2580.
188. Kaiser, A.; Liu, T.; Richtering, W.; Schmidt, A.M. Magnetic capsules and pickering emulsions stabilized by core-shell particles. *Langmuir* **2009**, *25*, 7335–7341.
189. Kaiser, A.; Schmidt, A.M. Phase behavior of polystyrene-brush-coated nanoparticles in cyclohexane. *J. Phys. Chem. B* **2008**, *112*, 1894–1898.
190. Xu, C.; Ohno, K.; Ladmiral, V.; Milkie, D.E.; Kikkawa, J.M.; Composto, R.J. Simultaneous block copolymer and magnetic nanoparticle assembly in nanocomposite films. *Macromolecules* **2009**, *42*, 1219–1228.
191. García, I.; Tercjak, A.; Gutierrez, J.; Rueda, L.; Mondragon, I. Nanostructuration *via* solvent vapor exposure of poly(2-vinyl pyridine-b-methyl methacrylate) nanocomposites using modified magnetic nanoparticles. *J. Phys. Chem. C* **2008**, *112*, 14343–14347.
192. Lei, Z.; Ren, N.; Li, Y.; Li, N.; Mu, B. Fe_3O_4/SiO_2-g-PSStNa polymer nanocomposite microspheres (PNCMs) from a surface-initiated atom transfer radical polymerization (SI-ATRP) Approach for pectinase immobilization. *J. Agric. Food Chem.* **2009**, *57*, 1544–1549.
193. Lattuada, M.; Hatton, T.A. Preparation and controlled self-assembly of Janus magnetic nanoparticles. *J. Am. Chem. Soc.* **2007**, *129*, 12878–12889.
194. Huggins, M.L. Theory of solutions of high polymers. *J. Am. Chem. Soc.* **2002**, *64*, 1712–1719.
195. Edwards, E.W.; Chanana, M.; Wang, D. Capping gold nanoparticles with stimuli-responsive polymers to cross water-oil interfaces: In-depth insight to the trans-interfacial activity of nanoparticles. *J. Phys. Chem. C* **2008**, *112*, 15207–15219.
196. Wakamatsu, H.; Yamamoto, K.; Nakao, A.; Aoyagi, T. Preparation and characterization of temperature-responsive magnetite nanoparticles conjugated with N-isopropylacrylamide-based functional copolymer. *J. Magn. Magn. Mater.* **2006**, *302*, 327–333.
197. Isojima, T.; Lattuada, M.; Vander Sande, J.B.; Hatton, T.A. Reversible clustering of pH- and temperature-responsive Janus magnetic nanoparticles. *ACS Nano* **2008**, *2*, 1799–1806.
198. Hicks, J.F.; Zamborini, F.P.; Osisek, A.J.; Murray, R.W. The Dynamics of electron self-exchange between nanoparticles. *J. Am. Chem. Soc.* **2001**, *123*, 7048–7053.

199. Hicks, J.F.; Miles, D.T.; Murray, R.W. Quantized double-layer charging of highly monodisperse metal nanoparticles. *J. Am. Chem. Soc.* **2002**, *124*, 13322–13328.
200. Chen, S.; Murray, R.W. Electrochemical quantized capacitance charging of surface ensembles of gold nanoparticles. *J. Phys. Chem. B* **1999**, *103*, 9996–10000.
201. Hicks, J.F.; Zamborini, F.P.; Murray, R.W. Dynamics of electron transfers between electrodes and monolayers of nanoparticles. *J. Phys. Chem. B* **2002**, *106*, 7751–7757.
202. Chen, S.; Murray, R.W.; Feldberg, S.W. Quantized capacitance charging of monolayer-protected Au clusters. *J. Phys. Chem. B* **1998**, *102*, 9898–9907.
203. Daniel, M.-C.; Astruc, D. Gold nanoparticles: Assembly, supramolecular chemistry, quantum-size-related properties, and applications toward biology, catalysis, and nanotechnology. *Chem. Rev.* **2003**, *104*, 293–346.
204. Shipway, A.N.; Katz, E.; Willner, I. Nanoparticle arrays on surfaces for electronic, optical, and sensor applications. *ChemPhysChem* **2000**, *1*, 18–52.
205. Vossmeyer, T.; Guse, B.; Besnard, I.; Bauer, R.E.; Müllen, K.; Yasuda, A. Gold nanoparticle/polyphenylene dendrimer composite films: Preparation and vapor-sensing properties. *Adv. Mater.* **2002**, *14*, 238–242.
206. Storhoff, J.J.; Elghanian, R.; Mucic, R.C.; Mirkin, C.A.; Letsinger, R.L. One-Pot colorimetric differentiation of polynucleotides with single base imperfections using gold nanoparticle probes. *J. Am. Chem. Soc.* **1998**, *120*, 1959–1964.
207. Wohltjen, H.; Snow, A.W. Colloidal metal−insulator−metal ensemble chemiresistor sensor. *Anal. Chem.* **1998**, *70*, 2856–2859.
208. Demaille, C.; Brust, M.; Tsionsky, M.; Bard, A.J. Fabrication and characterization of self-assembled spherical gold ultramicroelectrodes. *Anal. Chem.* **1997**, *69*, 2323–2328.
209. Al-Rawashdeh, N.A.F.; Sandrock, M.L.; Seugling, C.J.; Foss, C.A. Visible region polarization spectroscopic studies of template-synthesized gold nanoparticles oriented in polyethylene. *J. Phys. Chem. B* **1998**, *102*, 361–371.
210. Evans, S.D.; Johnson, S.R.; Cheng, Y.L.; Shen, T. Vapour sensing using hybrid organic-inorganic nanostructured materials. *J. Mater. Chem.* **2000**, *10*, 183–188.
211. Maier, S.A.; Brongersma, M.L.; Kik, P.G.; Meltzer, S.; Requicha, A.A.G.; Atwater, H.A. Plasmonics–A route to nanoscale optical devices. *Adv. Mater.* **2001**, *13*, 1501–1505.
212. Shukla, R.; Bansal, V.; Chaudhary, M.; Basu, A.; Bhonde, R.R.; Sastry, M. Biocompatibility of gold nanoparticles and their endocytotic fate inside the cellular compartment: A microscopic overview. *Langmuir* **2005**, *21*, 10644–10654.
213. Connor, E.E.; Mwamuka, J.; Gole, A.; Murphy, C.J.; Wyatt, M.D. Gold nanoparticles are taken up by human cells but do not cause acute cytotoxicity. *Small* **2005**, *1*, 325–327.
214. Fujiwara, H.; Yanagida, S.; Kamat, P.V. Visible laser induced fusion and fragmentation of thionicotinamide-capped gold nanoparticles. *J. Phys. Chem. B* **1999**, *103*, 2589–2591.
215. Lowe, A.B.; Sumerlin, B.S.; Donovan, M.S.; McCormick, C.L. Facile preparation of transition metal nanoparticles stabilized by well-defined (co)polymers synthesized *via* aqueous reversible addition-fragmentation chain transfer polymerization. *J. Am. Chem. Soc.* **2002**, *124*, 11562–11563.

216. Roth, P.J.; Theato, P. Versatile synthesis of functional gold nanoparticles: Grafting polymers from and onto. *Chem. Mater.* **2008**, *20*, 1614–1621.
217. Wuelfing, W.P.; Gross, S.M.; Miles, D.T.; Murray, R.W. Nanometer gold clusters protected by surface-bound monolayers of thiolated poly(ethylene glycol) polymer electrolyte. *J. Am. Chem. Soc.* **1998**, *120*, 12696–12697.
218. Kim, D.; Park, S.; Lee, J.H.; Jeong, Y.Y.; Jon, S. Antibiofouling polymer-coated gold nanoparticles as a contrast agent for *in vivo* X-ray computed tomography imaging. *J. Am. Chem. Soc.* **2007**, *129*, 7661–7665.
219. Corbierre, M.K.; Cameron, N.S.; Sutton, M.; Mochrie, S.G.J.; Lurio, L.B.; Ruhm, A.; Lennox, R.B. Polymer-stabilized gold nanoparticles and their incorporation into polymer matrices. *J. Am. Chem. Soc.* **2001**, *123*, 10411–10412.
220. Corbierre, M.K.; Cameron, N.S.; Lennox, R.B. Polymer-stabilized gold nanoparticles with high grafting densities. *Langmuir* **2004**, *20*, 2867–2873.
221. Zubarev, E.R.; Xu, J.; Sayyad, A.; Gibson, J.D. Amphiphilic gold nanoparticles with V-shaped arms. *J. Am. Chem. Soc.* **2006**, *128*, 4958–4959.
222. Zhu, M.-Q.; Wang, L.-Q.; Exarhos, G.J.; Li, A.D.Q. Thermosensitive gold nanoparticles. *J. Am. Chem. Soc.* **2004**, *126*, 2656–2657.
223. Yusa, S.-i.; Fukuda, K.; Yamamoto, T.; Iwasaki, Y.; Watanabe, A.; Akiyoshi, K.; Morishima, Y. Salt effect on the heat-induced association behavior of gold nanoparticles coated with poly(N-isopropylacrylamide) prepared *via* reversible addition-fragmentation chain transfer (RAFT) radical polymerization. *Langmuir* **2007**, *23*, 12842–12848.
224. Mangeney, C.; Ferrage, F.; Aujard, I.; Marchi-Artzner, V.; Jullien, L.; Ouari, O.; Rekai, E.D.; Laschewsky, A.; Vikholm, I.; Sadowski, J.W. Synthesis and properties of water-soluble gold colloids covalently derivatized with neutral polymer monolayers. *J. Am. Chem. Soc.* **2002**, *124*, 5811–5821.
225. Luo, S.; Xu, J.; Zhang, Y.; Liu, S.; Wu, C. Double hydrophilic block copolymer monolayer protected hybrid gold nanoparticles and their shell cross-linking. *J. Phys. Chem. B* **2005**, *109*, 22159–22166.
226. Fustin, C.-A.; Colard, C.; Filali, M.; Guillet, P.; Duwez, A.-S.; Meier, M.A.R.; Schubert, U.S.; Gohy, J.-F. Tuning the hydrophilicity of gold nanoparticles templated in star block copolymers. *Langmuir* **2006**, *22*, 6690–6695.
227. Kang, Y.; Taton, T.A. Core/shell gold nanoparticles by self-assembly and crosslinking of micellar, block-copolymer shells. *Angew. Chem. Int. Ed.* **2005**, *44*, 409–412.
228. Li, D.; He, Q.; Cui, Y.; Li, J. Fabrication of pH-Responsive nanocomposites of gold nanoparticles/Poly(4-vinylpyridine). *Chem. Mater.* **2007**, *19*, 412–417.
229. Li, D.; He, Q.; Yang, Y.; Mohwald, H.; Li, J. Two-Stage pH response of poly(4-vinylpyridine) grafted gold nanoparticles. *Macromolecules* **2008**, *41*, 7254–7256.
230. Ohno, K.; Koh, K.; Tsujii, Y.; Fukuda, T. Synthesis of gold nanoparticles coated with well-defined, high-density polymer brushes by surface-initiated living radical polymerization. *Macromolecules* **2002**, *35*, 8989–8993.
231. Ohno, K.; Koh, K.; Tsujii, Y.; Fukuda, T. Fabrication of ordered arrays of gold nanoparticles coated with high-density polymer brushes. *Angew. Chem., Int. Ed.* **2003**, *42*, 2751–2754.

232. Li, D.; Cui, Y.; Wang, K.; He, Q.; Yan, X.; Li, J. Thermosensitive nanostructures comprising gold nanoparticles grafted with block copolymers. *Adv. Funct. Mater.* **2007**, *17*, 3134–3140.
233. Kim, D.J.; Kang, S.M.; Kong, B.; Kim, W.J.; Paik, H.J.; Choi, H.; Choi, I.S. Formation of thermoresponsive gold nanoparticle/PNIPAAm hybrids by surface-initiated, atom transfer radical polymerization in aqueous media. *Macromol. Chem. Phys.* **2005**, *206*, 1941–1946.
234. Duan, H.; Kuang, M.; Zhang, G.; Wang, D.; Kurth, D.G.; Mohwald, H. pH-Responsive capsules derived from nanocrystal templating. *Langmuir* **2005**, *21*, 11495–11499.
235. The thermosensitive mechanism is commonly interpreted as a transformation in the state of the copolymer chains with “expire” and “inspire” water behavior, involving two formation processes of intermolecular and intramolecular hydrogen bonds. Below the LCST the so-called “inspire” water process is taking place which involves the formation of intermolecular hydrogen bonds which promotes the absorption of a quantity of water by the polymer chains causing the polymer chains to swell. Above the LCST, the formation of intramolecular hydrogen bonds will exclude the polymer-bonded water, causing the polymer chains to shrink forcing the water molecules out of the polymer space; the “expire” water process. The alternative “inspiration” and “expiration” of water imitates the “breathing” process [236].
236. Li, D.; He, Q.; Cui, Y.; Wang, K.; Zhang, X.; Li, J. Thermosensitive copolymer networks modify gold nanoparticles for nanocomposite entrapment. *Chem. Eur. J.* **2007**, *13*, 2224–2229.
237. Mandal, T.K.; Fleming, M.S.; Walt, D.R. Preparation of polymer coated gold nanoparticles by surface-confined living radical polymerization at ambient temperature. *Nano Lett.* **2002**, *2*, 3–7.
238. Nuβ, S.; Böttcher, H.; Wurm, H.; Hallensleben, M.L. Gold nanoparticles with covalently attached polymer chains. *Angew. Chem. Int. Ed.* **2001**, *40*, 4016–4018.
239. Wang, B.; Li, B.; Zhao, B.; Li, C.Y. Amphiphilic Janus gold nanoparticles *via* combining "solid-state grafting-to" and "grafting-from" methods. *J. Am. Chem. Soc.* **2008**, *130*, 11594–11595.
240. Watson, K.J.; Zhu, J.; Nguyen, S.T.; Mirkin, C.A. Hybrid nanoparticles with block copolymer shell structures. *J. Am. Chem. Soc.* **1999**, *121*, 462–463.
241. Jordan, R.; West, N.; Ulman, A.; Chou, Y.-M.; Nuyken, O. Nanocomposites by surface-initiated living cationic polymerization of 2-oxazolines on functionalized gold nanoparticles. *Macromolecules* **2001**, *34*, 1606–1611.
242. Raula, J.; Shan, J.; Nuopponen, M.; Niskanen, A.; Jiang, H.; Kauppinen, E.I.; Tenhu, H. Synthesis of gold nanoparticles grafted with a thermoresponsive polymer by surface-induced reversible-addition-fragmentation chain-transfer polymerization. *Langmuir* **2003**, *19*, 3499–3504.
243. Carrot, G.; Gal, F.; Cremona, C.; Vinas, J.; Perez, H. Polymer-grafted-platinum nanoparticles: From three-dimensional small-angle neutron scattering study to tunable two-dimensional array formation. *Langmuir* **2009**, *25*, 471–478.
244. Chen, Y.; Rosenzweig, Z. Luminescent CdSe quantum dot doped stabilized micelles. *Nano Lett.* **2002**, *2*, 1299–1302.
245. Murray, C.B.; Kagan, C.R.; Bawendi, M.G. Self-organization of CdSe nanocrystallites into three-dimensional quantum dot superlattices. *Science* **1995**, *270*, 1335–1338.
246. Wang, S.; Mamedova, N.; Kotov, N.A.; Chen, W.; Studer, J. Antigen/Antibody immunocomplex from CdTe nanoparticle bioconjugates. *Nano Lett.* **2002**, *2*, 817–822.

247. Myung, N.; Ding, Z.; Bard, A.J. Electrogenerated chemiluminescence of CdSe nanocrystals. *Nano Lett.* **2002**, *2*, 1315–1319.
248. Mattoussi, H.; Radzilowski, L.H.; Dabbousi, B.O.; Fogg, D.E.; Schrock, R.R.; Thomas, E.L.; Rubner, M.F.; Bawendi, M.G. Composite thin films of CdSe nanocrystals and a surface passivating/electron transporting block copolymer: Correlations between film microstructure by transmission electron microscopy and electroluminescence. *J. Appl. Phys.* **1999**, *86*, 4390–4399.
249. Alivisatos, A.P. Semiconductor clusters, nanocrystals, and quantum dots. *Science* **1996**, *271*, 933–937.
250. Michalet, X.; Pinaud, F.F.; Bentolila, L.A.; Tsay, J.M.; Doose, S.; Li, J.J.; Sundaresan, G.; Wu, A.M.; Gambhir, S.S.; Weiss, S. Quantum dots for live cells, *in vivo* imaging, and diagnostics. *Science* **2005**, *307*, 538–544.
251. Lee, J.; Sundar, V.C.; Heine, J.R.; Bawendi, M.G.; Jensen, K.F. Full color emission from II-VI semiconductor quantum dot-polymer composites. *Adv. Mater.* **2000**, *12*, 1102–1105.
252. Chan, W.C.W.; Nie, S. Quantum dot bioconjugates for ultrasensitive nonisotopic detection. *Science* **1998**, *281*, 2016–2018.
253. Medintz, I.L.; Uyeda, H.T.; Goldman, E.R.; Mattoussi, H. Quantum dot bioconjugates for imaging, labelling and sensing. *Nat. Mater.* **2005**, *4*, 435–446.
254. Carrot, G.; Scholz, S.M.; Plummer, C.J.G.; Hilborn, J.G.; Hedrick, J.L. Synthesis and characterization of nanoscopic entities based on poly(caprolactone)-grafted cadmium sulfide nanoparticles. *Chem. Mater.* **1999**, *11*, 3571–3577.
255. Ye, J.; Hou, Y.; Zhang, G.; Wu, C. Temperature-Induced aggregation of poly(N-isopropylacrylamide)-stabilized CdS quantum dots in water. *Langmuir* **2008**, *24*, 2727–2731.
256. Xu, J.; Wang, J.; Mitchell, M.; Mukherjee, P.; Jeffries-El, M.; Petrich, J.W.; Lin, Z. Organic-inorganic nanocomposites *via* directly grafting conjugated polymers onto quantum dots. *J. Am. Chem. Soc.* **2007**, *129*, 12828–12833.
257. Zorn, M.; Bae, W.K.; Kwak, J.; Lee, H.; Lee, C.; Zentel, R.; Char, K. Quantum dot-block copolymer hybrids with improved properties and their application to quantum dot light-emitting devices. *ACS Nano* **2009**, *3*, 1063–1068.
258. Farmer, S.C.; Patten, T.E. Photoluminescent polymer/quantum dot composite nanoparticles. *Chem. Mater.* **2001**, *13*, 3920–3926.
259. Sill, K.; Emrick, T. Nitroxide-mediated radical polymerization from CdSe nanoparticles. *Chem. Mater.* **2004**, *16*, 1240–1243.
260. Skaff, H.; Emrick, T. Reversible addition fragmentation chain transfer (RAFT) polymerization from unprotected cadmium selenide nanoparticles. *Angew. Chem. Int. Ed.* **2004**, *43*, 5383–5386.
261. Esteves, A.C.C.; Bombalski, L.; Trindade, T.; Matyjaszewski, K.; Barros-Timmons, A. Polymer grafting from CdS quantum dots *via* AGET ATRP in miniemulsion. *Small* **2007**, *3*, 1230–1236.
262. Skaff, H.; Ilker, M.F.; Coughlin, E.B.; Emrick, T. Preparation of cadmium selenide−Polyolefin composites from functional phosphine oxides and ruthenium-based metathesis. *J. Am. Chem. Soc.* **2002**, *124*, 5729–5733.
263. Tanke, R.S.; Kauzlarich, S.M.; Patten, T.E.; Pettigrew, K.A.; Murphy, D.L.; Thompson, M.E.; Lee, H.W.H. Synthesis of germanium nanoclusters with irreversibly attached functional groups: Acetals, Alcohols, esters, and polymers. *Chem. Mater.* **2003**, *15*, 1682–1689.

264. Zhou, L.; Gao, C.; Xu, W.; Wang, X.; Xu, Y. Enhanced biocompatibility and biostability of CdTe quantum dots by facile surface-initiated dendritic polymerization. *Biomacromolecules* **2009**, *10*, 1865–1874.

265. Zhou, L.; Gao, C.; Xu, W. Amphibious polymer-functionalized CdTe quantum dots: Synthesis, thermo-responsive self-assembly, and photoluminescent properties. *J. Mater. Chem.* **2009**, *19*, 5655–5664.

2

Chemistry of Ammonothermal Synthesis

Theresia M. M. Richter and Rainer Niewa *

Institut für Anorganische Chemie, Universität Stuttgart, Pfaffenwaldring 55, Stuttgart 70569, Germany; E-Mail: richter@iac.uni-stuttgart.de

* Author to whom correspondence should be addressed; E-Mail: niewa@iac.uni-stuttgart.de

---

Abstract: Ammonothermal synthesis is a method for synthesis and crystal growth suitable for a large range of chemically different materials, such as nitrides (e.g., GaN, AlN), amides (e.g., $LiNH_2$, $Zn(NH_2)_2$), imides (e.g., $Th(NH)_2$), ammoniates (e.g., $Ga(NH_3)_3F_3$, $[Al(NH_3)_6]I_3 \cdot NH_3$) and non-nitrogen compounds like hydroxides, hydrogen sulfides and polychalcogenides (e.g., NaOH, LiHS, CaS, Cs_2Te_5). In particular, large scale production of high quality crystals is possible, due to comparatively simple scalability of the experimental set-up. The ammonothermal method is defined as employing a heterogeneous reaction in ammonia as one homogenous fluid close to or in supercritical state. Three types of milieus may be applied during ammonothermal synthesis: ammonobasic, ammononeutral or ammonoacidic, evoked by the used starting materials and mineralizers, strongly influencing the obtained products. There is little known about the dissolution and materials transport processes or the deposition mechanisms during ammonothermal crystal growth. However, the initial results indicate the possible nature of different intermediate species present in the respective milieus.

Keywords: high pressure; high temperature; group III nitride semiconductors; ammonothermal synthesis

Nomenclature:

- *A* Alkali or alkaline-earth metal, *A*(1) = alkali metal, *A*(2) = alkaline-earth metal
- *B* Further metal, might be main group metal, transition or rare-earth metal
- *M* Transition metal, excluding Sc, Y, La
- *R* Rare-earth metal, Sc, Y, La-Lu
- *X* Halide
- *E* Other main group element

---

1. Introduction

Ammonothermal synthesis has gained increasing research interest over the last 20 years. The first ammonothermal syntheses were carried out in analogy to the hydrothermal synthesis of oxides in the 1960s by Juza and Jacobs and pursued by Jacobs and co-workers over the following decades [1–3]. Since ammonia resembles water in its physical properties, water was replaced by ammonia in order to obtain amides, imides and nitrides in place of hydroxides and oxides. Using ammonia instead of nitrogen for nitridation allows less harsh temperature and pressure conditions, since ammonia is more reactive. Furthermore, an abundance of further materials like hydroxides, chalcogenides and hydrogen sulfides can be also obtained from supercritical ammonia.

Both hydrothermal and ammonothermal synthesis are part of the large group of solvothermal methods. The use of the term solvothermal is not defined unambiguously. Definitions vary from "*any heterogenous chemical reaction in the presence of a solvent (whether aqueous or non-aqueous) above room temperature and at pressure greater than 1 atm in a closed system*" [4] to sub- or supercritical conditions of the solvent [5]. Rabenau's definition for hydrothermal conditions is frequently used "*an aqueous medium over 100 C and 1 bar*" [6] and transferred to solvothermal conditions in general by adjusting 373 K to "*the boiling point of the solvent*" [7]. The characteristics of ammonothermal synthesis and its products are mainly influenced by high pressure and temperature and usually syntheses are carried out under supercritial conditions. Above the critical point, solvents exist as homogeneous supercritical fluids, where gas and liquid phase can no more be distinguished and consequently the properties of the different phases converge. However, the properties of a supercritical fluid in its proper meaning and those of a fluid, with only one physical quantity above the critical state and the other one slightly below it, are hardly distinguishable. Hence, we define ammonothermal synthesis as a reaction in ammonia as one homogenous fluid next to or under supercritical conditions.

Various solvents showing distinct different chemical properties are used for solvothermal syntheses. This comprises (i) polar protic solvents like H_2O, NH_3, HF, HCl, HBr; (ii) polar non-protic solvents like tetrahydrofuran and (iii) non-polar solvents like benzene, xylene and CO_2. Table 1 gives some examples for solvents used in solvothermal techniques and the obtained products. Over recent years, the research interest in solvothermal methods has greatly increased, due the constantly raising demand of crystal growth of functional materials, like, for example, semiconductors ($CuInSe_2$ [8], InAs [9], GaN [10], AlN [11] and ZnO [12]), piezoelectrics (α-quartz [13], $GaPO_4$ [14]), electrodes for lithium batteries

($Li_{1-x}Mn_2O_{4-y}$ [15]), magnetic and catalytic materials ($La_{1-x}Ca/Sr/BaMnO_3$ [16]) and fine dielectric ceramics ($BaTiO_3$ [17]).

Table 1. Examples for solvothermal solvents and obtained products (see text).

	Solvent	Examples for products	References
(i)	H_2O	ZnO, α-quartz, α-Al_2O_3, $GaPO_4$	[12–14,18]
	H_2O + isopropyl alcohol	$BaTiO_3$	[17]
	$H_2O + C_2H_3Cl_3$	Diamond	[19]
	NH_3	GaN, AlN, Cu_3N, Cs_2S_2, NaOH	[10,11,20–22]
	HCl, HBr, HI	BiSCl, BiTeBr, SbSeI	[23]
	Ethanol	$Li_{1-x}Mn_2O_{4-y}$	[15]
	Benzyl alcohol	$La_{1-x}Ca/Sr/BaMnO_3$	[16]
(ii)	Ethylendiamine	Cu_7Te_4, $CuInSe_2$	[8,24]
	Diethylamine	$CuInSe_2$	[8]
	THF	β-MnS	[25]
(iii)	C_6H_6	Se, *c*/*h*-BN, γ-MnS	[25,26]
	Xylene	InAs	[9]
	Toluene	$CuCr_2Se_4$	[27]
	CO_2	Poly vinyl chloride	[28]
	Br_2	SbSBr	[23]

The choice of the solvent considerably influences the obtained product. Even different modifications occur depending on the solvent. For the reaction of $MnCl_2 \cdot 4\,H_2O$ with thiourea, $SC(NH_2)_2$, to MnS, for example, the use of tetrahydrofuran (THF) as solvent leads to metastable β-MnS, with benzene as solvent exclusively metastable γ-MnS and with water or ethylenediamine the stable α-MnS occur. Thus, the crystallization of the different modifications depends on the formed complexes of the solvent with the substrate, e.g., $[Mn(H_2O)_6]^{2+}$. The knowledge of the chemical nature of the dissolved species is the crucial information to understand the formation mechanism of MnS [25]. Those observations manifest the importance of the complex intermediates present in the solvent for the formed product and for the structure of the product.

The best explored solvothermal method is the hydrothermal synthesis, owing to its commercial application for the synthesis of oxides and hydroxides. Nowadays, the hydrothermal method is used, for example, in the industrial synthesis of over 3000 t α-quartz single crystals per year, due to their piezoelectric properties and of Al_2O_3 from more than 90 million t of bauxite per year, for processing to metal [4].

Already in 1839, Bunsen carried out experiments with liquids at high temperatures and high pressures (473 K, 150 MPa) [29]. For this purpose he used sealed glass tubes with integrated mercury manometer. Later, in 1848, he succeeded in the growth of $BaCO_3$ and $SrCO_3$ millimeter-long crystals from an aqueous solution at 473 K and 1.52 MPa using NH_4Cl as mineralizer [30]. Similar experiments for recrystallization of apophyllites were apparently conducted by Wöhler earlier [30]. Already in 1845 Schafhäutl obtained for the first time micro crystalline quartz crystals from hydrothermal conditions [31].

In 1851 Sénarmont laid the foundation for hydrothermal mechanisms in geology. He already realized the importance of pressure and temperature *...pressure to maintain the gaseous reactants in a forced dissolution and temperature to favorite certain combinations or decompositions...* for the formation of the minerals and synthesized a great number of natural minerals under high temperature and high pressure conditions, in order to elucidate the natural formation conditions. Using a similar experimental setup as Bunsen did, he applied high temperatures and high pressures in sealed glass tubes. To avoid explosion of the sealed glass tubes containing the reactants and the solvent (water), he placed them inside water filled autoclaves [32].

The ammonothermal method was established inter alia in order to synthesize high quality crystals of amides for XRD structure determination and of deuteroamides for neutron diffraction [3]. The first compounds obtained from ammonothermal conditions (up to 507 MPa and 823 K) were binary amides ($Be(NH_2)_2$, $Mg(NH_2)_2$ [2]) and binary nitrides (Be_3N_2 [1]), later also imides ($Th(NH)_2$ [33]), ternary compounds and other than nitrogen containing materials were synthesized [3].

Over the last twenty years the research interest in the ammonothermal method has increased considerably, since it is one of the few techniques leading to group III bulk nitrides. In comparison to other techniques, the ammonothermal method allows crystal growth on native substrate and growth of the initial native substrate itself in a very high quality. Increasing commercial efforts are directed to the ammonothermal growth of GaN and AlN on native substrates. GaN and AlN are semiconductors with wide band gaps used as base for optoelectronic and electronic devices (e.g., light-emitting diodes (LEDs), high electron mobility transistors, lasers with high optical storage capacity). The growth of group III nitrides as bulk crystal is much more difficult than as micro crystalline powder, but necessary for the use in optoelectronic and electronic devices [34,35].

There are several other methods known for the synthesis of bulk GaN and AlN. Up to now the most popular technique is hydride or halide vapor phase epitaxy, which is used for commercial purposes. Vapor-phase GaCl or $GaCl_3$ obtained from metallic Ga and gaseous HCl is carried in a gas flow and reacts with gaseous NH_3 to form GaN. The formed GaN deposits usually on a non-native substrate like sapphire, silicon or GaAs. At ambient pressure, growth temperatures of 1000–1100 °C are typical and growth rates of up to 0.5 mm/h are reported. Wafers are cut from the formed bulk material and subsequently polished. They can be used in devices or as a substrate for further syntheses [36].

Metal organic vapor phase epitaxy allows the growth of thin epitaxial layers used for devices. Metalorganic precursors (e.g., triethyl or trimethyl-gallium) react with NH_3 forming GaN on a substrate. Growth rates vary from 1–2 μm/h for thin layers and up to 50 μm/h for bulk growth [36].

The advantages of devices made from ammonothermal GaN are the lower defect, strain and bowing level due to the growth on native substrate, the possibility of high scalability and no tilt boundaries, which occur even at GaN grown on a native seed by hydride/halide vapor phase epitaxy. If the substrate shows tilt boundaries, even if it is a native substrate but grown itself on a foreign substrate, the tilt boundaries will appear in the grown material, too [10,37–42]. High quality of the substrate is very important, since it allows the production of more devices per substrate and devices made of low-defect-density GaN or AlN promise longer working periods at higher power without breakdown [43]. Solvothermal methods in general are known for their reliable and scalable process, which permits the efficient growth of a great number of crystals within one synthesis, e.g., 1400 quartz crystals, 1700 g each [36,44].

Slow growth rates were one of the disadvantages of the ammonothermal method in the past. According to the latest research results this problem seems to overcome. Recently, researchers increased growth rates for GaN by ammonothermal method from 24–106 μm/d [45] to 250 (*c*-plane) and 300 μm/d (*m*-plane) in the presence of NH_4F, used as so-called mineralizer [46]. Even *m*-plane growth rates of up to 40 μm/h (=960 μm/d) with rates of 10–30 μm/h (=240–720 μm/d) for all planes were reported recently [39].

2. Ammonia as Solvent

Ammonia and water show quite similar properties (see Table 2). Hence, the parallel from supercritical water to supercritical ammonia, drawn and realized by Juza and Jacobs, who established the ammonothermal synthesis, is not surprising [1,2]. Ammonia is a non-aqueous ionizing solvent, that means pure ammonia has a low specific electrical conductivity and dissolved electrolytes are partially or completely dissociated [47]. Ammonia was the first ionizing solvent, except for water, which has been studied thoroughly. Already at the beginning of the 20th century, inter alia Franklin, Kraus and Bronn studied the ammono-system of compounds, e.g., metal solutions in liquid ammonia [48–50]. Still, supercritical and liquid ammonia is a far less explored solvent than water. Compared to water ammonia is less protic and polar, at ambient pressure and temperature it is gaseous and shows much higher pressures at high temperatures [51]. Liquid ammonia has only small solubilities for many inorganic compounds, which yields often poorly crystallized or amorphous solids on precipitation, due to inhibited nucleation and crystal growth. Using supercritical ammonia rather than liquid ammonia overcomes this problem: Under ammonothermal conditions (supercritical fluid above a pressure of 11.3 MPa and a temperature of 405.2 K [52] see Figures 1 and 2 [53]) the solubility is high enough to dissolve a large number of inorganic compounds. To dissolve a solid the relative permittivity of the solvent has to be higher than the lattice energy of the solid. Due to the rising relative permittivity of solvents with increasing density, which occurs at higher pressure, the solubility of mostly ionic solids increases with increasing pressure [3,54]. Water has a higher relative permittivity than ammonia; therefore, solids dissolve better in water than in ammonia at similar conditions (see Table 2).

Table 2. Comparison of selected properties of ammonia and water [47,52,55].

	Water	Ammonia
$T_{crit.}$/K	647.65	405.2
$p_{crit.}$/MPa	22.1	11.3
ϵ_r	78.3 (298 K)	16.9 (298 K)
Autoprotolysis	$2H_2O \rightleftarrows H_3O^+ + OH^-$	$2NH_3 \rightleftarrows NH_4^+ + NH_2^-$
Ionic product	10^{-14} (298 K)	10^{-32} (239 K)
pk_B	15.7	4.75
Proton affinity/eV	−7.9	−9.2

Ammonia has a higher proton affinity than water ($E_{pa}(NH_3)$ = –9.2 eV; $E_{pa}(H_2O)$ = –7.9 eV [47]; the energy released when a proton is attached to the molecule in the gas phase). Hence, ammonia is a more basic solvent than water (see $pk_B(NH_3) = 4.8$ and $pk_B(H_2O) = 15.7$ [47]). Both water and ammonia show autoprotolysis since they possess at least one non-bonding pair of electrons and acidic hydrogen atoms, leading to formation of H_3O^+ or NH_4^+ cations and OH^- or NH_2^- anions, respectively. Due to the endothermic nature of the autoprotolysis reactions, the magnitude of autoprotolysis increases with increasing temperature. In the same way as water is a suitable solvent for the synthesis of oxides, ammonia is a unique and excellent solvent for nitride synthesis. Additionally, liquid and supercritical ammonia find application as solvent for water sensible compounds, which do not necessarily contain nitrogen, e.g., sulfides, hydrogen sulfides [3,56,57] and hydroxides [22].

Figure 1. Pressure-temperature phase diagram of ammonia. The dashed line is extrapolated [58].

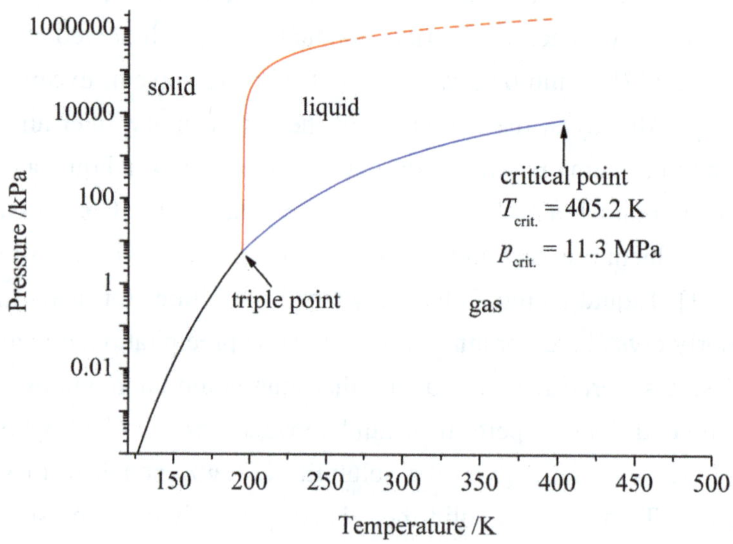

Figure 2. Pressure-temperature diagram of ammonia in dependence on the filling degree of the reaction vessel (in %). The inset shows an enlarged view in the range around the critical point ($T_{crit.}$ 405.2 K, $p_{crit.}$ = 11.3 K) [53].

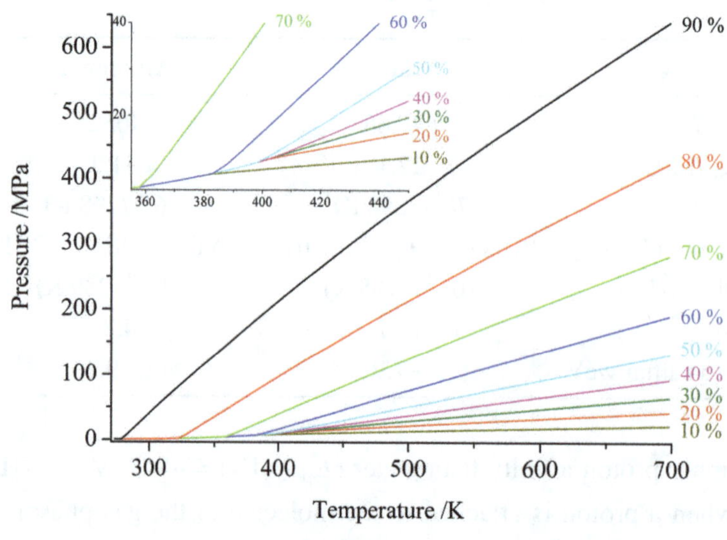

3. Technical Details for Ammonothermal Reactions

There are different installations for the realization of ammonothermal synthesis, starting with the materials and construction of the reaction vessel, the inner shape, liner and geometry of the reaction vessel via the filling of the reaction vessel with ammonia to the heating of the reaction vessel. In general, two types of ammonothermal syntheses can be distinguished, covering different aims: the research on fundamental questions and the research on application of the method for crystal growth. The former research branch investigates new compounds, new synthesis routes for compounds and the processes during crystal growth. The latter investigates and applies the conditions for an optimal crystal growth, including low defect and crack concentration, high growth rates and large crystal size, aiming at a commercial application. The different objectives require different laboratory installations, not only of size but also of type.

The reaction vessel is referred to as autoclave, since most of the installations use steel autoclaves and even the techniques with metallic capsules and glass tubes as reaction containers usually use steel autoclaves for mechanical stabilization of the capsule.

Figure 3. Flow chart of an ammonothermal reaction setup for the simultaneous pressure and temperature control during reaction using two pumps. The heat exchanger is used for pre-cooling one pump and the supply pipe [59,60].

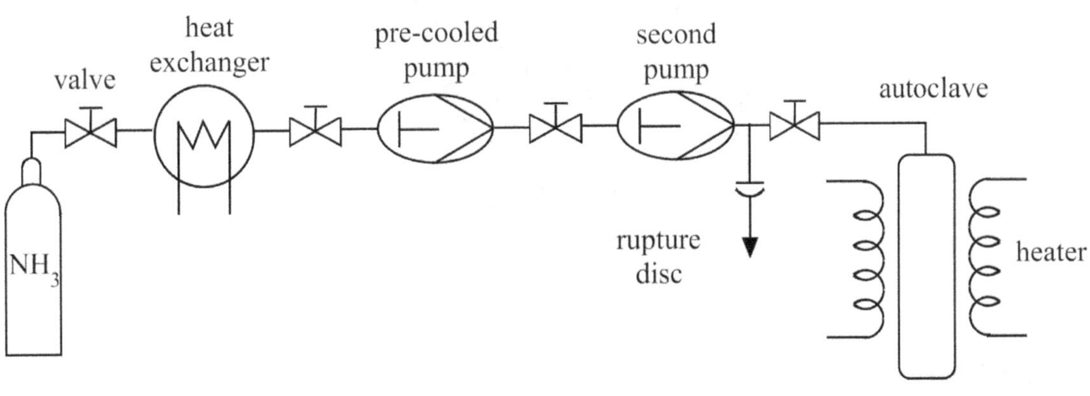

Different approaches find application to fill the autoclave with ammonia: Condensing ammonia into the autoclave by cooling, adding solid ammonia or filling the autoclave sequentially with ammonia by high pressure pumps. Condensing ammonia into the autoclave historically is realized by the help of a tensieudiometer, an apparatus for the simultaneous measurement of pressure and volume, developed by Hüttig [61]. The autoclave, already loaded with the solid reactants, can be connected to the tensieudiometer and is cooled down below the boiling point of ammonia ($T \leq 239.79$ K at ambient pressure [52]) and ammonia condenses in the autoclave. Since volume and pressure in the tensieudiometer, which realizes a closed system including the autoclave, are known an exact amount of ammonia can be filled inside the autoclave. To cool the autoclave down, usually the autoclave body is placed in a cooling bath of acetone or ethanol and dry ice. Adding solid ammonia, solidified by cooling in liquid nitrogen is an other option, which is necessary if glass tubes are used, since the glass tubes are sealed at high temperatures after filling. Liquid ammonia would evaporate too fast at those temperatures. Both filling methods only allow limited control of the pressure during synthesis, by the amount of added ammonia before the synthesis and the applied temperature during synthesis. Pressure control during synthesis is possible using high pressure pumps, which compress ammonia prior to reaction and control the pressure during synthesis. A system of two high pressure pumps with a pre-cooled compressor and supply pipe (258–263 K) guarantees the condensation of the gaseous ammonia, using only one pump allows the system to reach pressures of 5–30 MPa, since the pressure increases with increasing temperature (see Figure 3). The use of a second pump permits the variation of the pressure during synthesis and pressures of up to 450 MPa [3,59,60].

Concerning the temperature at least two different temperature zones inside the reaction chamber have to be achieved in order to obtain a temperature gradient. The temperature gradient contributes via convection to the mass transport of the dissolved species during synthesis and is essential for the crystallization process (see section *Crystallization Process*). The temperature gradient can be applied by different heaters around or integrated into the autoclave, by not heating or even cooling one part of the autoclave. Autoclave models with integrated cooling were also reported, see Figure 4 [3,39].

Figure 4. Flow chart of an ammonothermal autoclave with integrated cooling and external heating [3].

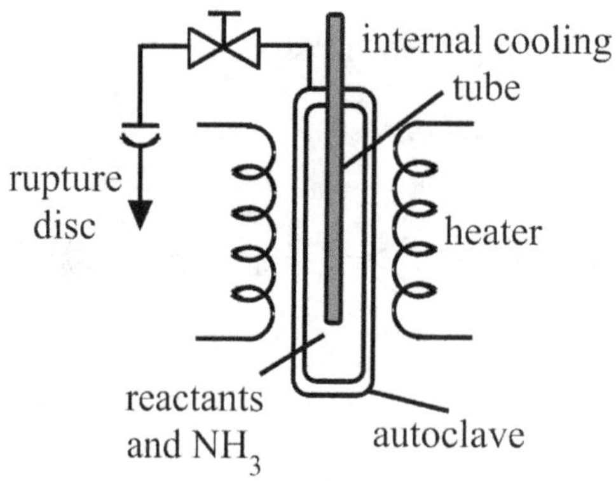

3.1. Reaction Vessels

The first reaction vessels for supercritical syntheses were sealed thick-walled glass tubes with integrated mercury manometer, which withstand temperatures of up to 473 K and pressures of up to 15 MPa [29]. Sénarmont was the first to apply a counter-pressure to avoid explosion of the sealed glass tubes containing the reactants, by placing them inside water filled autoclaves [32]. Similar counter pressure systems are still in use for ammonothermal synthesis. Sealed glass, silver or gold ampoules are used as reaction vessels for ammonothermal synthesis, containing the starting materials and supercritical ammonia. To avoid explosion, the ampoules are placed in a steel autoclave, connected to a high pressure pump, which produces a counter-pressure of nitrogen inside the autoclave (see Figure 5). The counter-pressure should only be slightly lower than the pressure inside the ampoule. However, in order to simplify the handling the counter-pressure is usually higher than the pressure inside the ampoule. For supercritical conditions, the autoclave containing the ampoule is placed in a heater. In this way, pressures and temperatures of up to 280 MPa and 470 K for glass ampoules are realized, metal ampoules enable the application of higher temperatures [3]. Recently, a similar construction was used for large scale

GaN crystal growth in a vessel with internal heating. The reaction is carried out in a welded capsule, containing the starting materials and supercritical ammonia. The capsule is coated directly by a heater, followed by a ceramic shell and an externally-cooled steel shell. The inner heater reaches temperatures of up to 1023 K at 600 MPa, while the temperature of the outer steel shell remains below 473 K, due to the insulating ceramic shell. The use of an internal heating allows the application of higher temperatures and higher pressures compared to autoclaves made of nickel based superalloys, additionally the use of conventional steel is less expensive in production and processing [39]. Nickel based superalloys, for example, Inconel 750 [6], Vacumelt ATS 340 [3], Inconel 718 [59] and Rene 41 [36] are used for the construction of high pressure autoclaves with external heating. Those superalloys almost resist corrosion in ammonobasic milieu up to temperatures of $\sim$873 K at $\geq$3000 MPa (Inconel 750) and can stand even $\leq$1123 K at $\leq$150 MPa (Rene 41) [62]. Autoclaves made of those alloys are loaded directly with the starting materials and ammonia and are heated up with an external direct heater. This explains the milder working conditions.

Figure 5. Flow chart of an ammonothermal reaction setup using a glass or metal ampoule filled with ammonia and the reactants placed in a steel autoclave connected to a high pressure pump, applying a counter-pressure of nitrogen [3].

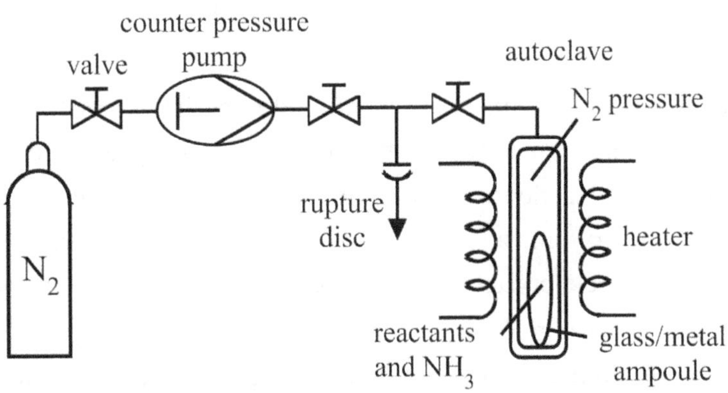

Depending on the reaction conditions, the use of a liner may be necessary. Ammonobasic conditions are usually less corrosive than ammonoacidic conditions. Thus, a nickel based superalloy as autoclave material is often sufficient. However, in ammonobasic conditions, nickel based alloys can also be affected and a supplementary liner can be useful. We have observed the formation of nickel compounds from the autoclave material at $T \geq 840$ K with zinc or gallium as reactant. Indium metal used as reactant passed the autoclave material by diffusion along the grain boundaries and leads to the destruction of the autoclave. For ammonoacidic conditions, liners are used to protect the products from metal impurities originating from the autoclave material. Especially chromium and nickel impurities in the compounds or even chromium and nickel based compounds are observed. Precious metals are used as liner materials, particularly silver, gold and platinum, although the choice of liner material restricts the choice of the mineralizer. However, the formation of a nitride layer on the surface of the autoclave material at $T \sim 800$ K in supercritical ammonia during synthesis, resulting in a hard surface of nitrides, seems to protect the autoclave material in some cases to a certain amount from corrosion [3].

According to the use of the autoclave, the inner geometry is constructed differently. Ammonothermal syntheses on a fundamental research level, investigating new compounds, synthesis routes and processes do not obligatory require special inner shapes of the autoclave. Only the starting materials, including mineralizer, are loaded into the autoclave in one of the two temperature zones, subsequently the formed products, the crystallization spot and the dependence on temperature and pressure are investigated. However, ammonothermal syntheses exploring the crystal growth usually use seed crystals and a feedstock of micro crystalline powder, both containing the same metal (mostly GaN or AlN but also metal amides, metal halides or pure metal) and placed in different temperature zones of the reaction vessel. The micro crystalline powder is diluted with the help of a mineralizer and transported to the seed crystals, where it deposits. In this way, large crystals are obtained. The epitaxial growth on the seed crystal can be heteroepitaxial on foreign substrate or homoepitaxial on native substrate. This method is suitable for large scale commercial synthesis of AlN and GaN [63]. Often a baffle is used, in order to optimize the heat and mass transport inside the autoclave, since it is capable of controlling the flow pattern in the system. Different baffle designs are used, depending on the transport direction in the system.

4. Crystallization Process

The ammonothermal synthesis corresponds, in its thermodynamical fundaments and many chemical aspects, to the well studied Chemical Vapor Transport. The definition of the Chemical Vapor Transport by Binnewies *et al.* applies on a fundamental level: "textita condensed phase, typically a solid, is volatilized in the presence of a [...] transport agent, and deposits elsewhere, usually in the form of crystals" [64]. For the ammonothermal method this means: A solid phase is dissolved in supercritical ammonia initiated by chemical reaction with the added mineralizer forming complex ions, the complex ions are transported into the crystallization zone and deposit there under reformation of the solid as micro crystalline powder or single crystals. Crucial for the transport is the solubility of the compound ($\geq$3% [65]) and the presence of a gradient. In principle, the gradient may concern any state variable leading to a difference in solubility. Typically, temperature gradients are used in crystal growth due the most simple realization by applying two or more different heaters realizing different temperatures and consequently producing two or more different temperature zones inside the reaction vessel.

There are currently many research efforts in maximizing growth rates and crystal sizes and in minimizing defects and strain/bowing levels. However, there appears only very little work done in order to understand the reasons for the achieved improvements. Knowledge and understanding of the physical and chemical processes during ammonothermal crystal growth allows a controlled and efficient improvement of the crystal growth process. Due to the influence of various parameters, this process is very complex. Two groups of parameters can be distinguished: Chemical parameters like the solvent, the chemical nature and the concentration of nutrient and mineralizer, and thermodynamical parameters like the pressure, the temperature in dissolution and crystallization zone and the temperature gradient contribute to the crystal growth [5]. Over the last years, there have been some initial studies carried out clarifying the processes during ammonothermal synthesis and especially the formation mechanism of GaN from supercritical ammonia. The research efforts cover different points of both groups of parameters, yet the influence of most parameters are not well examined and require further intense research efforts.

By isolation of intermediate compounds, evaluation of crystallization and dissolution zone, and determination of solubility of Ga species, important information for the formation mechanism of GaN from supercritical ammonia are obtained. Additionally, spectroscopic data of the intermediate compounds were collected representing the base for future *in-situ* measurements for further exploration of the GaN formation mechanism [66,67]. Special optical cells for *in-situ* monitoring during ammonothermal process were developed [59,60,68]. A combination of the *in-situ* monitoring technology and preliminary work on intermediate compounds allows to obtain further information for a complete unravelment of the ammonothermal synthesis. In order to understand the role of applied temperature, temperature gradient and baffle design for an optimized transport and consequently for the crystal growth first theoretical 3-D simulations for the ammonothermal process were furnished, revealing a strong dependence of the flow pattern by the baffle shape. A positive inclination of the baffle leads to an upward directed jet stream through the baffle opening and vice versa. The baffle shape not only influences the flow pattern but also the temperature profile. These information may be used to control specifically the crystallization and the mass transport from dissolution to crystallization zone, and to avoid parasitic crystal growth [69].

4.1. Thermodynamical Parameters

Knowledge about thermodynamical parameters (temperatures, pressure and reaction time) during ammonothermal crystal growth is crucial to adjust the growth conditions. High pressure allows us to decrease the temperature so that comparatively mild conditions can be applied. Depending on the desired product, temperature and pressure have to be adapted. In general, high temperature and low pressure favors the formation of binary nitrides and low temperatures and high pressures the stabilization of ternary amides, ternary ammoniates or mixed ternary compounds. Regarding the temperature, different aspects have to be respected: the dissolution temperature, the growth temperature and the temperature gradient. Generally speaking, increasing temperature and pressure yields a higher solubility of the starting materials, but the nature of the reactants and the products' solubility (negative or positive temperature dependance) have to be considered. For the transport of the mobile species, a gradient, typically in temperature, is necessary, determining the transport of the mobile species in solution and the kinetics. The transport can be performed dominated by convection or by diffusion, with convection by a gradient being much faster. A vertical gradient, typically meaning a vertical position of the reaction vessel, favors the transport via convection, while a horizontal position minimizes almost entirely the transport to diffusion, which leads to much slower transport rates. However, higher transport rates often yields higher defect densities. Thus, the kinetics have to be adapted according to the advantages of a high growth rate and the disadvantages of increased defect densities [5,7]. A temperature gradient can easily be applied by different heaters, by the lack of a heater on some parts of the reaction vessel or by an induced cooling.

The influence of the thermal decomposition of ammonia obeying reaction Equation (1) during ammonothermal synthesis at temperatures up to 873 K is expected to be small. H_2 formed from the dissociation or as by-product of chemical reactions during dissolution or recrystallization has been shown to diffuse through the autoclave wall [3,70]. Jacobs and Schmidt concluded that equilibrium Equation (1) has to lie more on the left side, as they could not detect any N_2 after reaction and their autoclaves would not have stand pressures as high as the extrapolated values [3].

$$2\,NH_3 \rightleftarrows N_2 + 3\,H_2 \tag{1}$$

4.2. Chemical Parameters

Chemical parameters have previously been studied for solvothermal methods using other solvents than ammonia. We will limit ourselves to few examples: For the basic hydrothermal synthesis of wurtzite-type ZnO (starting materials: $Zn(NO_3)_{2(aq)}$, NH_4OH in H_2O at ambient pressure and 373 K) the mechanism starts with dissolution of the ZnO nutrient by means of a mineralizer forming different soluble intermediate species $[ZnOOH]^-$, $[Zn(OH)_4]^{2-}$ and $[ZnO_2]^{2-}$, which occur in different concentrations depending on OH^- concentration and temperature and the solid ϵ-$Zn(OH)_2$, which crystallizes as solid intermediate [44]. *h*-ZnO was shown to form via the crystalline intermediate ϵ-$Zn(OH)_2$, where the conversion takes place interior solid ϵ-$Zn(OH)_2$. Simultaneously, on the outer surface layers exposed to the solution, *h*-ZnO is formed directly from ϵ-$Zn(OH)_2$, or exchanged with solution. ϵ-$Zn(OH)_2$ crystallizes from soluble species ($[Zn(OH)_{4-n}(H_2O)_n]^{2-n}$; $n = 0, 1$) [12]. A possible ZnO formation mechanism for the growth on a ZnO seed crystal is shown in reaction Equation (2), while reaction Equation (3) shows the solid-solid transformation from spontaneously nucleated ϵ-$Zn(OH)_2$ to *h*-ZnO.

$$\begin{aligned} \mathrm{ZnO}\ (nutrient) + 2\,\mathrm{OH^-} + \mathrm{H_2O} &\rightleftarrows [\mathrm{Zn(OH)_4}]^{2-} \rightleftarrows \mathrm{Zn(OH)_2} + 2\,\mathrm{OH^-} \\ &\rightleftarrows [\mathrm{ZnO_2}]^{2-} + 2\,\mathrm{H_2O} \rightleftarrows \mathrm{ZnO}\ (crystal) + 2\,\mathrm{OH^-} + \mathrm{H_2O} \end{aligned} \tag{2}$$

$$\mathrm{Zn(OH)_2} \rightleftarrows \mathrm{ZnO} + \mathrm{H_2O} \tag{3}$$

Extensive research has also been done on the mechanism during hydrothermal growth of zeolites. It has been shown, e.g., that the dissolved species taking part in the crystallization process are small and of simple structure. The growth from larger and more complex species is too time consuming and has a higher defect probability due to different docking possibilities on the crystal surface. A high defect concentration may slow down or even entirely stop the crystal growth [71]. $[Si(OH)_3O]^-$ and $[Al(OH)_4]^-$ monomers are taken to represent the building species in zeolite crystal growth [71,72].

During ammonothermal synthesis, various ionic species may occur. The nature of these species is strongly related to the acidity of the system. Three types of milieus arise, evoked by the used starting materials: ammonoacidic (NH_4^+ present), ammononeutral (e.g., NO_3^{2-} present) and ammonobasic (NH_2^- present). The milieu determining ions (NH_4^+ and NH_2^-) derive from the acidic or basic mineralizer and the autoprotolysis of ammonia. Mineralizers added to the reactants are applied to establish a defined milieu. Sometimes a tiny amount of a co-mineralizer (e.g., alkali metal halides) is added to activate the reaction. Although the role of mineralizers in chemical reactions is not always known, they are used in ammonothermal synthesis [66,67] as well as in various further methods (gas phase transport reactions [64], hydrothermal synthesis [4], synthesis in liquid ammonia [54,73,74]). Similar applies to the so-called co-mineralizers. The mineralizers fulfill different functions in the growth process: to convey the solubility of the starting material by inducing the formation of soluble and consequently mobile species, to enable and enhance the formation of new chemical bonds and compounds, and finally to detach from the desired compound to be able to act like a catalyst (especially in case of growth of binary compounds). The importance of the choice of the mineralizer can be seen in the fact that they determine the transportation direction of the product in the gradient within the autoclave (see paragraph *Intermediate Species controlling Solubility and Growth Rates*).

4.2.1. Ammonobasic Systems

Typically, alkali metal amides $A(1)NH_2$, alkali metals or alkali metal azides are use as ammonobasic mineralizers. Ammonia reacts with alkali metals and alkali metal azides according to reactions Equations (4) and (5) to form NH_2^-, such that the solution turns basic. The same applies for alkaline-earth metals and some rare-earth metals. However, alkali metal amides and azides are easier to handle due to their powder shape and lower reactivity compared to the metal bulk material. Alkali metal amides are favorable if no formation of hydrogen or additional nitrogen pressure is desired.

$$2\,\mathrm{NH_3} + 2\,A(1) \rightarrow 2\,\mathrm{NH_2^-} + \mathrm{H_2} + 2\,A(1)^+ \tag{4}$$

$$2\,\mathrm{NH_3} + 3\,\mathrm{N_3^-} \rightarrow 3\,\mathrm{NH_2^-} + 4\,\mathrm{N_2} \tag{5}$$

During ammonothermal synthesis, in analogy to the experience from hydrothermal research, it is expected that the metal dissolves by help of the mineralizer, forming intermediate species of the type $[B(NH_2)_n]^{m-}$, possibly also as complex imides or amide imides. Those complex anions are transported in the supercritical ammonia from the dissolution to the crystallization zone, where the crystallization takes place.

Ammonobasic mineralizers like KNH_2 are known from syntheses of binary metal amides in liquid ammonia for a long time. The latter binary amides are formed from metal salt solutions in liquid ammonia with KNH_2 as a little soluble precipitate. An example for this synthesis is the formation of $Cd(NH_2)_2$ from $Cd(SCN)_2$ in liquid ammonia with KNH_2 as mineralizer. The application of alkali metal amides as mineralizers only work, if there are no ternary alkali metal amides formed. Often with small amounts of mineralizer binary amides, with large amounts ternary amides are obtained. Beryllium and zinc always form ternary amides [74].

There is a large number of ternary amides of two metals synthesized in supercritical ammonia from the metals (see paragraph *Ternary Amides*) at comparatively high pressures and low temperatures. Ternary amides decompose to imides and further to nitrides, under release of ammonia. Before the formation of pure nitrides is completed, metal hydrides and mixed hydride nitrides may be obtained as shown in the systems Th/H/N [33], Zr/H/N [75] and Ce/H/N [76], especially at temperatures below the formation temperature for pure nitrides.

It is also known that various bulk binary nitrides are formed from the metal in supercritical ammonia using an ammonobasic mineralizer (see paragraph *Nitrides*), from technological point of view, the most interesting examples currently being the semiconductors GaN [77] and AlN [78]. Recent research in ammonobasic conditions indicates that the binary nitrides are formed via ternary amides functioning as intermediate species [67,79] following the assumed reaction Equation (6). $K[Ga(NH_2)_4]$ and $Na[Ga(NH_2)_4]$ convert to *h*-GaN at higher temperatures [80,81]. For GaN synthesis from Ga with $LiNH_2$ in supercritical ammonia, $Li[Ga(NH_2)_4]$ was proposed as possible intermediate compound [79] and was indeed recently found to crystallize in two modifications in the hot zone of the autoclave [67]. In the system $Ga/Na/NH_3$ two ternary amides are known, namely $Na[Ga(NH_2)_4]$ [81] and $Na_2[Ga(NH_2)_4]NH_2$ [67,82] to form under ammonothermal conditions. The former predominates at high pressures, whereas the latter is obtained mainly at lower pressures. With potassium only one solid compound $K[Ga(NH_2)_4]$ [80] and an intriguing liquid "$KGa(NH)_n \cdot xNH_3$" were reported [65]. Decomposition of $Na[Al(NH_2)_4]$ under release of ammonia at temperatures below 373 K is known to proceed via a liquid compound. An intermediate "$NaAl(NH_2)_2(NH)$" was proposed according to volumetric measurements of released ammonia. Further heating leads to a mixture of AlN and $NaNH_2$ [81]. Knowledge about those intermediate compounds is not only interesting for the formation mechanism, but also for understanding the temperature dependance of the solubility of GaN in supercritical ammonia.

$$A(1)_n\mathrm{Ga(NH_2)}_{3+n} \rightleftarrows \mathrm{GaN} + nA(1)\mathrm{NH_2} + 2\mathrm{NH_3} \qquad (6)$$

4.2.2. Ammonoacidic Systems

Ammonothermal synthesis in ammonoacidic milieu is a powerful method to obtain ammoniates, nitrides and even amides. The effect of ammonacidic milieu is due to the presence and increased concentration of ammonium ions NH_4^+. Ammonoacidic milieu is well known from syntheses in liquid ammonia [54,73], where, e.g., NH_4I is used as ammonoacidic catalyst for the precipitation of $Mg(NH_2)_2$ from magnesium according to the following reaction [73]:

$$\begin{aligned} \mathrm{Mg} + 2\,\mathrm{NH_4^+} &\rightleftarrows \mathrm{Mg^{2+}} + 2\,\mathrm{NH_3} + \mathrm{H_2} \\ \mathrm{Mg^{2+}} + 4\,\mathrm{NH_3} &\rightleftarrows \mathrm{Mg(NH_2)_2} + 2\,\mathrm{NH_4^+} \end{aligned} \qquad (7)$$

Typical ammonoacidic mineralizers comprise ammonium halides NH_4X (X = F, Cl, Br, I) and metal halides BX_n (X = F, Cl, Br, I; B = for example, Al, Ga, Fe, Zn).

Several aluminum halide ammoniates are known $[Al(NH_3)_5X]X_2$ (X = F [83], Cl, Br, I [84]), obtained from different synthetic approaches. Ammoniates of aluminum halides are supposed to represent intermediate compounds in the ammonoacidic AlN synthesis. Analogously, there are ammoniates from gallium halides with different amounts of ammonia molecules per gallium kown. In GaN hydride or halide vapor phase epitaxy these occur as volatile precursors, accomplishing the gallium transport to the crystallization zone.

In liquid ammonia gallium halides form hexammoniates such as $[Ga(NH_3)_6]Br_3 \cdot NH_3$ and $[Ga(NH_3)_6]I_3 \cdot NH_3$ [66,85]. Those ammoniates contain octahedrally surrounded Ga^{3+} forming positively charged complex ions $[Ga(NH_3)_6]^{3+}$, which may represent the mobile Ga-containing species in solution. At higher temperature and pressure gallium metal reacts with ammonium halides NH_4x (X = F, Cl) to the ammoniates $[Ga(NH_3)_5Cl]Cl_2$ or $Ga(NH_3)_3F_3$ and GaN [66]. Obtaining $[Ga(NH_3)_5Cl]Cl_2$ and $Ga(NH_3)_3F_3$ during ammonothermal crystal growth of GaN next to GaN confirms the assumption of their role as intermediate species in the GaN synthesis from supercritical ammonia. $[Ga(NH_3)_5Cl]Cl_2$ contains octahedral units $[Ga(NH_3)_5Cl]^{2+}$, which may again represent the Ga-transporting species in solution. In solid $Ga(NH_3)_3F_3$ gallium occurs in two different species, namely $[Ga(NH_3)_4F_2]^+$ and $[Ga(NH_3)_2F_4]^-$.

Thus, in solution cationic $[Ga(NH_3)_4F_2]^+$ and anionic $[Ga(NH_3)_2F_4]^-$ complex ions may be present. This observation may be the key to the enravelment of solubility and crystallization processes during GaN crystal growth.

4.2.3. Intermediate Species controlling Solubility and Growth Rates

There is little known about the solubility of group III nitrides in supercritical ammonia. We believe it to strongly depend on the present intermediate compounds and thus on mineralizer, temperature and pressure. In the system Ga/NH_3 a retrograde solubility for GaN in supercritical ammonia using KNH_2 as mineralizer was found, meaning that the solubility of GaN decreases with increasing temperature. This results from experiments, where a Ga nutrient was placed at the midpoint of an autoclave, presenting a temperature gradient with a hot and a cold zone, and observing the crystallization spot are shown. At $T \geq 723$ K, GaN crystallizes in the hot zone [65,67,86]. An equivalent behavior was observed for AlN, which can be formed from Al and KNH_2 probably via the intermediate $K[Ga(NH_2)_4]$ [78]. This observation might be attributed to dynamic temperature dependant equilibria of different potassium amides and imides of gallium as intermediate species. $K[Ga(NH_2)_4]$, with tetrahedrally coordinated Ga, is known from synthesis in liquid ammonia [80]. At 853 K and 100 MPa we obtain from Ga and KNH_2 pure *h*-GaN. At 733 K and 50 MPa the same starting materials yield a liquid with the composition "$KGa(NH_2)_2(NH)$", which was earlier reported as "$KGa(NH)_n \cdot xNH_3$" [65]. This liquid is supposed to behave similarly to the intermediate "$NaAl(NH_2)_2(NH)$", which occurs during thermal decomposition of $Na[Al(NH_2)_4]$ under release of ammonia (see paragraph *Ammonobasic Systems*) [81]. The coordination of gallium in this liquid is not yet known. Initial results assume a similar behavior for the system $Na/Ga/NH_3$, where the compounds $Na[Ga(NH_2)_4]$ [81] and $Na_2[Ga(NH_2)_4]NH_2$ [67,82] are known and the existence of an equivalent liquid phase is possible.

In ammonoacidic milieu, negative and positive temperature dependance of the solubilities are observed. For the temperature range 473–823 K GaN shows a positive temperature dependance of the solubility in ammonoacidic milieu, using NH_4X (X = Cl, Br, I) as mineralizer [87,88]. This is observed for the syntheses of *c*-GaN and *h*-GaN with Ga metal or GaN as nutrient [88]. Additionally, there are two examples of GaN manifesting a negative solubility in ammonoacidic milieu reported: A change in the temperature dependance of the solubility of GaN from positive to negative, using NH_4Cl has been noticed at temperatures above 923 K with pressures of 110 MPa [62]. Also, a negative solubility has been revealed in the temperature range 823–923 K using NH_4F as mineralizer [46]. Up to now NH_4F and NH_4Cl are the only acidic mineralizers known to evoke a negative solubility for GaN and consequently the crystallization in the hot zone.

Different growth rates for the negatively charged $(000\bar{1})$ *N*-face and the positively charged (0001) Ga-face of the GaN seed crystal were observed. It is assumed that the nature of the intermediate species determines the predominant growth on one of the faces. Thus, the existence of positively charged complex ions such as $[Ga(NH_3)_6]^{3+}$ and $[Ga(NH_3)_5Cl]^{2+}$ may explain a higher growth rate on the negatively charged face (see Figure 6a) [87]. Furthermore, a growth on both *c*-faces, the negatively charged *N*-face and the positively charged Ga-face, is observed using NH_4F as mineralizer [46]. The presence of both cationic $[Ga(NH_3)_4F_2]^+$ and anionic $[Ga(NH_3)_2F_4]^-$ complex ions as intermediates in solution, which should deposit each on the opposite charged face of the GaN seed, may explain this phenomenon (see Figure 6b). Consequently, the solubility, the crystallization zone and the deposition on the seed crystal of group III nitrides dependent strongly on the formed intermediate compounds.

Figure 6. Schematic crystallization of GaN on a seed crystal under ammonoacidic conditions. (**a**) by cationic $[Ga(NH_3)_4F_2]^+$ and anionic $[Ga(NH_3)_2F_4]^-$ intermediates, (using NH_4F as mineralizer), manifesting a negative temperature dependance of the solubility of GaN; (**b**) by cationic $[Ga(NH_3)_6]^{3+}$ intermediates (using NH_4Br or NH_4I as mineralizer), manifesting a positive temperature dependance of the solubility of GaN.

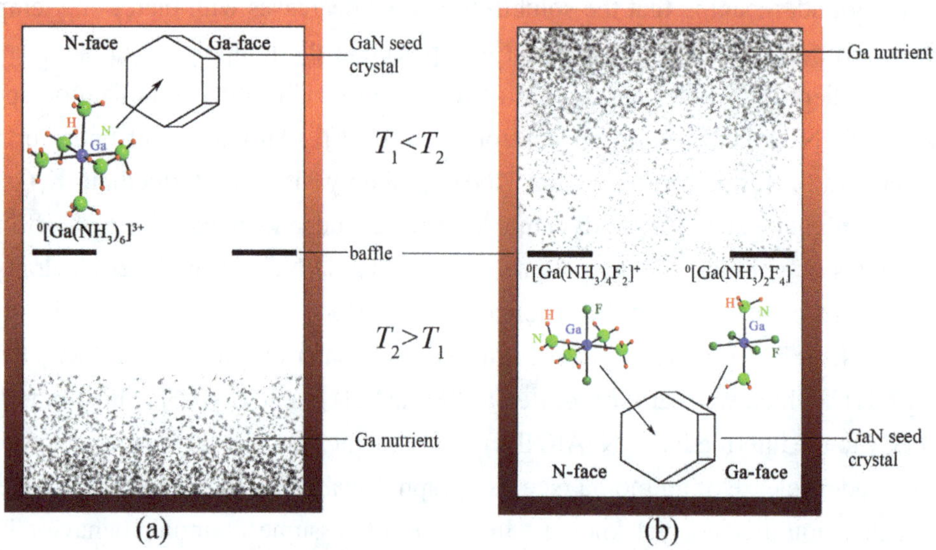

5. Compounds from Ammonothermal Synthesis

5.1. Ammoniates of Metal Halides

Ammoniates of metal halides can be grown from metal halides or from the metal and ammonium halides in supercritical ammonia (see Table 3). Such ammoniates of metal halides are suggested to represent crystallized intermediate compounds during ammonoacidic III-nitride growth [66], since ammoniates are expected to have a high solubility in supercritical ammonia (see paragraph *Ammonoacidic Systems*). In liquid and supercritical ammonia, one typically obtains the ammoniates with highest ammonia content known, for example, $[Al(NH_3)_6]I_3 \cdot NH_3$ [89], $[Al(NH_3)_5Cl]Cl_2$, $[Al(NH_3)_5Br]Br_2$, $[Al(NH_3)_5I]I_2$ [84], $[Ga(NH_3)_6]I_3 \cdot NH_3$, $[Ga(NH_3)_6]Br_3 \cdot NH_3$, $[Ga(NH_3)_5Cl]Cl_2$, $Ga(NH_3)_3F_3$ [66], $[Fe(NH_3)_6]I_2$ and $[Mn(NH_3)_6]I_2$ [90].

Other synthesis methods leading to metal ammoniates are reaction of metal halides with gaseous ammonia or reaction of metals with ammonia donors (e.g., NH_4HF_2, NH_4Cl, NH_4Br) producing compounds such as $Mg(NH_3)_2Cl_2$, $Mg(NH_3)_2Br_2$, $Mg(NH_3)_2I_2$ [91], $[Zn(NH_3)_4]Br_2$, $[Zn(NH_3)_4]I_2$ [92], $Ga(NH_3)_2F_3$ [93], $Fe(NH_3)_2Cl_2$, $[Fe(NH_3)_6]Cl_2$, $[Fe(NH_3)_6]Br_2$ [94–96] and $[B(NH_3)_5Cl]Cl_2$ with B = Al [84], Cr [97], Co [97,98], Rh [97,99], Ru [97], Os [97]. Further ammoniates may be obtained by thermal decomposition of higher ammoniates. The step-wise ammonia release on heating of $[Fe(NH_3)_6]Cl_2$ via $Fe(NH_3)_2Cl_2$ and amorphous $Fe(NH_3)Cl_2$ to ϵ-Fe_3N_{1+x} illustrates the formation of different metal halide ammoniates by thermal decomposition [94–96].

Table 3. Conditions for the ammonothermal crystal growth of metal halide ammoniates and of ammoniates of metal amides.

Compound	Reactants + mineralizer	*T*/K	*p*/MPa	*t*/d	References
$Al(NH_3)_2F_3$	$AlN + NH_4F$	673	–	3	[83]
$[Al(NH_3)_5Cl]Cl_2$	$AlCl_3$	603	–	3–6	[84]
$[Al(NH_3)_5Br]Br_2$	$AlBr_3$	623	–	3–6	[84]
$[Al(NH_3)_5I]I_2$	AlI_3	673	–	3–6	[84]
$[Al(NH_3)_6]I_3 \cdot NH_3$	$Al + NH_4I$	393	9	1	[89]
$Ga(NH_3)_3F_3$	$Ga + NH_4F$	753	238	3	[66]
$[Ga(NH_3)_5Cl]Cl_2$	$Ga + NH_4Cl$	853	95	1	[66]
$[Ga(NH_3)_6]Br_3 \cdot NH_3$	$GaBr_3$	197–373	$\leq$6	–	[66]
$[Ga(NH_3)_6]I_3 \cdot NH_3$	GaI_3	197–373	$\leq$6	–	[66]
$[Mn(NH_3)_6]I_2$	$Mn + I_2$	673–873	600	$\leq$7	[90]
$[Fe(NH_3)_6]I_2$	$Fe + I_2$	673–873	600	$\leq$7	[90]
$Cs_3La(NH_2)_6 \cdot NH_3$	$Cs + La$	490–570	400–600	31–103	[100]
$Cs_4La(NH_2)_7 \cdot NH_3$	$Cs + La$	490–570	400–600	31–103	[100]
$BaAl_2(NH_2)_8 \cdot 2\,NH_3$	$Al + BaAl_2$	823	245	29	[101]
$InF_2(NH_2) \cdot NH_3$	$InN + NH_4F$	673	220	1	[83]

5.2. Binary Amides and Deuteroamides

Several binary amides were obtained from the metals dissolved in liquid ammonia at ambient temperature. In 1891, Joannis had already discovered liquid ammonia as useful solvent to obtain single crystals of metal amides. He obtained colorless $NaNH_2$ crystals from a solution of Na in liquid ammonia [102]. Europium, ytterbium, alkali and alkaline-earth metals dissolve in liquid ammonia at ambient temperature and form intensely blue or bronze colored solutions, if in higher concentration [103]. The solutions are metastable and react to metal amides in form of colorless crystals, hydrogen and a colorless solution [104]. The formation of amides can be enhanced by higher temperatures, higher pressures, exposure to light or addition of a catalyst, e.g., elemental platinum or Fe(II)-compounds like iron oxide [49,73,74,105,106].

5.2.1. Alkali Metal Amides

The alkali metal amides $LiNH_2$ [107], $NaNH_2$ [108], KNH_2, $RbNH_2$ and $CsNH_2$ [73] can be obtained from the metals in liquid ammonia at ambient temperature. The presence of a catalyst (platinum net) or exposure to light enhances the reaction rate. The reaction duration depends on the solubility of the metal in liquid ammonia, which rises with increasing atomic weight of the alkali metal. Higher temperatures accelerate the reactions considerably. For KNH_2, $RbNH_2$ and $CsNH_2$ the reaction can be carried out within a few hours even at low temperatures ($T \leq 273$ K), usually yielding micro crystalline powders [106]. The formation of $LiNH_2$ from lithium metal and liquid ammonia at ambient temperature takes 8 days or more [73,107]. At 400 K and 20 MPa the reaction can be carried out within one day without any catalyst [109], after seven days at 583 K and 71 MPa crystals were obtained [1]. $NaNH_2$ shows similar behavior [108]. To shorten the reaction time and to grow alkali metal amide crystals, it is favorable to work under ammonothermal conditions (see Table 4) [1].

Table 4. Conditions for the ammonothermal synthesis of binary amides and deuteroamides (synthesized using ND_3).

Compound	Reactants + mineralizer	*T*/K	*p*/MPa	*t*/d	Sample	References
$LiND_2$	Li	473	304	–	m.c.	[110]
$NaNH_2$	Na	393	≤10	14	s.c.	[111]
$NaND_2$	Na	423–473	405	8	m.c.	[111]
KND_2	K	320	≤10	4	m.c.	[112]
$CsNH_2$	Cs	423	180	2	m.c.	[113]
$CsND_2$	Cs	423	180	2	m.c.	[113]
$Be(NH_2)_2$	Be	633	253	5	s.c.	[2]
$Be(NH_2)_2$	Be + NaN_3	643	355	20	s.c.	[114]
$Mg(NH_2)_2$	Mg	613–653	10	2	m.c.	[115]
$Mg(NH_2)_2$	Mg + NaN_3	523	253	2–4	s.c.	[1,2]
$Mg(NH_2)_2$	Mg_3N_2	633–648	1	≤7	m.c.	[115]
$Mg(NH_2)_2$	Mg + $NaNH_2$	623–653	212–345	2–4	s.c.	[115]
$Ca(NH_2)_2$	Ca	370	6	14	s.c.	[116]
$Sr(NH_2)_2$	Sr + K	625	550	7	s.c.	[117]
$Sr(ND_2)_2$	Sr	625	550	9	m.c.	[117]
$Ba(NH_2)_2$	Ba	533	324	3	s.c.	[1]
$Ba(NH_2)_2$	Ba	398	≤20	120	s.c.	[118]
$Mn(NH_2)_2$	Mn + $Na_2[Mn(NH_2)_4]$	393	10	10	s.c.	[119]
$Zn(NH_2)_2$	Zn + $Na_2[Zn(NH_2)_4] \cdot 0 \cdot 5\,NH_3$	523	380	60	s.c.	[119]
$La(NH_2)_3$	La + KNH_2	623	405	6	s.c.	[120]
$Sm(NH_2)_3$	Sm	403–493	200–500	–	–	[121]
$Eu(NH_2)_2$	Eu + K	523–673	500–557	7–9	s.c.	[122]
$Eu(NH_2)_2$	Eu	323	≥0.9	3	m.c.	[123]
$Yb(NH_2)_3$	Yb	453	507	32	m.c.	[123]

s.c. means single crystal, m.c. micro crystalline.

Although crystals of the heavier alkali metal amides were obtained from liquid ammonia, the atomic positions of hydrogen could not be determined with X-ray diffraction. This stems back to the low scattering contribution of hydrogen in combination with these metals. Neutron diffraction of the corresponding deuteroamides allows determination of the atomic position for hydrogen. Under ammonothermal conditions, deuteroamides can be grown with small amounts of expensive ND_3, while getting fast a comparatively large yield per synthesis. The exact atomic position provided the base for the interpretation of the interesting electrostatic interactions between the protons and the cations, due to an asymmetric distribution of the charge on NH_2^- [110,111,117,124,125]. An increasing interaction with rising charge density of the cation was observed, *i.e.*, for the amides of the lighter alkali and alkaline-earth metals (Li, Na, Be, Mg) those interactions are stronger than for the heavier ones. In the amides of Li, Na, Be and Mg, the anions form the motif of a cubic closed packing with the cations occupying tetrahedral holes. The resulting structures show an occupation of specific tetrahedral holes within the anion substructure by the cations and can even lead to layered structures similar to LiOH [126]. In lithium amide, Li occupies alternatingly 3/4 of the tetrahedral holes within one layer and 1/4 within the next. Due to strong proton–cation interactions, the anion is hindered in vibration. For the heavier metals the coordination number increases to six for K, Rb, Ca and Sr and eight for Cs. The interatomic interactions decrease resulting in the realization of several modifications, depending on the vibrational and rotational freedom of the amide ions influenced by temperature, as can be seen in the case of KNH_2 [3,124,125,127,128].

Alkali metal amides are important starting materials in ammonothermal synthesis, due to their application as ammonobasic mineralizers. $NaNH_2$ and KNH_2 are the most used ones, since they are less reactive than $RbNH_2$ and $CsNH_2$, but show a higher solubility as $LiNH_2$. The solubility of alkali metal amides in liquid ammonia increases with increasing atomic weight. $LiNH_2$ is only very poorly soluble in liquid ammonia, $NaNH_2$ is poorly soluble (0.144 g/100 g NH_3 at 253 K), KNH_2 is well soluble (65.8 g/100 g NH_3 at 241 K), $RbNH_2$ is very soluble (several hundred grams/100 g NH_3 at 241 K) and $CsNH_2$ is little less soluble as $RbNH_2$ [73,129].

5.2.2. Alkaline-Earth Metal Amides

The alkaline-earth metal amides $Mg(NH_2)_2$, $Ca(NH_2)_2$, $Sr(NH_2)_2$ and $Ba(NH_2)_2$ are obtained from the metals in liquid ammonia at ambient temperature with reaction times of two days ($Ba(NH_2)_2$), eight days ($Sr(NH_2)_2$), four months ($Ca(NH_2)_2$) [130] and up to 1.5–2 years ($Mg(NH_2)_2$) [115]. $Mg(NH_2)_2$ and $Ba(NH_2)_2$ crystals, produced in this way, were not suitable for X-ray diffraction and no $Be(NH_2)_2$ could be obtained. For the heavier alkaline-earth metals (Ca, Sr, Ba) temperatures and pressures somewhat below the critical point of ammonia are sufficient to obtain well crystallized amides and deuteroamides. For the synthesis of $Ca(NH_2)_2$, $Ca(ND_2)_2$, $Sr(NH_2)_2$ and $Sr(ND_2)_2$ 370 K and 6 MPa were applied for two weeks starting from the metal and ammonia [116], $Ba(NH_2)_2$ crystals were obtained at 398 K and $\leq$20 MPa after four months [118]. Nevertheless, working under ammonothermal conditions reduces the reaction time considerably and micro crystalline $Ba(NH_2)_2$ and $Sr(NH_2)_2$ can be obtained from the respective metals in supercritical ammonia after 2 days at 573 K and 60 MPa. Increasing the reaction conditions for $Mg(NH_2)_2$ from $\sim$0.09 MPa and ambient temperature to $\sim$10 MPa and 613–653 K reduces the reaction time from 1.5–2 years to two days and forms a micro crystalline product suitable for powder X-ray diffraction [115]. $Mg(NH_2)_2$ crystals were obtained from Mg metal under ammonobasic conditions (using $NaNH_2$ or NaN_3 as mineralizers) at 523–653 K and 10–345 MPa [1,2,115]. The reaction of Mg_3N_2 at elevated temperatures of 633–648 K and NH_3 pressure of 1 MPa also yields micro crystalline $Mg(NH_2)_2$ [115]. First synthesis of $Be(NH_2)_2$ proceeded under ammonothermal conditions: Single crystals were obtained from the reaction of beryllium metal and ammonia with NaN_3 as mineralizer at 643 K and 355 MPa after 20 days [114].

5.2.3. Lanthanum Amide, Samarium Amide, Europium Amide and Ytterbium Amide

Apart from $Yb(NH_2)_3$ [123] and $Sm(NH_2)_3$ [121], $La(NH_2)_3$ [120] is the only binary trivalent rare-earth metal amide so far. Colorless single crystals are obtained from the reaction of lanthanum metal and potassium (molar ratio 80:1) in supercritical ammonia at 623 K and 405 MPa after six days. The use of a mineralizer is crucial for the crystal growth, without only micro crystalline powder is produced. Using NH_4I as mineralizers leads to smaller crystals than presented for the reaction in presence of the potassium mineralizer [120,131]. $Sm(NH_2)_3$ was obtained from samarium metal and potassium (molar ratio 23:1) in supercritical ammonia at 403–493 K and 200–500 MPa as micro crystalline powder [121]. A solution of europium in liquid ammonia already forms micro crystalline $Eu^{II}(NH_2)_2$ at 323 K and ≥0.9 MPa. Increasing the temperature and pressure to 523–673 K and 500–557 MPa and using optionally potassium as mineralizer leads to dark red crystals. The amount of mineralizer is crucial for the formation of the binary amide, since two ternary amides, namely $K[Eu^{II}(NH_2)_3]$ and $K_3[Eu^{III}(NH_2)_6]$, can be obtained in the same temperature and pressure range [122,123]. There are two binary ytterbium amides known, $Yb(NH_2)_2$ and $Yb(NH_2)_3$, which were so far only obtained as micro crystalline powders. $Yb(NH_2)_2$ was prepared from ytterbium metal in liquid ammonia at ambient temperature within some hours and forms with ammonia $Yb(NH_2)_3$, already before all metal has reacted to $Yb(NH_2)_2$ [123]. Nearly pure $Yb(NH_2)_3$ was obtained from ammonothermal conditions (453 K, 507 MPa, 32 days), however, no successful crystal structure determination for $Yb(NH_2)_3$ was presented so far [123].

5.2.4. Transition Metal, Group III and Group IV Metal Amides

There are various synthetic methods leading to binary transition metal amides. Some transition metal amides precipitate from metal salt solutions in liquid ammonia in presence of KNH_2, e.g., $Cd(NH_2)_2$ can be obtained from $Cd(SCN)_2$ [74,132]. Earlier, binary metal amides were obtained from the reaction from the metal with gaseous or liquid ammonia or of a metal ethyl compound with ammonia, e.g., $Zn(NH_2)_2$ [133]. As a result of these synthesis techniques conducted at comparably low temperatures, the products are often amorphous or poorly crystallized. Under ammonothermal conditions, such poorly crystallized or amorphous binary amides can be recrystallized in form of single crystals or well crystallized powders. Alternatively, they yield from the metal in presence of a mineralizer under ammonothermal conditions: Amorphous zinc amide can be recrystallized at 723 K and 30.4 MPa [1]. Zinc amide was first crystallized under ammonothermal conditions, subsequently the crystal structure was determined. Crystals were obtained from zinc powder and ammonia with $Na_2[Zn(NH_2)_4]\cdot 0.5\ NH_3$ as mineralizer at 523 K and 380 MPa [119]. The analogous reaction of Mn and ammonia at 393 K and 10 MPa (slightly lower than the critical pressure) using $Na_2[Mn(NH_2)_4]$ as mineralizer yielded $Mn(NH_2)_2$ [119].

Binary group III amides are potential precursors during ammonothermal synthesis of group III nitrides. However, the correctness of the reported formula $B(NH_2)_3$ (B = B, Al, Ga, In) was questioned or could not be reproduced. The lack of crystal structure data for these compounds hinders an unequivocal chemical assignment. $B(NH_2)_3$ obtained from BCl_3 or BBr_3 with NH_4Cl in liquid ammonia [134] is suspected to represent polymeric $B(NH)_{3/2}$ (see paragraph *Imides, Nitride Imides, Nitride Amides and*

Amide Azides) [135]. The reaction of $AlBr_3$ and KNH_2 in liquid ammonia did not yield $Al(NH_2)_3$, but a mixed polymeric aluminum amide imide $[Al(NH_2)(NH)]_n$ (see paragraph *Imides, Nitride Imides, Nitride Amides and Amide Azides*) [135–137]. The reaction of $A(1)[Ga(NH_2)_4]$ with NH_4Cl in liquid ammonia at 237 K yields an amorphous product, with the proposed formula $Ga(NH_2)_3$. Due to the lack of diffraction data no crystal structures are proposed, although the composition was determined by chemical analysis [80]. Polymeric gallium imide $[Ga(NH_{3/2}]_n$ was reported from the reaction of $[Ga(NMe_2)_3]_2$ with liquid ammonia in reflux for 8 h (see paragraph *Imides, Nitride Imides, Nitride Amides and Amide Azides*) [135].

$In(NH_2)_3$ crystals were obtained from KNH_2 and InI_3 or $K_3In(NH_2)_6$ in liquid ammonia after one day. Several ternary alkali metal indium amides are known from snythesis in liquid ammonia, e.g., $Li_3In(NH_2)_6$, $K_2In(NH_2)_5$, $K_3In(NH_2)_6$ [138]. However, since no structural information on these compounds is available, the compositions might be questioned.

$Si(NH_2)_4$ is obtained from $SiCl_4$ in liquid ammonia at $T \leq 273$ K and decomposes at $T \geq 373$ K to $Si(NH)_2$ [139,140]. Again, no crystal structure data for $Si(NH_2)_4$ and $Si(NH)_2$ are reported due to the lack of crystalline products. The chemical nature of the compounds for now remains to be proven.

5.3. Ternary Amides

There were various ternary amides of the formulas $A(1)^I_n A(1)^I_m(NH_2)_{n+m}$, $A(1)^I_n A(2)^{II}_m/M^{II}_m/R^{II}_m(NH_2)_{n+2\cdot m}$, $A(1)^I_n B^{III}/R^{III}(NH_2)_{n+3}$ ($A(1)$ = alkali metal, $A(2)$ = alkaline-earth metal, M = transition metal, R = rare-eart metal, B = main group metal; n = 1, 2, 3; m = 1, 3, 7) obtained from liquid or supercritical ammonia. Usually, they contain two different metals, from which at least one is an alkali or alkaline-earth metal. Examples for ternary amides were reported from syntheses in liquid ammonia, e.g., $A(1)[Al(NH_2)_4]$ (with $A(1)$ = Na, K, Cs) [81,141], $Na_2Al(NH_2)_5$ [82], $Na[Ga(NH_2)_4]$ [81], $KBe(NH_2)_3$ [142], $RbBe(NH_2)_3$ [142], $Li[Al(NH_2)_4]$ [143], α-$K[Al(NH_2)_4]$ and β-$K[Al(NH_2)_4]$ [144]. By increasing temperature and pressure the list of ternary amides was extended and the structure determination often was possible (see Tables 5–7), since crystal growth was realized only under ammonothermal conditions, e.g., for $Na_2[Ga(NH_2)_4]NH_2$ [67,82]. Such compounds are expected to exhibit significantly higher solubilities in supercritical NH_3. Therefore, they are suggested as intermediates formed from the reactant and the mineralizer during nitride crystal growth of, for example, GaN. The isolation and crystallization of such compounds sheds light on the formation mechanism of the nitride crystal growth.

For the preparation of ternary alkali and alkaline-earth metal amides some tendencies has been established, which can often also be applied to other ternary amides. Thus, it is recommendable to use a surplus of the less soluble metal. In general, alkali metals possess a higher solubility in ammonia than alkaline-earth metals. The same applies for amides of the heavier metals compared to their lighter homologues. The thermal stability of ternary amides decreases with increasing charge density of the cations. At higher temperatures nitrides of the $A(2)$, M, R or B metal and binary amides of the alkali metals are formed. The water and oxygen sensitivity of the ternary amides increases with increasing atomic weight of the metal. In the case of ternary alkali metal alkaline-earth metal amides they depend more of the alkali metal than of the alkaline-earth metal [3].

Table 5. Conditions for the ammonothermal synthesis of alkali metal alkaline-earth metal ternary amides and coordination numbers of the metal atoms by amide ions.

Compound	*CN A*(1), *A*(1)/*A*(2) by NH_2^-	Reactants	*T*/K	*p*/MPa	*t*/d	References
$K_2Li(NH_2)_3$	6, 4	K + Li 2:1	333	70	60	[145,146]
$KLi(NH_2)_2$	[a], 4	K + Li 1:1	333–473	70–210	4	[145]
$KLi_3(NH_2)_4$	8, 4	K ≤ Li	333–473	70–210	4	[145]
$KLi_7(NH_2)_8$	8, 4	K ≤ Li	333–473	70–210	4	[145]
$K_2[Mg(NH_2)_4]$	7, 4	K + Mg	423	200	3	[147]
$Rb_2[Mg(NH_2)_4]$	7, 4	Rb + Mg	423	200	3	[147]
$Cs[Mg(NH_2)_4]$	9/11, 4	Cs + Mg	415	200	2	[148]
$NaCa(NH_2)_3$	6, 6	Na + Ca 5:1–1:2	740–773	500	60	[149,150]
$KCa(NH_2)_3$	6, 6	K + Ca 1:1	573	500	20	[151]
$RbCa(NH_2)_3$	8, 6	Rb + Ca 1:2–3:1	573	500	17	[152]
$CsCa(NH_2)_3$	8, 6	Cs + Ca	573–773	500–600	10–35	[153]
$KSr(NH_2)_3$	6, 6	K + Sr 1:1	570	500	7	[149]
$RbSr(NH_2)_3$	6, 6	Rb + Sr 3:1–1:1	540–573	800	8	[150]
$CsSr(NH_2)_3$	8, 6	Cs + Sr	573–673	500–600	14–55	[153]
$KBa(NH_2)_3$	6, 6	K + Ba 3:1–1:1	540–573	500	7	[150]
$RbBa(NH_2)_3$	6, 6	Rb + Ba 3:1–1:1	540–573	500	7	[150]
$CsBa(NH_2)_3$	8, 6	Cs + Ba	473	500	11	[154]
$Na_2Sr_3(NH_2)_8$	6, 6	Na + Sr 1:2	570	500	4	[149]

[a] Coordination number uncertain cf. [3].

Table 6. Conditions for the ammonothermal synthesis of ternary amides containing rare-earth metals and coordination numbers of the metal atoms by amide ions.

Compound	*CN A*(1), *R* by NH_2^-	Reactants	*T*/K	*p*/MPa	*t*/d	References
$Na_3[La(NH_2)_6]$	6, 6	Na + La 1:1	523	507	30	[131]
$KLa_2(NH_2)_7$	6, 8	K + La 1:2	623	507	6	[155]
$K_3[La(NH_2)_6]$	6, 6	K + La	473	405	–	[156]
$RbLa_2(NH_2)_7$	[a]					[3]
$Rb_3[La(NH_2)_6]$	6, 6	Rb + La 3:1	573	400–450	6–12	[157]
$CsLa_2(NH_2)_7$	9, 8	Cs + La	470–570	400–600	3–100	[3,158]
$Na_3[Ce(NH_2)_6]$	[a]					[3]
$KCe_2(NH_2)_7$	6, 8	K + Ce 1:1	455	400–500	5–10	[76]
$K_3[Ce(NH_2)_6]$	6, 6	K + Ce 3:1	455	400–500	5–10	[76]
$Cs_3Ce_2(NH_2)_9$	12, 6	Cs + Ce	490	600	21	[159]
$Cs_3Ce_2(NH_2)_9$	12, 6	Cs + Ce	490	200	21	[159]
$Na_3[Nd(NH_2)_6]$	6, 6	[a]				[3]
$KNd_2(NH_2)_7$	6, 8	[a]				[3]

Table 6. *Cont.*

Compound	*CN A*(1), *R* by NH_2^-	Reactants	*T*/K	*p*/MPa	*t*/d	References
$K_3[Nd(NH_2)_6]$	6, 6	[a]				[3]
$Rb_3[Nd(NH_2)_6]$	6, 6	Rb + Nd 3:1	573	400	7	[157]
$Cs_3Nd_2(NH_2)_9]$	12, 6	Cs + Nd	430–530	300	7–100	[159]
$Na_3[Sm(NH_2)_6]$	6, 6	[a]				[3]
$KSm_2(NH_2)_7$	6, 8	K + Sm 1:2	403–493	200–500	–	[121]
$K_3[Sm(NH_2)]_6$	6, 6	K + Sm 3:1	403–493	200–500	–	[121]
$Cs_3Sm_2(NH_2)_9$	12, 6	Cs + Sm	470	600	60	[159]
$KEu(NH_2)_3$	6, 6	K + Eu 1:1	573	500	3	[122]
$RbEu(NH_2)_3$	6, 6	Rb + Eu 1:1-2:1	540–73	500	28	[150]
$Rb_3[Eu(NH_2)_6]$	10, 6	Rb + Eu 10:1	423	500	40	[160]
$CsEu(NH_2)_3$	6, 6	Cs + Eu	573	500–600	9–14	[153]
$K_3[Eu(NH_2)_6]$	6, 6	K + Eu 12:1	573	500	3	[122,149]
$Na_2Eu_3(NH_2)_8$	6, 6	Na + Eu 1:1	570	500	8	[149]
$NaGd(NH_2)_4$	4, 6	$NaNH_2$ + Gd 1:1	493	507	20	[161]
$Na_3[Gd(NH_2)_6]$	6, 6	$NaNH_2$ + Gd 3:1	573	304	51	[3,161]
$K_3[Gd(NH_2)_6]$	6, 6	[a]				[3]
$Cs_3Gd_2(NH_2)_9$	12, 6	Cs + Gd	440	600	160	[162]
$NaY(NH_2)_4$	6, 6	Na + Y 1:1	523	507	7	[163]
$KY(NH_2)_4$	6, 6	K + Y 1:4–6	485–505	600	22	[164]
$RbY(NH_2)_4$	11, 6	Rb + Y 1:4–6	485–505	600	22	[164]
$Na_3[Y(NH_2)_6]$	6, 6	Na + Y 3:1	523	507	7	[163]
$Cs_3Y_2(NH_2)_9$	12, 6	Cs + Y	490	600	70	[162]
$Rb_3[Y(NH_2)_6]$	10, 6	Rb + Y 3:1	473	500	14	[157,160]
$K_3[Y(NH_2)_6]$	10, 6	K + Y 3:1	473	500	14	[157,160]
$NaYb(NH_2)_4$	4, 6	Na + Yb 1:1	413–463	507	14	[163]
$Na_3[Yb(NH_2)_6]$	6, 6	Na + Yb 3:1	453	608	8	[165]
$KYb(NH_2)_7$	[a]					[3,121]
$K_3[Yb(NH_2)_6]$	10, 6	K + Yb 3:1	473	500	14	[157,160]
$Rb_3[Yb(NH_2)_6]$	8, 6	Rb + Yb 3:1	473	500	14	[157,160]
$Cs_3Yb_2(NH_2)_9$	12, 6	Cs + Yb	450	600	160	[162]

[a] No further information in literature.

Table 7. Conditions for the ammonothermal synthesis of ternary amides of main group and transition metals and coordination numbers of the metal atoms by amide ions.

Compound	*CN A*(1), *B*/*M* by NH_2^-	Reactants	*T*/K	*p*/MPa	*t*/d	References
$Li[Ga(NH_2)_4]$	4, 4	Ga + $LiNH_2$	673	250	3	[67]
$Li[Ga(NH_2)_4]$	4, 4	Ga + $LiNH_2$	673	250	3	[67]
$Na_2[Ga(NH_2)_4]NH_2$	4, 4	Ga + $NaNH_2$	853	130	2	[67,82]
$Na[Ga(NH_2)_4]$	4, 4	Na + Ga	853	130	2	[67,81]
$Rb[Al(NH_2)_4]$	12, 4	Rb + Al	393–473	80–120	20	[166]
$Cs[Al(NH_2)_4]$	12, 4	Cs + Al	423–473	120–600	15	[166]
$Na_2[Mn(NH_2)_4]$	4/6, 4	Na + Mn	373	10	–	[167]
$K_2[Zn(NH_2)_4]$	7, 4	Zn + KNH_2	720	249	2	[168]

Both the nature of the $A(1)$ and the $A(2)$, M, R or B metal have an impact on the crystal structure, the number of compounds formed and the composition of the ternary amide. The charge density of both cations is crucial for the realized structure. Increasing the size of the alkali metal leads in the case of ternary lanthanum amides to different structures for $Na_3[La(NH_2)_6]$ [131], $K_3[La(NH_2)_6]$ [156] and $Rb_3[La(NH_2)_6]$ [157]: In $Na_3[La(NH_2)_6]$ all cations occupy octahedral sites in a close packing of amide ions with ABC stacking. In $K_3[La(NH_2)_6]$ both types of cations occupy octahedral sites in a close packing of amide ions with ABC stacking, one layer of octahedral voids is fully occupied by K cations, in the next layer one third of the octahedral voids is filled with La atoms. For $K_3[R(NH_2)_6]$ (R = Y, La, Ce, Nd, Sm, Eu, Gd, Yb) two structure types were observed, one for rare-earth metals from La to Gd and one for Y and Yb. The first group crystallizes in $C2/m$ with all cations in octahedral holes of a close packing of amide groups, occupying layers of octahedral holes alternately. The second group consists of $K_3[Y(NH_2)_6]$ and $K_3[Yb(NH_2)_6]$, which are isotypes of $Rb_3[Eu(NH_2)_6]$ and $Rb_3[Y(NH_2)_6]$ [160]. Those compounds crystallize in $R32$. R^{3+} cations occupy octahedral holes in a close packing of amide groups, while A^+ cations are surrounded by a larger number of amide groups. In $Rb_3[La(NH_2)_6]$ [157] again the cations are surrounded octahedrally, realizing layers. Perhaps surprisingly, there is no such compound as $Cs_3[La(NH_2)_6]$ known, but two amide ammoniates $Cs_3La(NH_2)_6 \cdot NH_3$ and $Cs_4La(NH_2)_7 \cdot NH_3$ [100], where cesium and amide ions together with ammonia molecules form the motif of a cubic close packing with lanthanum cations in octahedral voids, exclusively formed by amide groups. A change of the structure for compounds of the same composition can be observed with changing charge density of both cation types, as can be seen from alkali metal rare-earth metal amides. Thus, compounds with small charge density realize close packings of amide ions with all cations situated in octahedral holes. With increasing size of the alkali metal, layered structures of the cations are preferred and the rare-earth cations form structural units with the amide groups similar to the binary amides. In the cesium amides cesium is surrounded by twelve amide ions, the R^{3+} ions occupy also octahedral sites in a close packing. $A(1)$-rich phases manifest lower coordination numbers at $A(2)$, M, R or B metal, as can be seen in $K_3[La(NH_2)_6]$ [156] and $KLa_2(NH_2)_7$ with CN(La) of 8 and 6 respectively [155]. In Table 5 the coordination numbers of the cations by amide ions in ternary alkali metal alkaline-earth metal amides are listed and the change of the surrounding depending on the size of the metals can be compared.

Four different types of ternary alkali metal amides can be distinguished according to their crystal structures: The first group consists of the isotypic compounds $NaCa(NH_2)_3$ [149,150], $KSr(NH_2)_3$ [149], $KBa(NH_2)_3$ [150], $RbBa(NH_2)_3$ [150], $KEu(NH_2)_3$ [122], $RbEu(NH_2)_3$ [150] and $NaY(NH_2)_4$ [163]. The amide groups arrange in the motif of a cubic closed packing, the cations occupy octahedral holes. $Na_2Sr_3(NH_2)_8$ [149], $Na_2Eu_3(NH_2)_8$ [149], $Na_3La(NH_2)_6$ [131] and $Na_3Y(NH_2)_6$ [163] crystallize in the same structure, irrespective of the different composition. This is possible, since in $Na_3R(NH_2)_6$ with R = La, Ce, Nd, Sm, Gd, Y the sites are occupied completely and ordered, whereas for all other compounds a statistical occupation of the sites with different cations is observed [148,150]. Compared to the ternary amides of the smaller alkaline-earth metals, those compounds are more ionic.

In the second group the difference of the radii of the two types of cations is larger, which results in different coordination numbers. Those ternary amides crystallize in the space groups $C2/c$ ($RbCa(NH_2)_3$ [152]) and $P2_1/c$ ($KCa(NH_2)_3$ [151]). Both structures are related by a close structural relationship, since $P2_1/c$ is a translationengleiche maximal subgroup of $C2/c$. The change of the space group is a result of the influence of the radii of the alkali metals. For potassium a distorted octahedral surrounding is realized, however rubidium and cesium require higher coordination numbers. This group also contains the compounds $CsCa(NH_2)_3$, $CsSr(NH_2)_3$ and $CsEu(NH_2)_3$ [153]. The resulting crystal structures are described as distorted hexagonal perovskites, where cesium and NH_2^- form the motif of a hexagonal close packing. Ca occupies face-sharing octahedral voids exclusively surrounded by amide ions. In this way, one-dimensional chains of $^1_\infty[Ca(NH_2)_6^{4-}]$ result [153]. The change from group one to group two may surprise: One could expect the same crystal structure for $KCa(NH_2)_3$ and $KEu(NH_2)_3$, because the binary amides $Ca(NH_2)_2$ and $Eu(NH_2)_2$ are isotypes [149].

The third group includes $K[Be(NH_2)_3]$ and $Rb[Be(NH_2)_3]$, where nearly trigonal planar complex anions $[Be(NH_2)_3]^-$ appear. Up to now, no amide of this group was obtained from ammonothermal conditions [142].

Finally, the fourth group contains isolated tetrahedral complex anions of the type $[B(NH_2)_4]^{2-}$. Examples, obtained from supercritical ammonia are one modification of dimorphic $K_2[Zn(NH_2)_4]$ [168,169] and $CsMg(NH_2)_4$ [148]. The divalent cation is surrounded by four amide groups, whereas the monovalent cations show coordination with larger numbers of amide groups. Various ternary amides of this type containing two metals were synthesized from liquid ammonia, e.g., $Na_2[Mn(NH_2)_4]$ [169], $Li[Al(NH_2)_4]$ [143,170], $Na[Al(NH_2)_4]$ [81], $Na[Ga(NH_2)_4]$ [81,141], $K_2[Mn(NH_2)_4]$ [171] and $Rb_2[Zn(NH_2)_4]$ [171].

There was a remarkable change of the valence state of Eu from +2 to +3 observed, when a large surplus of KNH_2 was used in ammonothermal synthesis of ternary amides. Apparently, an increased NH_2^- concentration provokes formation of $K_3[Eu^{III}(NH_2)_6]$ rather than $KEu^{II}(NH_2)_3$ [122]. Divalent europium is known to behave similarly to divalent strontium and thus it is not surprising that $KEu^{II}(NH_2)_3$ is an isotype of $KSr^{II}(NH_2)_3$ [149]. Trivalent europium behaves rather similar to the trivalent lanthanides: $K_3[Eu^{III}(NH_2)_6]$ is an isotype of $K_3[La^{III}(NH_2)_6]$ [156]. The same applies for the respective ternary rubidium europium amide. Due to the lower solubility of $NaNH_2$ and the resulting lower NH_2^- concentration in this system a product with lower sodium content $Na_2Eu_3^{II}(NH_2)_8$ was obtained in this system [149].

5.4. Imides, Nitride Imides, Nitride Amides and Amide Azides

Certain imides and nitrides can be formed via thermal decomposition of amides under release of ammonia. From *in-situ* powder X-ray diffraction it is known that during decomposition compounds containing N^{3-}, NH_2^- and NH^{2-} ions may occur. Examples are Li_2NH, CaNH, MgNH [73,114] and $Th_2N_2(NH)$ [33]. There are binary imides of lithium, beryllium, magnesium, calcium, strontium and barium known [73,114]. A few imides and nitride imides were obtained from ammonothermal conditions, but only in the form of micro crystalline powders, see Table 8. From the thermal degradation of ternary rare-earth metal amides the formation of $Yb_{0.66}(NH)$ [123], $Ce_3(NH)_3N$ [76] and $La_{0.667}NH$ [172] was reported.

Table 8. Conditions for the ammonothermal synthesis of micro crystalline binary imides, nitride imides and amide azides.

Compound	Reactants mineralizer	*T*/K	*p*/MPa	*t*/d	References
MgNH	Mg_3N_2	773	≥5	7	[115]
$Th(NH)_2$	Th + Li/Na/K	573	608	29	[33]
Th_2N_2NH	Th	823	507	2	[1]
$ThN(NH_2)$	ThNJ + $A(1)NH_2$	573	405	2	[33]
$Th_3N_2(NH)_3$	ThNJ + $A(1)NH_2$	623	608	27	[33]
Si_2N_2NH	Si + KNH_2	873	600	5	[173]
$ZrN(NH_2)$	ZrNI + KN_3	633	507	10	[75]
$Cs_2(NH_2)N_3$	Cs + Y	463–493	500–600	21–26	[174]

The reaction of Mg_3N_2 with NH_3 at $T \geq 773$ K and $p \geq 5$ MPa yields MgNH with $Mg(NH_2)_2$ and Mg_3N_2 impurities after one week. This finding may be explained by the following equations [115]:

$$Mg_3N_2 + NH_3 \rightleftarrows 3MgNH \tag{8}$$

$$MgNH + NH_3 \rightleftarrows Mg(NH_2)_2 \tag{9}$$

MgNH can also be obtained by thermal decomposition of $Mg(NH_2)_2$ at 513–638 K. At 633 K a micro crystalline powder suitable for X-ray diffraction is produced [115]. $Th(NH)_2$ is formed from thorium metal and ammonia at 613 K and 304 MPa after 4 days as micro crystalline powder [1]. At higher temperatures of 823 K and 507 MPa a micro crystalline nitride imide with the composition Th_2N_2NH is formed [1].

A cesium amide azide $Cs_2(NH_2)N_3$ is reported from the reaction of Cs and ammonia in presence of Y at 463–493 K and 500–600 MPa. The appearance of azide ions may be regarded somewhat surprising under the applied conditions, as the nitrogen of the ammonia would become oxidized during formation. Jacobs et al. explained this effect by a drastically reduced volume of the obtained $Cs_2(NH_2)N_3$ compared to the volume of $CsNH_2$ and NH_3, assuming release of hydrogen by diffusion through the autoclave wall and perhaps also by formation of yttrium hydride. Following reaction Equation (10) a volume reduction of ~29% occurs when $Cs_2(NH_2)N_3$ is formed. In addition, the formation of the mixed amide azide in contrast to two compounds ($Cs(NH_2)$ + CsN_3) is favorable due to a smaller volume of ~3% [174].

$$2\,CsNH_2 + 2\,NH_3 \rightleftarrows Cs_2(NH_2)N_3 + 4\,H_2 \tag{10}$$

However, in the past several cyanamides (or carbodiimides, respectively), formed by reaction with unintenional carbon impurities, were initially misinterpreted as representing azide compounds [175,176]. Since any carbon impurity under ammonothermal conditions in presence of alkali or alkaline-earth metals will readily form cyanamide ions [177], it may be regarded as likely that this compound rather represents an amide carbodiimide.

Group III metals manifest a strong tendency to form polymeric imides from synthesis in liquid ammonia. Thus, the reaction of BX_3 (X = Cl, Br) and NH_4Cl in liquid ammonia yields a polymeric solid with the approximate formula $[B(NH)_{3/2}]_n$ [135]. Earlier, a boron amide $B(NH_2)_3$ was discussed to form under similar reaction conditions [134], but this material might be identical to the polymeric $[B(NH)_{3/2}]_n$. In a further report boron imide was obtained from thermal decomposition of $B_2S_3 \cdot 6\,NH_3$ at $T \geq 378$ K [178]. A mixed aluminum amide imide $AlNH_2(NH)$ was reported as intermediate species during AlN synthesis from aluminum hydride AlH_3 and liquid ammonia at $T \leq 273$ K. However, the formula was only determined by chemical analysis and no crystal structure determination was reported. A pure aluminum amide $Al(NH_2)_3$ is reported to occur at lower temperatures ($T \leq 223$ K) from the same starting materials and to decompose to AlN via $AlNH_2(NH)$ at $T \geq 243$ K [137]. The metathesis reaction of $AlBr_3$ with KNH_2 in liquid ammonia and the reaction of $H_3Al \cdot NMe_3$ with liquid ammonia at ambient temperature yields polymeric $[AlNH_2(NH)]_n$ [135,136]. It is very likely that the reported $AlNH_2(NH)$ is the same polymeric compound [135]. Synthesis of polymeric gallium imide $[Ga(NH_{3/2}]_n$ was reported from the reaction of $[Ga(NMe_2)_3]_2$ with liquid ammonia in reflux for 8 h, the composition was deduced from elemental anaylsis and IR spectroscopy. Thermal decomposition of $[Ga(NH_{3/2}]_n$ yields a mixture of nanosized *c*-GaN and *h*-GaN [135]. Reaction of silicon with KNH_2 in supercritical ammonia at 873 K and 600 MPa yields a silicon nitride imide Si_2N_2NH, which is an intermediate compound during synthesis of Si_3N_4 by thermal decomposition of $Si(NH)_2$. Crystal growth and subsequent structure determination of Si_2N_2NH succeeded after ammonothermal synthesis at 873 K and 600 MPa. In Si_2N_2NH silicon is coordinated tetrahedrally by four N. The SiN_4 tetrahedra are connected via corners, resulting in layers of Si–N hexagons, which are linked by imide groups in prependicular direction [173]. Aditionnaly, in the system Si/NH_3, silicon imide $Si(NH)_2$ and silicon amide $Si(NH_2)_4$ are reported (see paragraph *Transition Metal, Group III and Group IV Metal Amides*). $Si(NH)_2$ is obtained from $SiCl_4$ in liquid ammonia at 353–363 K, however, up to now no crystal structure data is reported and the chemical nature of the compound is still unvertain. $Si(NH)_2$ serves as starting material for Si_3N_4 synthesis by thermal decomposition [179].

Table 9. Conditions for the ammonothermal synthesis of metal hydrides and hydride nitrides.

Compound	Reactants Mineralizer	*T*/K	*p*/MPa	*t*/d	References
ScH_2	$Sc + NH_4I$	773	≥11.3		[161]
$ZrH_{0.6}N$	$ZrN(NH_2)$	633	600		[75]
CeH_x ($x \leq 3$)	Ce	395	400–500		[76]
ThH_2	Th + Na/K	473	500	7	[33]
$ThHN_{1.23}$	$Th + NH_4I$	573	355	10	[33]

5.5. Metal Hydrides and Nitride Hydrides

In some systems, even metal hydrides can be synthesized prior the formation of the metal nitride is completed (see Table 9). This applies especially for the lighter lanthanides at temperatures below the formation temperature for the pure nitride. For example, CeH_x with $x \leq 3$ was obtained next to CeN when pure cerium metal reacts with ammonia at 395 K and 400–500 MPa. The amount of hydrogen in CeH_x ($x \leq 3$) decreases with increasing temperature, simultaneously the amount of CeN increases. At 475 K only pure CeN is obtained [76]. $ZrH_{0.6}N$ was obtained from $ZrN(NH_2)$ and NH_3 at 633 K and

600 MPa [75]. Only one compound is known from scandium under ammonothermal conditions, namely ScH_2. Scandium metal does not show any reaction up to 673 K. Using NH_4I as mineralizer ScH_2 was obtained from scandium metal at 773 K [161].

5.6. Nitrides

By the ammonothermal method nitride single crystals can be obtained at comparatively low temperatures $T \leq 700$ K [3] as can be seen in Table 10. The synthesis can be carried out in ammonobasic (e.g., EuN from Eu and K [122]), neutral (e.g., Cu_3N from $[Cu(NH_3)_4]NO_3$ [20]) or ammonoacidic milieu (e.g., GaN + NH_4I [62]). The formation process of the nitrides is not yet well understood (compare section *Crystallization Process*).

Table 10. Conditions for the ammonothermal synthesis of binary nitrides.

Compound	Reactants + Mineralizer	T/K	p/MPa	t/d	Sample	References
c-Be_3N_2	Be	673	20.3	7	m.c.	[2]
YN	Y + NH_4I	623	61	7	m.c.	[163,174]
EuN	Eu + K 40:1	673	500	7	s.c.	[122]
LaN	$K_3[La(NH_2)_6]$ + KNH_2	650	500	10	s.c.	[3]
LaN	La + Na	523–773	300–507	10	s.c.	[3,131]
CeN	Ce + Cs	490	200	12	–	[159]
SmN	Sm + K	433–453	500	10–30	–	[121]
GdN	Gd + NH_4I	523	507	23	–	[161]
GaN	Ga + $LiNH_2$/K	823	500	≥7	s.c.	[77]
GaN	GaN + NH_4I	≤1123	≤150	–	s.c.	[62]
GaN	GaN + KNH_2 + KI	673	240	7	s.c.	[180]
c-GaN	$[Ga(NH)_{3/2}]_n$ + NH_4I	753	–	2–3	s.c.	[181]
GaN:Mn	Ga + Mn + KNH_2	723–823	400–500	3–10	s.c.	[182]
GaN:Cr	Ga + $CrBr_3$ + KNH_2	723–823	400–500	3–10	s.c.	[182]
GaN:Fe	Ga + Fe + $LiNH_2$	723–773	400–500	3–10	s.c.	[182]
AlN	Al + K	723–873	200	1–18	s.c	[78]
AlN	Al + NH_4Cl	723	–	2	s.c	[11]
InN	In + KNH_2	723	–	–	m.c.	[43]
$Fe_{4-x}Ni_xN$	Fe + FeI_2	733–853	600–800	–	s.c.	[183–185]
Θ-Mn_6N_{5+x}	MnI_2 + $NaNH_2$	673–723	600	60–120	s.c.	[167]
η-Mn_3N_2	Mn + I_2 or MnI_2 + $NaNH_2$	673–873	600	30	s.c.	[167,186]
η-Mn_3N_2	Mn + K/Rb	673–873	≤600	35	s.c.	[186]
ϵ-Mn_4N	Mn + GaN $A(1)NH_2$	723–823	400–500	3–10	m.c.	[182]
Ni_3N	$[Ni(NH_3)_6]Cl_2$ + $NaNH_2$	523	200	7	s.c.	[187]
Cu_3N	$[Cu(NH_3)_4]NO_3$	623–853	≥600	–	s.c.	[20]
Cu_3Pd_xN $x = 0.020/0.989$	$[Cu(NH_3)_4]NO_3$ + Cu + $[Pd(NH_3)_4](NO_3)_2$	723	600	7	s.c.	[188]

s.c. means single crystal, m.c. micro crystalline.

5.6.1. Alkaline-Earth Metal Nitrides

The first binary nitride synthesized by the ammonothermal method was Be_3N_2, which was obtained as pure cubic phase in 1966 by Juza and Jacobs from beryllium in supercritical ammonia at 673 K and 20.3 MPa [2]. In this way, the usual synthesis temperatures for beryllium nitride could be drastically reduced, since the reaction of beryllium to cubic beryllium nitride in ammonia flow requires temperatures of 1173–1373 K [189] and in nitrogen flow 1600 K [190]. Similar ammonothermal reaction conditions should apply for Mg_3N_2, although we are not aware of any publication describing a synthesis of Mg_3N_2 from supercritical ammonia. However, Mg_3N_2 decomposes at 633–648 K and 1 MPa in liquid ammonia to $Mg(NH_2)_2$ and at $T \geq 773$ K and $p \geq 5$ MPa to MgNH [115] (see paragraph *Imides, Nitride Imides, Nitride Amides and Amide Azides*). α-Be_3N_2 crystallizes in the anti-bixbyite structure like Mg_3N_2, α-Ca_3N_2 [190], Cd_3N_2 [191] and Zn_3N_2 [192]. It is not yet established, if the nitrides of strontium Sr_3N_2 and barium Ba_3N_2 can be synthesized, although the ternary nitrides $CaMg_2N_2$ and $SrMg_2N_2$ were reported [193]. The anti-bixbyite structure manifests the motif of a cubic closest packing of the nitrogen atoms, with the metal atoms occupying 75% of the tetrahedral holes in an ordered manner. Nitrogen is surrounded by distorted metal atom octahedra [190]. α-Be_3N_2, β-Be_3N_2 [194], Mg_3N_2, α-Ca_3N_2 [190] and Zn_3N_2 [195] are known to represent semiconductors. A decrease of the band gap value (obtained from experimental data) with increasing atomic weight is observed (α-Be_3N_2: direct band gap of 3.8 eV [196], Mg_3N_2: 2.80 eV, Ca_3N_2: 1.55 eV [190] and Zn_3N_2 probably direct band gap of 1.01–1.25 eV [195,197]). Ammonothermal synthesis of nitrides with anti-bixbyite structure was only published for α-Be_3N_2, nevertheless it is a promising method for bulk growth, especially since most of the other established synthesize routes only yield micro crystalline powders or films and no bulk material, which is necessary for many applications as semiconducting material. Additionally, due to its setup as transport growth, the ammonothermal method could lower the oxygen concentration, which is crucial for nitride semiconductors.

5.6.2. Group III and IV Nitrides

The nitrides of group III metals are considered very promising materials for optical devices in the short-wavelength region, high-frequency high-power electronics and fast-speed communication. Furthermore, they prove high physical and chemical endurance, which makes them attractive for applications in various settings. The use in optic and electronic devices stems back to their semiconducting nature. For example, GaN shows a wide direct bandgap of 3.39 eV [198], an electron mobility of $\sim$1500 cm^2/Vs [36], a large critical breakdown electric field of $\sim$3 MV/cm and a saturation velocity of $\nu \leq 19$–10^6 cm/s. AlN and GaN find applications as blue lasers (e.g., for high capacity optical storage in CD-ROMs), in light emitting diodes, high electron mobility transistors and other devices [34,43]. The band gap for the group III nitrides decreases with increasing atomic weight of the metal constituent: AlN $E_g = 6.2$ eV at 295 K [199], GaN $E_g = 3.39$ eV [198] and InN $E_g = 0.7$–0.8 eV [200,201]. *h*-AlN, *h*-GaN and *h*-InN crystallize in the wurtzite-type structure (space group $P6_3mc$), where anions and cations are surrounded tetrahedrally by the respective counter ions.

For the growth of GaN or AlN for optoelectronic devices a feedstock setup is used, as described in paragraph *Reaction Vessels*(see Figure 6). This method allows the homo- and heteroepitaxial growth of large single crystals. Both ammonobasic and ammonoacidic conditions are used for large scale syntheses, in which AlN [78] requires the highest growth temperatures and InN the lowest [43]. The ammonobasic method enables the growth of GaN wafers up to two inches in diameter [202] or bulk single crystals of 10 mm^2 by 1 mm thick [43] and dislocation density below $5 \cdot 10^4$ cm^2 [203]. Maximal growth rates of up to 40 µm/h (=960 µm/d) with rates of 10–30 µm/h (=240–720 µm/d) for all planes were reported [39]. *c*-GaN in a sphalerite structure (space group $F\bar{4}3m$) is also known and can be obtained from $[Ga(NH)_{3/2}]_x$ and NH_4I under ammonothermal conditions (753 K) [181]. Indium nitride, like its lighter homologues, is an intriguing semiconducting material. Although, up to know research efforts are hindered by indium metal incorporations in as-grown InN, produced by decomposition of the thermodynamic instable InN, especially at temperatures above 723 K [204]. Similar structural defects provoked by formation of In particles are encountered in $In_xGa_{1-x}N$ solid solutions ($0 \leq x \leq 0.5$) [205], which are currently used in light-emitting and laser diodes. Their broad range of the direct band gap of E_g = 0.77–1.75 eV allows to cover a spectral region from near-infrared to near-ultraviolet [206]. Due to the lower thermal stability of InN and its solid solutions with AlN and GaN, lower temperatures in synthesis and crystal growth are required. InN is supposed to be metastable at $T \geq 258$ K, kinetic constraints are believed to inhibit the decomposition at ambient conditions [207]. To our best knowledge, only once a synthesis of InN from ammonothermal conditions has been reported and has not yet been confirmed (see Table 10) [43]. However, the low temperatures and high pressures, which can be applied with the ammonothermal method, may be useful to stabilizes InN and its solid solutions without incorporation of In metal.

Doping of group III nitride semiconductors is used to enhance the emission range and *n*-type conductivity. The most frequently used donor dopant is silicon, but also rare-earth metal doping like with erbium was applied. On the other hand, unintentional doping, usually by oxygen contaminations, leads to reduced carrier concentrations and lower transparency [208]. Erbium doped GaN has shown emission of two erbium induced narrow green lines at 537 and 558 nm in the range of the visible spectrum, with the former one peaking at 300 K [209]. Doping of GaN with erbium under ammonothermal conditions manifests an infrared emission at 2.029 eV, due to intra-4*f* transitions, although no green transition was observed, probably due to the low erbium concentration (1×10^{18} cm^{-3} for the Ga-polar growth, 1×10^{17} cm^{-3} for the N-polar growth). Additionally, erbium is discussed to work as getter for oxygen, crystallizing as erbium oxide nitride in the dissolution zone, thus lowering the oxygen concentration in the crystallization zone and consequently in the crystallized GaN. A lower concentration of Er and a higher impurity level (usually oxygen) in the nitrogen polar growth was observed, which is not surprising considering the different polarities of the seed crystal faces and the soluble species [210]. Furthermore, transition metals were used as dopants for GaN under ammonothermal conditions, yielding GaN:Mn (max. 1% Mn), GaN:Fe (max. 0.03% Fe), ferromagnetic GaN:Cr (max. 0.02% Cr) [182], GaN:Zn and GaN:Mg [211].

Cubic tin nitride Sn_3N_4 has been reported to form from SnI_4 or $SnBr_4$ with KNH_2 in liquid (but not supercritical) ammonia at 243 K followed by annealing in vacuo at 573 K [212].

5.6.3. Rare-Earth Metal Nitrides

Several rare-earth metal nitrides were obtained under ammonothermal conditions, namely YN [163,174], EuN [122], LaN [3,131], CeN [159], SmN [121] and GdN [161]. The reation conditions cover ammonoacidic and ammonobasic milieu, a temperature range from 433 (SmN) to 673 K (EuN) and pressures from 200 (CeN) to 507 MPa (EuN, LaN, SmN, GdN) (see Table 10). ScN, YN, LaN and GdN attract increasing research interest, since they are strongly suspected to be semiconductors, for ScN it even seems unequivocal [213]. Rare-earth metal nitrides cover a large range of properties, especially concerning their metallic or insulating character, CeN to GdN are considered half-metallic, TbN to HoN insulating and ErN to YbN metallic. GdN seems to play a special role, showing ferromagnetic properties with the Curie temperature in the range of T_C = 58–69 K. The intricate physical and chemical properties of those nitrides stem back mainly to their partly filled $4f$ shells [214,215]. They crystallize in the rocksalt structure, with a six-fold coordination of R^{3+} and N^{3-} ions.

5.6.4. Transition Metal Nitrides

Transition metal nitrides manifest a vast number of interesting physical and chemical properties such as high hardness, mechanical strength, high melting point, magnetic and semiconducting properties [216]. For example, η-Mn_3N_2 exhibits an antiferromagnetic spin structure [167], ϵ-Mn_4N has a non-collinear ferrimagnetic spin-structure with a Curie temperature in the range of T_C = 738–748 K [217,218], γ'-Fe_4N is a ferromagnetic conductor with high thermal stability [219,220] and Cu_3N is a diamagnetic semiconductor with an optical band gap of 0.8–1.9 eV [219,221]. The ammonothermal method so far yielded the following transition metal nitrides: Θ-Mn_6N_{5+x} [167], η-Mn_3N_2 [167,186], ϵ-Mn_4N [182], $Fe_{4-x}Ni_xN$ [183–185], Ni_3N [187] and Cu_3N [20]. From a supercritical ammonia-methanol mixture at 443–563 K and 16 MPa some nano crystalline nitrides were obtained, namely Cr_2N, Co_2N, Fe_4N, Cu_3N and Ni_3N [222].

The synthesis conditions for the above mentioned transition metal nitrides vary remarkably. The early $3d$-transition metals are thermodynamically more stable than the later ones, which is manifested in the synthesis of the nitrides of Ti to Ni from the metals in supercritical nitrogen at $\sim$1800 K and 10,000 MPa. Yet, no copper nitrides were obtained from similar conditions [223], although the ammonothermal method in neutral conditions yields up to 10 mm·10 mm·2 mm Cu_3N crystals [20]. The neutral milieu is applied by using $[Cu(NH_3)_4](NO_3)_2$ as mineralizer in the molar ratio 1:1 with Cu metal at 620–850 K and 300–800 MPa. The reaction proceeds by the comproportion of $[Cu(NH_3)_4](NO_3)_2$ with Cu metal to $[Cu(NH_3)_3]NO_3$, which subsequently forms Cu_3N via $[Cu(NH_3)_2]NO_3$ [224]. The presence of $[Cu(NH_3)_2]NO_3$ crystals next to Cu_3N in the product confirms this mechanism. The formation of pure Cu_3N without any presence of oxides may surprise given the proposed reaction Equation (11) of nitrate to form water and elemental nitrogen.

$$3\,[Cu(NH_3)_4]NO_3 \rightarrow Cu_3N + 9\,H_2O + 4\,N_2 + 6\,NH_3 \qquad (11)$$

However, the oxygen may be catched by the autoclave metal surface. Cu_3N crystallizes in the anti-ReO_3-type structure, with nitrogen octahedrally surrounded by six Cu atoms and Cu forming collinear bonds with two nearest nitrogen atoms. The structure offers vacant sites, which can be occupied, for example, by Cu (forming Cu_4N) or by Pd (e.g., forming Cu_3Pd_xN [188]). Like Cu_3N the ternary nitride Cu_3Pd_xN ($x = 0.020/0.989$) is metastable at ambient temperature and can be obtained from Cu and $[Cu(NH_3)_2]NO_3$ in presence of $[Pd(NH_3)_4](NO_3)_2$ under ammonothermal conditions at 723 K and 600 MPa [188]. In contrast to semiconducting Cu_3N, the Pd doped variant exhibits metallic or semimetallic behavior [225].

Three different manganese nitrides could be obtained under ammonothermal conditions so far: Θ-Mn_6N_{5+x} and η-Mn_3N_2 from MnI_2 and $NaNH_2$ at 600 MPa and in temperature gradients from 673 K to 723 K and 673 K 873 K, respectively [167], ϵ-Mn_4N as by-product next to GaN:Mn during doping of GaN with Mn at 723–823 K and 400–500 MPa in ammonobasic milieu [182]. In η-Mn_3N_2 the Mn atoms form an *fcc* substructure with the N atoms occupying octahedral sites in such a way that perpendicular to [001] two fully occupied layers are followed by an empty one. Occupying all octahedral holes would lead to a rocksalt-type structure, like in the early transition and rare-earth metal nitrides [186]. Θ-Mn_6N_{5+x} realizes such a defect rocksalt structure, however, suffering a tetragonal distortion due to magnetostriction at ambient temperatures [226]. The structure of ϵ-Mn_4N also shows an *fcc* substructure of Mn atoms. Here the nitrogen atoms only occupy 1/4 of the octahedral voids in an ordered manner forming an inverse perovskite structure. Similar to the case of Cu_3N, one type of manganese atoms is coordinated linearly by two nitrogen atoms, a second type is exclusively surrounded by Mn of the first type [217].

γ'-Fe_4N an isotype of ϵ-Mn_4N, crystallizing in the inverse perovskite structure [183]. However, γ'-Fe_4N obtained from supercritical ammonia [185] contained about 5 wt.% Ni ($Fe_{4-x}Ni_xN$) [183]. The nickel most likely origins from the autoclave wall, since the nickel based alloys used for autoclave manufacturing are corroded to a certain degree under ammonoacidic conditions. $Fe_{4-x}Ni_xN$ was obtained from Fe and FeI_2 at 733–853 K and 600–800 MPa.

Ni_3N was synthesized ammonothermally from $[Ni(NH_3)_6]Cl_2$ and $NaNH_2$ at 523 K and 200 MPa in 7 days [187]. Ni_3N crystallizes in the ϵ-Fe_3N-type structure, with the Ni atoms realizing the motif of a hexagonal closed packing and the N atoms occupying corner-sharing octahedra. The Ni atoms are coordinated by two N atoms.

5.7. Non-Nitrogen Compounds

Solvothermal methods allow crystal growth at comparatively low temperatures. Thus, oxides, carbonates, fluorides, sulfates and sulfide minerals such as proustite Ag_3AsS_3 can be grown under hydrothermal conditions [6]. However, use of water as solvent is prohibited in chemical synthesis and crystal growth if the target compound is water sensitive or forms stable solid hydrates. The ammonothermal technique in some cases proves superior even for formation of hydroxides, hydroxide hydrates, sulfides and hydrogen sulfides.

Alkali metal hydroxides, except LiOH, are difficult to grow as crystals suitable for single crystal X-ray diffraction, since they exist in different modifications in the range of ambient temperature to their melting point. Thus, the crystal growth from the melt is inhibited, while during growth from acqueous solution hydroxide hydrates are formed. The ammonothermal method allows the growth of single crystals of NaOH, KOH, RbOH [22] and CsOH [227] at conditions below the transition temperature of the modification, which is stable at ambient conditions (see Table 11). Some alkali metal hydroxide hydrates were synthesized from ammonothermal synthesis as well. NaOH [22], $K(H_2O)OH$, $Rb(H_2O)OH$ [228] and $Cs(H_2O)OH$ [227] crystals were obtained by recrystallization of the micro crystalline substance in supercritical ammonia. The reaction of alkali metal hydroxide hydrates $A(1)(H_2O)OH$ ($A(1)$ = K, Rb, Cs) with alkali metal amides $A(1)NH_2$ ($A(1)$ = K, Rb, Cs) yields crystals of KOH, RbOH and CsOH [227]. This synthesis furnished the first crystallographic data of CsOH. $Ca(OH)_2$ can be obtained at 723 K and 69–207 MPa in one day from CaO and NH_4I [57].

Table 11. Conditions for the ammonothermal synthesis of non-nitrogen compound crystals.

Compound	Reactants + mineralizer	*T*/K	*p*/MPa	*t*/d	References
NaOH	NaOH	523–473	≤600	10	[22]
KOH	$K(H_2O)OH + KNH_2$	≤423	≤600	~10	[22]
RbOH	$Rb(H_2O)OH + RbNH_2$	≤365	≤600	~10	[22]
CsOH	$Cs(H_2O)OH + CsNH_2$	460	300	5	[227]
CaOH	$CaO + NH_4I$	723	69–207	1	[57,229]
$K(H_2O)OH$	$K(H_2O)OH$	390–410	500	7–10	[228]
$Rb(H_2O)OH$	$Rb(H_2O)OH$	400–420	180	6	[228]
$Cs(H_2O)OH$	$Cs(H_2O)OH$	450	250	8	[227]
LiHS	$LiNH_2 + H_2S$	300–370	–	–	[3,230]
KHS	$KNH_2 + H_2S$	393	≤30	7	[3,231]
CaS	$CaS + NH_4I$	573–673	69–207	1	[57,229]
SrS	$SrS + NH_4I$	573–673	69–207	1	[57,229]
CdS	$CdS + NH_4I$	573–673	69–207	1	[57,229]
CuS	$CuI + CaS + NH_4I$	643	69–207	1	[229]
Cu_7S_4	$CuI + CaS + NH_4I$	643	69–207	1	[229]
$CaCu_2S_2$	$CuI + CaS + NH_4I$	573–673	69–207	1	[57,229]
$NH_4Cu_4S_3$	$CuI + CaS + NH_4I$	573–673	69–207	1	[229]
Na_2S_2	a				[3,232]
K_2S_2	a				[3,232]
Rb_2S_2	a				[3,232]
Cs_2S_2	Cs + Se	573	200–300	–	[21]
$Na_2S_3 \cdot NH_3$	Na + S	300–320	200	–	[233]
K_2S_3	K + S	423	50	–	[234]
Rb_2S_3	Rb + S	670	300	–	[235]
Cs_2S_3	Cs + S	370	50	–	[235]
K_2S_5	a				[3,232]
Rb_2S_5	Rb + S	450	5	–	[236]
Cs_2S_5	Cs + S	323–373	10–200	–	[237]
Cs_2Se	Cs + Se	573	200	–	[3,238]
Na_2Se_2	a				[3,232]

Table 11. *Cont.*

Compound	Reactants + mineralizer	*T*/K	*p*/MPa	*t*/d	References
K_2Se_2	[a]				[3,232]
Rb_2Se_2	[a]				[3,232]
$Cs_2Se_2{\cdot}xNH_3$	[a]				[3,232]
K_2Se_3	K + Se	423	50	–	[234]
Rb_2Se_3	Rb + Se	600	100	–	[235]
Cs_2Se_3	Cs + Se	570	300	–	[235]
Rb_2Se_5	Rb + S	450	500	–	[236]
K_2Te_3	[a]				[3,232]
Rb_2Te_3	Rb + Te	500	100	–	[239]
Cs_2Te_3	Cs + Te	500	100	–	[239]
Rb_2Te_5	Rb + Te	473	200	7	[56]
Cs_2Te_5	Cs + Te	473	200	90	[240]
$K_2[CN_2]$	Ga + K	853	90	–	[177]

[a] No further information in literature.

For the synthesis of alkali metal hydrogen sulfides and selenides the common methods (e.g., molten metals or hydrogen sulfide hydrates in H_2S stream, precipitation from a saturated solution) usually yield micro crystalline products. The application of liquid or supercritical ammonia in combination with H_2S on alkali metal amides leads to well crystallized alkali metal hydrogen sulfide suitable for single crystal XRD, for example, LiHS and KHS (see Table 11) [3].

Micro crystalline metal sulfides CaS, SrS and CdS can be recrystallized in supercritical ammonia, using NH_4I as mineralizer, at 573–673 K and 69–207 MPa in one day [57]. CuS, Cu_7S_4 and $NH_4Cu_4S_3$ were obtained at 573–673 K and 69–207 MPa as by-products during the synthesis of $CaCu_2S_2$ from CaS, CuI, NH_4I in supercritical ammonia [229] (for details see Table 11).

Reactions of alkali metals with elemental chalcogens in supercritical or subcritical ammonia (e.g., Cs_2S_5 [237]) lead to various polychalcogenides of the types $A(1)_2E_2$ ($A(1)$ = Na, K, Rb, Cs; E = S, Se, Te) and $A(1)_2E_5$ ($A(1)$ = Na, K, Rb, Cs; E = S, Se) [3], namely, Na_2S_3 [233], K_2S_3 [234], Rb_2S_3 [235], Cs_2S_3 [235], Cs_2S_5 [237], K_2Se_3 [234], Rb_2Se_3 [235], Cs_2Se_3 [235], Rb_2Te_3 [239], Cs_2Te_3 [239], Rb_2Te_5 [56], Cs_2Te_5 [240]. Also, Cs_2Se [3,238] and Cs_2S_2 [21] form under ammonothermal conditions and two alkali metal sulfide or selenide ammoniates were reported $Na_2S_3 \cdot NH_3$ [233] and $Cs_2Se_2{\cdot}xNH_3$ [3,232]. For the reaction conditions see Table 11. $Na_2S_3 \cdot NH_3$ [233], Cs_2S_5 [237] and Rb_2S_5 [236] were synthesized under subcritical conditions, but are mentioned for reasons of completeness. Although, some of those compounds were already known prior the introduction of ammonothermal synthesis in this field, a large number of structure determinations suceeded only thanks to single crystals grown from supercritical ammonia. In all mentioned polychalcogenides anions of the type E_n^{2-} (E = S, Se, Te, n = 2–5) occur. There is a significant reduction of the distances between the anions of the type E_n^{2-} (E = S, Se, Te, n = 2–6) in the alkali metal polychalcogenides osberved with increasing atomic weight of the chalcogen atom. This leads to extended anionic networks, e.g., in $RbTe_5$ and $CsTe_5$ [56].

Smaller complex anions occur in K_2S_3, K_2Se_3 [234], Rb_2S_3, Rb_2Se_3, Cs_2S_3, Cs_2Se_3 [235] and Cs_2Te_3 [239]. These compounds crystallize isoytpically in the K_2S_3-type structure and contain kinked $[S_3]^{2-}$ and $[Se_3]^{2-}$ anions, respectively. The alkali metal cations occupy 1/2 of trigonal prisms formed by four or five $[S_3]^{2-}$ or $[Se_3]^{2-}$ groups, resulting in a coordination number 1 + 6 for the cations, with one shorter (S/Se atom above plane of prism) and six larger distances (S/Se atoms in corners of the polyhedron) [234]. Interestingly, although Cs_2Te_3 crystallizes in the K_2S_3-type structure, Rb_2Te_3 crystallizes with the K_2Te_3-type structure [239]. Still, both structures contain Te_3^{2-} polyanions. The reaction of alkali or alkaline-earth metals with carbon impurities under ammonothermal conditions leading to carbodiimides or cyanamides has already been discussed in paragraph *Imides, Nitride Imides, Nitride Amides and Amide Azides*. In this way $K_2[CN_2]$ was obtained as by-product during the synthesis of GaN from Ga and K metal with unintentional carbon impurities at 853 K and 90 MPa [177]. From a supercritical ammonia-methanol mixture at 443–563 K and 16 MPa some nano crystalline oxides were obtained, namely Al_2O_3, TiO_2 and Ga_2O_3 [222].

6. Conclusions

The ammonothermal synthesis is a useful technique suitable for various applications: for commercial large scale productions just as well as for small scale fundamental research. Its advantages for commercial applications are obviously the synthesis of bulk material even of difficult to crystallize compounds (for example, AlN and GaN), the high purity and low defect concentration of the products, due to the inherent chemical material transport, the comparatively mild conditions and the high scalability. For small scale fundamental research the method offers the crystal growth and stabilization of various materials, which are difficultly to crystallize by other synthesis routes, the decrease of reaction time compared to other methods and the access to new materials, which are not accessible by other methods. Nevertheless, the method has its limitations in temperature and pressure. The maximum temperature and pressure vary strongly, depending on the used reaction vessel design and material. Additionally, the corrosion of the autoclave material affects the purity of the products and the working lifetime of the autoclave. The use of a suitable liner material (e.g., precious metals, ceramics) can reduce the corrosion considerably and new constructions with inner heating and counter pressure allow the application of higher temperatures and pressures. Recent research reports indicate the suitability of the ammonothermal method in particular for commercial GaN crystal growth, due to competitive growth rates, high quality and high scalability. However, it is crucial for both the synthesis of novel materials and the improvement of the growth of large crystals to develop a fundamental comprehension of the physical and chemical processes during ammonothermal crystal growth. Knowledge about the thermodynamical and the chemical parameters will permit a specific and well-directed growth of novel functional materials.

The preparative possibilities of ammonothermal synthesis, more than fifty years after the beginning of development, are still far from exhausted. Tasks and future directions are manifold, starting from further exploration of new chemical compounds and materials synthesized by this technique including solid solutions and doped systems with potential for various applications. For this task a deeper understanding of the influence of starting materials, *i.e.*, nutrients for crystal growth as well as the chemistry of the diverse mineralizers and co-mineralizers is badly needed. There is only little knowledge about dissolved species of the various constituents eventually forming the desired products under ammonobasic, ammononeutral and ammonoacidic conditions, depending on temperature, pressure and concentrations. This directly relates to the fundamental physical properties of the ammonia-based solutions of the mineralizers and dissolved materials, which will significantly deviate from those of pure ammonia, e.g., in terms of pressure-temperature relations. Similarly, more complex solutions, like, for example, buffer systems modifying the properties of the solvent with direct influence on the chemistry within the solution. Combined solvent systems as, e.g., ammonia-alcohol mixtures specially tailored for the desired products are a further completely open field, where scientists in the field just have scratched the surface of possibilities. The information on physical properties of ammonia-based solvents and chemistry in solution during the ammonothermal process will clearly lead to new directions of exploratory chemistry. Additionally, it will aid the improved mass-production of better semiconductor materials in sense of lower impurity and defect levels, enhanced growth rates of different crystallographic faces or even improved reactor design, due to a better understanding of dissolution, transport and material deposition processes including transport direction in the gradient and solubility of the desired material within the ammono-based solution. The latter point may lead to new concepts and combinations of nutrient, mineralizer, co-mineralizer and more complex solvent combinations. In terms of technical improvements, we will surely soon see novel liner materials and concepts as well as autoclaves for higher pressures or higher temperatures to be developed, in order to broaden the approachable range in synthetic chemistry and enhance the crystal growth process for commercial products. These commercial products in future may not only be restricted to *h*-GaN crystals, but be broadened to AlN, GaN-based solid solutions, further nitride-based materials and possibly even compounds not containing nitrogen. All in all, we see great prospects for solvothermal reactions in general, but for ammonothermal synthesis, crystal growth and materials production in particular.

Acknowledgments

This work was funded by the Deutsche Forschungsgemeinschaft (DFG) within the frame of the research group FOR1600 "Chemie und Technologie der Ammonothermal-Synthese von Nitriden". We thank all cooperation partners from FOR1600, in particular the group of Eberhard Schlücker and Nicolas Alt (Friedrich-Alexander Universität Erlangen-Nürnberg), without which our experimental work in this field would not be possible.

References

1. Juza, R.; Jacobs, H.; Gerke, H. Ammonothermalsynthese von Metallamiden und Metallnitriden. *Ber. Bunsenges. Phys. Chem.* **1966**, *70*, 1103–1105.
2. Juza, R.; Jacobs, H. Ammonothermal synthesis of magnesium and beryllium amides. *Angew. Chem. Int. Ed.* **1966**, *5*, 247.

3. Jacobs, H.; Schmidt, D. High-pressure ammonolysis in solid-state chemistry. *Curr. Top. Mater. Sci.* **1982**, *8*, 387–427.

4. Byrappa, K.; Yoshimura, M. *Handbook of Hydrothermal Technology a Technology for Crystal Growth and Materials Processing*; Noyes Publications/Wiliam Andrew Publishing, LLC: Park Ridge, NY, USA, 2001.

5. Largeteau, A.; Darracq, S.; Goglio, G.; Demazeau, G. Solvothermal crystal growth of functional materials. *High Press. Res.* **2008**, *28*, 503–508.

6. Rabenau, A. The role of hydrothermal synthesis in preparative chemistry. *Angew. Chem. Int. Ed.* **1985**, *24*, 1026–1040.

7. Demazeau, G. Solvothermal reactions: An original route for the synthesis of novel materials. *J. Mater. Sci.* **2007**, *43*, 2104–2114.

8. Li, B.B.; Xie, Y.; Huang, J.; Qian, Y. Synthesis by a solvothermal route and characterization of $CuInSe_2$ nanowhiskers and nanoparticles. *Adv. Mater.* **1999**, *11*, 1456–1459.

9. Li, Y.D.; Duan, X.F.; Qian, Y.T.; Yang, L.; Ji, M.R.; Li, C.W. Solvothermal co-reduction route to the nanocrystalline III–V semiconductor InAs. *J. Am. Chem. Soc.* **1997**, *119*, 7869–7870.

10. Dwiliński, R.; Wysmolek, A.; Baranowski, J.; Kamińska, M.; Doradziński, R.; Jacobs, H. GaN synthesis by ammonothermal method. *Acta Phys. Pol. A* **1995**, *88*, 833–836.

11. Lan, Y.C.; Chen, X.L.; Cao, Y.G.; Xu, Y.P.; Xun, L.; Xu, T.; Liang, J.K. Low-temperature synthesis and photoluminescence of AlN. *J. Cryst. Growth* **1999**, *207*, 247–250.

12. Nicholas, N.J.; Franks, G.V.; Ducker, W.A. The mechanism for hydrothermal growth of zinc oxide. *CrystEngComm* **2012**, *14*, 1232–1240.

13. Nacken, R. Hydrothermale Mineralsynthese zur Züchtung von Quarzkristallen. *Chem. Ztg.* **1950**, *74*, 745–749.

14. Gleichmann, H.; Richert, H.; Hergt, R.; Barz, R.U.; Grassl, M. Hydrothermal liquid phase epitaxy of gallium orthophosphate on quartz crystal substrates. *Cryst. Res. Technol.* **2001**, *36*, 1181–1188.

15. Li, W.J.; Shi, E.W.; Chen, Z.Z.; Zhen, Y.Q.; Yin, Z.W. Solvothermal synthesis of superfine $Li_{1-x}Mn_2O_{4-\sigma}$ powders. *J. Solid State Chem.* **2002**, *163*, 132–136.

16. Vázquez-Vázquez, C.; López-Quintela, A.M. Solvothermal synthesis and characterisation of $La_{1-x}A_xMnO_3$ nanoparticles. *J. Solid State Chem.* **2006**, *179*, 3229–3237.

17. Bocquet, J.F.; Chhor, K.; Pommier, C. Barium titanate powders synthesis from solvothermal reaction and supercritical treatment. *Mater. Chem. Phys.* **1999**, *57*, 273–280.

18. Suchanek, W. Hydrothermal synthesis of alpha alumina (α-Al_2O_3) powders: Study of the processing variables and growth mechanisms. *J. Am. Ceram. Soc.* **2010**, *93*, 399–412.

19. Korablov, S.; Yokosawa, K.; Korablov, D.; Tohji, K.; Yamasaki, N. Hydrothermal formation of diamond from chlorinated organic compounds. *Mater. Lett.* **2006**, *60*, 3041–3044.

20. Zachwieja, U.; Jacobs, H. Ammonothermalsynthese von Kupfernitrid, Cu_3N. *J. Less-Common Met.* **1990**, *161*, 175–184.

21. Böttcher, P. Zur Synthese und Struktur von Cs_2S_2. *J. Less-Common Met.* **1979**, *63*, 99–103.

22. Jacobs, H.; Kockelkorn, J.; Tacke, T. Hydroxide des Natriums, Kaliums und Rubidiums: Einkristallzüchtung und Röntgenographische Strukturbestimmung an der bei Raumtemperatur stabilen Modifikation. *Z. Anorg. Allg. Chem.* **1985** *531*, 119–124.

23. Popolitov, V.; Litvin, B.; Lobachev, A. Hydrothermal crystallization of semiconducting compounds of group $A^V B^{VI} C^{VII}$ (A^V:Sb, Bi; B^{VI}:S, Se, Te; C^{VII}:I, Br, Cl). *Phys. Status Solidi A* **1970**, *3*, K1–K4.

24. Li, B.; Xie, Y.; Huang, J.; Su, H.; Qian, Y. A Solvothermal route to nanocrystalline Cu_7Te_4 at low temperature. *J. Solid State Chem.* **1999**, *146*, 47–50.

25. Lu, J.; Qi, P.; Peng, Y.; Meng, Z.; Yang, Z. Metastable MnS crystallites through solvothermal synthesis. *Chem. Mater.* **2001**, *13*, 2169–2172.

26. Hao, X.; Yu, M.; Cui, D.; Xu, X.; Wang, Q.; Jiang, M. The effect of temperature on the synthesis of BN nanocrystals. *J. Cryst. Growth* **2002**, *241*, 124–128.

27. Ramesha, K.; Seshadri, R. Solvothermal preparation of ferromagnetic sub-micron spinel $CuCr_2Se_4$ particles. *Solid State Sci.* **2004**, *6*, 841–845.

28. Kendall, J.L.; Canelas, D.A.; Young, J.L.; deSimone, J.M. Polymerizations in supercritical carbon dioxide. *Chem. Rev.* **1999**, *99*, 543–564.

29. Bunsen, R. Über die Spannkraft einiger condensirten Gase. *Ann. Phys.* **1839**, *46*, 97–103.

30. Bunsen, R. Bemerkungen zu einigen Einwürfen gegen mehrere Ansichten chemisch-geologischen Erscheinungen in Island. *Liebigs Ann.* **1848**, *65*, 70–85.

31. Schafhäutl, A. Die neuesten geologischen Hypothesen und ihr Verhältnis zur Naturwissenschaft überhaupt. *Gelehrte Anzeigen München* **1845**, *20*, 577.

32. De Sénarmont, M. Sur la formation des minéraux par voie humide dans les gites métallifères concrétionnés. *Ann. Chim. Phys.* **1851**, *32*, 129–175.

33. Blunck, H.; Juza, R. Verbindungen des Thoriums mit Stickstoff und Wasserstoff. *Z. Anorg. Allg. Chem.* **1974**, *410*, 9–20.

34. Akasaki, I. Nitride semiconductors–Impact on the future world. *J. Cryst. Growth* **2002**, *237-239*, 905–911.

35. Zhu, D.; Wallis, D.; Humphreys, C. Prospects of III-nitride optoelectronics grown on Si. *Rep. Progr. Phys.* **2013**, *76*, 106501.

36. Ehrentraut, D.; Meissner, E.; Bockowski, M. *Technology of Gallium Nitride Crystal Growth*; Springer Materials Sciences: Berlin Heidelberg, Germany, 2010.

37. Denis, A.; Goglio, G.; Demazeau, G. Gallium nitride bulk crystal growth processes: A review. *Mater. Sci. Eng. R.* **2006**, *R50*, 167–194.

38. Dwiliński, R.; Doradziński, R.; Garczyński, J.; Sierzputowski, L.; Puchalski, A.; Kanbara, Y.; Yagi, K.; Minakuchi, H.; Hayashi, H. Bulk ammonothermal GaN. *J. Cryst. Growth* **2009**, *311*, 3015–3018.

39. Ehrentraut, D.; Pakalapati, R.T.; Kamber, D.S.; Jiang, W.; Pocius, D.W.; Downey, B.C.; Mclaurin, M.; D'Evelyn, P.M. High quality, low cost ammonothermal bulk GaN substrates. *Jpn. J. Appl. Phys.* **2013**, *52*, 1–4.

40. Kim, E. Next Generation LED-SORAA. In Proceedings of Strategies in Light: Santa Clara Convention, Silicon Valley, CA, USA, 2 August 2012.

41. Letts, E. Development of GaN wafers for solid-state lighting via the ammonothermal method. In Proceedings of 8th International Workshop on Bulk Nitride Semiconductors, Seeon, Germany, 1 October 2013.

42. D'Evelyn, P.M.; Park, D.S.; LeBoeuf, S.F.; Rowland, L.B.; Narang, K.J.; Hong, H.; Arthur, S.D.;

Sandvik, M. Gallium nitride crystals and wafers and method of making. US Patent 7,786,503 B2, 31 August 2010.

43. Wang, B.; Callahan, M. Ammonothermal synthesis of III-nitride crystals. *Cryst. Growth Des.* **2006**, *6*, 1227–1246.

44. Ehrentraut, D.; Sato, H.; Kagamitani, Y.; Sato, H.; Yoshikawa, A.; Fukuda, T. Solvothermal growth of ZnO. *Progr. Cryst. Growth Charact. Mater.* **2006**, *52*, 280–335.

45. Tomida, D.; Kagamitani, Y.; Bao, Q.; Hazu, K.; Sawayama, H.; Chichibu, S.F.; Yokoyama, C.; Fukuda, T.; Ishiguro, T. Enhanced growth rate for ammonothermal gallium nitride crystal growth using ammonium iodide mineralizer. *J. Cryst. Growth* **2012**, *353*, 59–62.

46. Bao, Q.; Saito, M.; Hazu, K.; Furusawa, K.; Kagamitani, Y.; Kayano, R.; Tomida, D.; Qiao, K.; Ishiguro, T.; Yokoyama, C.; *et al.* Ammonothermal crystal growth of GaN using an NH4F mineralizer. *Cryst. Growth Des.* **2013**, *13*, 4158–4161.

47. Holleman, A.; Wiberg, N.; Wiberg, E. *Lehrbuch der Anorganischen Chemie*, 102nd ed.; de Gruyter: Berlin, Germany, 2007.

48. Franklin, E.C.; Fernelius, W.C. *The Nitrogen System of Compounds*; Reinhold Publishing Corporation: New York, NY, USA, 1935.

49. Bronn, J. Über den Zustand der metallischen Lösungen. *Ann. Phys.* **1905**, *16*, 166–171.

50. Kraus, C.A. Solutions of metals in non-metallic solvents. VI. The conductance of the alkali metals in liquid ammonia. *J. Am. Chem. Soc.* **1921**, *43*, 749–770.

51. Haar, L.; Gallagher, J. Thermodynamic properties of ammonia. *J. Phys. Chem. Ref. Data* **1978**, *7*, 635–792.

52. Xiang, H.W. Vapor pressures, critical parameters, boiling points, and triple points of ammonia and trideuteroammonia. *J. Phys. Chem. Ref. Data* **2004**, *33*, 1005.

53. Lemmon, E.W.; McLinden, M.O.; Friend, D.G. Thermophysical properties of fluid systems, NIST standard reference data. Available online: http://webbook.nist.gov/chemistry/fluid/ (accessed on 09 December 2013).

54. Shatenshtein, A.I. A study of acid catalysis in liquid ammonia. *J. Am. Chem. Soc.* **1937**, *59*, 432–435.

55. Rondinini, S.; Longhi, P.; Mussini, P.R.; Mussini, T. Autoprotolysis constants in nonaqueous solvents and aqueous organic solvent mixtures. *Pure Appl. Chem.* **1987**, *59*, 1693–1702.

56. Böttcher, P.; Kretschmann, U. Darstellung und Kristallstruktur von Dirubidiumpentatellurid, Rb2Te5. *J. Less-Common Met.* **1983**, *95*, 81–91.

57. Purdy, A.P. Ammonothermal crystal growth of sulfide materials. *Chem. Mater.* **1998**, *10*, 692–694.

58. Glasser, L. Equations of state and phase diagrams of ammonia. *J. Chem. Educ.* **2009**, *86*, 1457–1458.

59. Alt, N.S.A.; Schlücker, E. Hochdruck-Sichtzelle für Untersuchungen des ammonothermalen Prozesses. *Chem. Ing. Tech.* **2011**, *83*, 280–285.

60. Alt, N.S.A.; Meissner, E.; Schlücker, E. Development of a novel in situ monitoring technology for ammonothermal reactors. *J. Cryst. Growth* **2012**, *350*, 2–4.

61. Hüttig, G.F. Apparat zur gleichzeitigen Druck- und Raummessung von Gasen (Tensi-eudiometer). *Z. Anorg. Allg. Chem.* **1920**, *114*, 161–173.

62. Yoshida, K.; Aoki, K.; Fukuda, T. High-temperature acidic ammonothermal method for GaN crystal growth. *J. Cryst. Growth* 2013, in press.

63. Callahan, M.J.; Chen, Q.S.C. Hydrothermal and Ammonothermal Growth of ZnO and GaN. In *Handbook of Crystal Growth*, 1st ed.; Dhanaraj, G., Byrappa, K., Prasad, V., Dudley, M., Eds.; Springer: Berlin/ Heidelberg, Germany, 2010; pp. 655–689.

64. Binnewies, M.; Glaum, R.; Schmidt, M.; Schmidt, P. *Chemical Vapor Transport Reactions*; Walter de Gruyter: Berlin, Germany & Boston, MA, USA, 2012.

65. Wang, B.; Callahan, M. Transport growth of GaN crystals by the ammonothermal technique using various nutrients. *J. Cryst. Growth* **2006**, *291*, 455–460.

66. Zhang, S.; Hintze, F.; Schnick, W.; Niewa, R. Intermediates in ammonothermal GaN crystal growth under ammonoacidic conditions. *Eur. J. Inorg. Chem.* **2013**, *2013*, 5387–5399.

67. Zhang, S.; Alt, N.S.A.; Schlücker, E.; Niewa, R. Novel alkali metal amidogallates as intermediates in ammonothermal GaN crystal growth. *J. Cryst. Growth* 2013, in press.

68. Alt, N.S.A.; Meissner, E.; Schlücker, E.; Frey, L. In situ monitoring technologies for ammonothermal reactors. *Phys. Status Solidi C* **2012**, *9*, 436–439.

69. Erlekampf, J.; Seebeck, J.; Savva, P.; Meissner, E.; Friedrich, J.; Alt, N.; Schlücker, E.; Frey, L. Numerical time-dependent 3D simulation of flow pattern and heat distribution in an ammonothermal system with various baffle shapes. *J. Cryst. Growth* 2014, in press.

70. Vogelsang, K.; Schröter, W.; Hoffmann, R.; Jacobs, H. Ein Beitrag zum Problem der Porenbildung. *Härterei Techn. Mitt.* **2002**, *57*, 42–48.

71. Cundy, C.S.; Cox, P.a. The hydrothermal synthesis of zeolites: Precursors, intermediates and reaction mechanism. *Microporous Mesoporous Mater.* **2005**, *82*, 1–78.

72. Moolenaar, R.J.; Evans, J.C.; McKeever, L.D. The structure of the aluminate ion in solutions at high pH. *J. Phys. Chem.* **1970**, *74*, 3629–3636.

73. Juza, R. Amides of the alkali and the alkaline earth metals. *Angew. Chem. Int. Ed.* **1964**, *3*, 471–481.

74. Juza, R. Über die Amide der 1. und 2. Gruppe des periodischen Systems. *Z. Anorg. Allg. Chem.* **1937**, *231*, 121–135.

75. Juza, R. Über ein Nitridamid des Zirkoniums. *Z. Anorg. Allg. Chem.* **1974**, *406*, 145–152.

76. Jacobs, H.; Kablitz, D. Untersuchung des Systems Kalium/Cer/Ammoniak. *Z. Anorg. Allg. Chem.* **1979**, *454*, 35–42.

77. Dwiliński, R.; Baranowski, J.; Kamińska, M. On GaN crystallization by ammonothermal method. *Acta Phys. Pol. A* **1996**, *90*, 763–766.

78. Peters, D. Ammonothermal synthesis of aluminum nitride. *J. Cryst. Growth* **1990**, *104*, 411–418.

79. Lan, Y.; Chen, X.L.; Xu, Y.; Cao, Y.; Huang, F. Syntheses and structure of nanocrystalline gallium nitride obtained from ammonothermal method using lithium metal as mineralizator. *Mater. Res. Bull.* **2000**, *35*, 2325–2330.

80. Guarino, R.; Rouxel, J. L'amidogallate de potassium $KGa(NH_2)_4$ et l'imidogallate $KGa(NH)_2$. L'obtention de l'amidure de gallium $Ga(NH_2)_3$. *Bull. Soc. Chim. Fr.* **1969**, *7*, 2284–2287.

81. Jacobs, H.; Nöcker, B. Neubestimmung von Struktur und Eigenschaften isotyper Natriumtetraamidometallate des Aluminiums und Galliums. *Z. Anorg. Allg. Chem.* **1993**, *619*,

381–386.

82. Molinié, P.; Brec, R.; Rouxel, J. Le pentaamidogallate de sodium: $Na_2Ga(NH_2)_5$. *C. R. Hebd. Seances Acad. Sci. C* **1972**, *274*, 1388–1391.

83. Ketchum, D.; Schimek, G.; Pennington, W.; Kolis, J. Synthesis of new group III fluoride–ammonia adducts in supercritical ammonia: Structures of $AlF_3(NH_3)_2$ and $InF_2(NH_2)(NH_3)$. *Inorg. Chim. Act.* **1999**, *294*, 200–206.

84. Jacobs, H.; Schröder, F.O. Penta-ammoniates of aluminium halide: The crystal structures of $AlX_3{\cdot}5NH_3$ with X = Cl , Br , I. *Z. Anorg. Allg. Chem.* **2002**, *628*, 951–955.

85. Johnson, W.C.; Parsons, J.B. Nitrogen compounds of gallium. *J. Phys. Chem.* **1932**, *36*, 2588–2594.

86. Wang, B.; Callahan, M.; Rakes, K.; Bouthillette, L.; Wang, S.Q.; Bliss, D.; Kolis, J. Ammonothermal growth of GaN crystals in alkaline solutions. *J. Cryst. Growth* **2006**, *287*, 376–380.

87. Ehrentraut, D.; Kagamitani, Y.; Yokoyama, C.; Fukuda, T. Physico-chemical features of the acid ammonothermal growth of GaN. *J. Cryst. Growth* **2008**, *310*, 891–895.

88. Purdy, A.P. Ammonothermal synthesis of cubic gallium nitride. *Chem. Mater.* **1999**, *11*, 1648–1651.

89. Peters, D.; Bock, J.; Jacobs, H. Hexaaminaluminiumiodidammoniakat-$[Al(NH_3)_6]I_3NH_3$-Darstellung und Kristallstruktur. *J. Less-Common Met.* **1989**, *154*, 243–250.

90. Jacobs, H.; Bock, J.; Stüve, C. Röntgenographische Strukturbestimmung und IR-spektroskopische Untersuchungen an Hexaamindiiodiden, $[M(NH_3)_6]I_2$, von Eisen und Mangan. *J. Less-Common Met.* **1987**, *134*, 207–214.

91. Leineweber, A.; Friedriszik, M.W.; Jacobs, H. Preparation and crystal structures of $Mg(NH_3)_2Cl_2$, $Mg(NH_3)_2Br_2$, and $Mg(NH_3)_2I_2$. *J. Solid State Chem.* **1999**, *234*, 229–234.

92. Essmann, R. Influence of coordination on N-H...X- hydrogen bonds. Part 1. $[Zn(NH_3)_4]Br_2$ and $[Zn(NH_3)_4]I_2$. *J. Mol. Struct.* **1995**, *356*, 201–206.

93. Meyer, G.; Roos, M. Zwei Galliumfluorid-Ammoniakate: $Ga(NH_3)F_3$ und $Ga(NH_3)_2F_3$. *Z. Anorg. Allg. Chem.* **1999**, *625*, 1129–1134.

94. Bremm, S.; Meyer, G. Reactivity of ammonium halides: Action of ammonium chloride and bromide on iron and iron(III) chloride and bromide. *Z. Anorg. Allg. Chem.* **2003**, *629*, 1875–1880.

95. Bremm, S.; Meyer, G.; Möller, A.; Amann, P.; Sobotka, B. Einwirkung von Ammoniumhalogeniden auf Eisen und Eisenhalogenide. *Z. Anorg. Allg. Chem.* **2002**, *628*, 2190.

96. Widenmeyer, M.; Hansen, T.C.; Meissner, E.; Niewa, R. Formation and decomposition of iron nitrides observed by in situ powder neutron diffraction and thermal analysis. *Z. Anorg. Allg. Chem.* 2014, in press.

97. Hambley, T.W.; Lay, P.A. Comparisons of π-bonding and hydrogen bonding in isomorphous compounds: $[M(NH_3)_4Cl]Cl_2$ (M = Cr, Co, Rh, Ir, Ru, Os). *Inorg. Chem.* **1986**, *25*, 4553–4558.

98. Podberezskaya, N.; Yudanova, T.; Magarill, S.; Ipatova, E.; Romanenko, G.; Pervukhina, N.; Borisov, S. Structures of crystals of inorganic coordination compounds with complex ions $[MA_5X]$ and $[MX_5A]$ containing neutral (A) and acid (X) ligands, at very high packing

densities. *J. Struct. Chem.* **1992**, *32*, 894–904.

99. Weishaupt, M.; Bezler, H.; Strähle, J. Darstellung und Kristallstruktur von $(NH_4)_2[V(NH_3)Cl_5]$. Die Kristallchemie der Salze $(NH_4)_2[V(NH_3)Cl_5]$, $[Rh(NH_3)Cl]Cl_2$ und M_2VXCl_5 mit M = K, NH_4, Rb, Cs und X = Cl, O. *Z. Anorg. Allg. Chem.* **1978**, *440*, 52–64.

100. Jacobs, H.; Schmidt, D.; Schmitz, D. Struktur und Eigenschaften der Caesiumamidolanthanat-monoammoniakate $Cs_3La(NH_2)_6{\cdot}NH_3$ und $Cs_4La(NH_2)_7{\cdot}NH_3$. *J. Less-Common Met.* **1981**, *81*, 121–133.

101. Pust, P.; Schmiechen, S.; Hintze, F.; Schnick, W. Ammonothermal synthesis and crystal structure of $BaAl_2(NH_2)_8{\cdot}2NH_3$. *Z. Anorg. Allg. Chem.* **2013**, *639*, 1185–1187.

102. Joannis, A. Sur l'amidure de sodium et sur un chlorure de disodammonium. *C. R. Hebd. Seances Acad. Sci. C* **1891**, *112*, 392–394.

103. Mammano, N.; Sienko, M. Low-temperature X-ray study of the compound tetraaminelithium. *105. J. Am. Chem. Soc.* **1968**, *90*, 6322–6324.

104. Kirschke, E.J.; Jolly, W.L. The reversibility of the reaction of alkali metals with liquid ammonia. *Inorg. Chem.* **1967**, *6*, 855–862.

105 Ruff, O.; Geisel, E. Über die Natur der sogenannten Metallammoniumverbindungen. *Ber. Dtsch. Chem. Ges.* **1906**, *39*, 828–843.

106. Juza, R.; Fasold, K.; Haeberle, C. Untersuchungen über die Amide der Alkalimetalle. *Z. Anorg. Allg. Chem.* **1937**, *234*, 75–85.

107. Juza, R.; Opp, K. Die Kristallstruktur des Lithiumamides. *Z. Anorg. Allg. Chem.* **1951**, *266*, 313–324.

108. Juza, R.; Weber, H.H.; Opp, K. Kristallstruktur des Natriumamids. *Z. Anorg. Allg. Chem.* **1956**, *284*, 73–82.

109. Jacobs, H.; Juza, R. Neubestimmung der Kristallstruktur des Lithiumamids. *Z. Anorg. Allg. Chem.* **1972**, *391*, 271–279.

110. Nagib, M.; Jacobs, H. Neutronenbeugung am Lithiumdeuteroamid. *Atomkernenergie* **1973**, *21*, 275–278.

111. Nagib, M.; Kistrup, H.; Jacobs, H. Neutronenbeugung am Natriumdeuteroamid. *Atomkernenergie* **1975**, *26*, 87–90.

112. Bohger, P.; Zeiske, T.; Jacobs, H. Neutronenbeugung an der Tieftemperaturmodifikation von Rubidiumdeuteroamid. *Z. Anorg. Allg. Chem.* **1998**, *624*, 364–366.

113. Nagib, M.; von Osten, E.; Jacobs, H. Röntgen- und Neutronenbeugung und Bestimmung der Wärmekapazität an Caesiumamid–$CsNH_2$–und Caesiumdeuteroamid–$CsND_2$–bei Temperaturen von 348 bis 33 K. *Atomkernenergie* **1983**, *43*, 47–54.

114. Jacobs, H.; Juza, R. Darstellung und Eigenschaften von Berylliumamid und -imid. *Z. Anorg. Allg. Chem.* **1969**, *370*, 248–253.

115. Jacobs, H.; Juza, R. Darstellung und Eigenschaften von Magnesiumamid und -imid. *Z. Anorg. Allg. Chem.* **1969**, *370*, 254–261.

116. Senker, J.; Jacobs, H.; Müller, M. Reorientational dynamics of amide ions in isotypic phases of strontium and calcium amide. 1. Neutron diffraction experiments. *J. Phys. Chem. B* **1998**, *102*, 931–940.

117. Nagib, M.; Jacobs, H.; Kistrup, H. Neutronenbeugung am Strontiumdeuteroamid, $Sr(ND_2)_2$,

bei Temperaturen von 31 bis 570 K. *Atomkernenergie* **1979**, *33*, 38–42.

118. Jacobs, H.; Hadenfeldt, C. Die Kristallstruktur von Bariumamid, $Ba(NH_2)_2$. *Z. Anorg. Allg. Chem.* **1975**, *418*, 132–140.

119. Fröhling, B.; Kreiner, G.; Jacobs, H. Synthesis and crystal structure of manganese(II) and zinc amides, $Mn(NH_2)_2$ and $Zn(NH_2)_2$. *Z. Anorg. Allg. Chem.* **1999**, *625*, 211–216.

120. Hadenfeldt, C.; Gieger, B.; Jacobs, H.Die Kristallstruktur von Lanthanamid, $La(NH_2)_3$. *Z. Anorg. Allg. Chem.* **1974**, *410*, 104–112.

121. Jacobs, H.; Kistrup, H. Über das System Kalium/Samarium/Ammoniak. *Z. Anorg. Allg. Chem.* **1977**, *435*, 127–136.

122. Jacobs, H.; Fink, U. Untersuchung des Systems Kalium/Europium/Ammoniak. *Z. Anorg. Allg. Chem.* **1978**, *438*, 151–159.

123. Hadenfeldt, C.; Jacobs, H.; Juza, R. Über die Amide des Europiums und Ytterbiums. *Z. Anorg. Allg. Chem.* **1970**, *379*, 144–156.

124. Nagib, M.; von Osten, E.; Jacobs, H. Neutronenbeugung an drei Modifikationen des Kaliumdeuteroamids KND_2. *Atomkernenergie* **1977**, *29*, 41–47.

125. Nagib, M.; Jacobs, H.; von Osten, E.Neutronenbeugung am Kaliumdeuteroamid KND_2 bei 31 K. *Atomkernenergie* **1977**, *29*, 303–304.

126. Dachs, H.Bestimmung der Lage des Wasserstoffs in LiOH durch Neutronenbeugung. *Z. Kristallogr.* **1956**, *112*, 60–67.

127. Juza, R.; Jacobs, H.; Klose, W. Die Kristallstrukturen der Tieftemperaturmodifikationen von Kalium- und Rubidiumamid. *Z. Anorg. Allg. Chem.* **1965**, *338*, 171–178.

128. Jacobs, H.; Nagib, M.; Osten, E.V. Einkristallzüchtung und Kristallchemie der Alkali- und Erdalkalimetallamide. *Acta Crystallogr. A* **1978**, *34*, 168.

129. Schenk, P.; Tulhoff, H. Das System Kaliumamid/Ammoniak. *Angew. Chem.* **1962**, *74*, 962.

130. Juza, R.; Schumacher, H. Zur Kenntnis der Erdalkalimetallamide. *Z. Anorg. Allg. Chem.* **1963**, *324*, 278–286.

131. Jacobs, H.; Scholze, H. Untersuchung des Systems $Na/La/NH_3$. *Z. Anorg. Allg. Chem.* **1976**, *427*, 8–16.

132. Juza, R.; Fasold, K.; Kuhn, W. Untersuchungen über Zink- und Cadmiumamid. *Z. Anorg. Allg. Chem.* **1937**, *234*, 86–96.

133. Fitzgerald, F.F. Reactions in liquid ammonia. Potassium ammonozincate, cuprous nitride and 110. an ammonobasic mercuric bromide. *J. Am. Chem. Soc.* **1907**, *29*, 656–665.

134. Joannis, M. Action du chlorure de bore sur le gaz ammoniac. *C. R. Hebd. Seances Acad. Sci.* **112. 1902**, *135*, 1106.

135. Janik, J.F.; Wells, R.L. Gallium imide, $\{Ga(NH)_{3/2}\}_n$, a new polymeric precursor for gallium nitride powders. *Chem. Mater.* **1996**, *8*, 2708–2711.

136. Wiberg, E.; May, A. Über die Umsetzung von Aluminiumwasserstoff mit Ammoniak u and Aminen. *Z. Naturforsch.* **1955**, *10b*, 229.

137. Semenenko, K.N.; Bulychev, B.M.; Shevlyagina, E.A. Aluminium hydride. *Russ. Chem. Rev. (Engl. Transl.)* **1966**, *35*, 649–658.

138. Purdy, A.P. Indium(III) amides and nitrides. *Inorg. Chem.* **1994**, *33*, 282–286.

139. Vigouroux, E.; Hugot, C. Silicon amide and imide. *C. R. Hebd. Seances Acad. Sci. C* **1903**, *136*, 1670–1672.

140. Glemser, O.; Naumann, P. Über den thermischen Abbau von Siliciumdiimid $Si(NH)_2$. *Z. Anorg. Allg. Chem.* **1959**, *298*, 134–141.

141. Molinié, P.; Brec, R.; Rouxel, J.; Herpin, P. Structures des amidoaluminates alcalins $MAl(NH_2)_4$ (M = Na, K, Cs). Structure de l'amidogallate de sodium $NaGa(NH_2)_4$. *Acta Crystallogr. B* **1973**, *29*, 925–934.

142. Drew, M.; Goulter, J.; Guémas-Brisseau, L.; Palvadeau, P.; Rouxel, J.; Herpin, P. Etude structurale d'amidobéryllates de rubidium et de potassium. *Acta Crystallogr. B* **1974**, *30*, 2579–2582.

143. Jacobs, H.; Jänichen, K. Lithiumaluminiumamid, $LiAl(NH_2)_4$, Darstellung, röntgenographische Untersuchung, Infrarotspektrum und thermische Zersetzung. *Z. Anorg. Allg. Chem.* **1985**, *531*, 125–139.

144. Tenten, A.; Jacobs, H. Strukturen und thermisches Verhalten von Kaliumtetraamidoaluminat, α- und β-$KAl(NH_2)_4$. *Z. Kristallogr.* **1989**, *186*, 289–291.

145. Jacobs, H.; Harbrecht, B. Substitution in layers of cations in lithium amide: Potassium trilithium amide, $KLi_3(NH_2)_4$, and potassium heptalithium amide, $KLi_7(NH_2)_8$. *Z. Anorg. Allg. Chem.* **1984**, *518*, 87–100.

146. Kraus, F.; Korber, N. $K_2Li(NH_2)_3$ and $K_2Na(NH_2)_3$–Synthesis and crystal structure of two crystal-chemically isotypic mixed-cationic amides. *J. Solid State Chem.* **2005**, *178*, 1241–1246.

147. Jacobs, H.; Kockelkorn, J. Darstellung und Eigenschaften der Amidomagnesate des Kaliums und Rubidiums $K_2[Mg(NH_2)_4]$- und $Rb_2[Mg(NH_2)_4]$-Verbindungen mit isolierten $[Mg(NH_2)_4]^{2-}$ Tetraedern. *J. Less-Common Met.* **1984**, *97*, 205–214.

148. Jacobs, H.; Birkenbeul, J.; Schmitz, D. Strukturverwandschaft des Dicaesiumamidomagnesats, $Cs[Mg(NH_2)_4]$, zum β-K_2SO_4-Typ. *J. Less-Common Met.* **1982**, *85*, 79–86.

149. Jacobs, H.; Fink, U. Über Natrium- und Kaliumamidometallate des Calciums, Strontiums und Europiums. *J. Less-Common Met.* **1979**, *63*, 273–286.

150. Jacobs, H.; Kockelkorn, J.; Birkenbeul, J. Struktur und Eigenschaften der ternären Metallamide $NaCa(NH_2)_3$, $KBa(NH_2)_3$, $RbBa(NH_2)_3$, $Rb(Eu(NH_2)_3$ und $RbSr(NH_2)_3$. *J. Less-Common Met.* **1982**, *87*, 215–224.

151. Jacobs, H.; Fink, U. Darstellung und Kristallstruktur von $KCa(NH_2)_3$. *Z. Anorg. Allg. Chem.* **1977**, *435*, 137–145.

152. Jacobs, H.; Kockelkorn, J. Darstellung und Kristallstruktur des Rubdiumcalciumamids, $RbCa(NH_2)_3$. *Z. Anorg. Allg. Chem.* **1979**, *456*, 147–154.

153. Jacobs, H.; Kockelkorn, J. ber Caesiumamidometallate ($CsM(NH_2)_3$) des Calciums, Strontiums und Europiums; Verbindungen mit der Struktur "Hexagonaler Perowskite". *J. Less-Common Met.* **1981**, *81*, 143–154.

154. Jacobs, H. Darstellung und Eigenschaften des Caesiumbariumamids, $CsBa(NH_2)_3$: Strukturverwandschaft zum NH_4CdCl_3-Typ. *J. Less-Common Met.* **1982**, *85*, 71–78.

155. Hadenfeldt, C.; Gieger, B.; Jacobs, H. Darstellung und Kristallstruktur von $KLa_2(NH_2)_7$. *Z. Anorg. Allg. Chem.* **1974**, *408*, 27–36.

156. Hadenfeldt, C.; Gieger, B.; Jacobs, H. Darstellung und Kristallstruktur von $K_3La(NH_2)_6$. *Z. Anorg. Allg. Chem.* **1974**, *403*, 319–326.

157. Jacobs, H.; Stüve, C. Rubidiumhexaamidolanthanat und -neodymat, $Rb_3[La(NH_2)_6]$ und $Rb_3[Nd(NH_2)_6]$; Strukturverwandtschaft zu $K_3[Cr(OH_6)]$ und K_4CdCl_6. *Z. Anorg. Allg. Chem.* **1987**, *546*, 42–47.

158. Jacobs,H.; Schmidt,D. Ü ber ein Caesiumheptaamidodilanthanat $CsLa_2(NH_2)_7$. *J. Less-Common Met.* **1981**, *78*, 51–59.

159. Jacobs, H.; Schmidt, D. Struktur und Eigenschaften von perowskitartigen Caesiumamidometallaten des Cers, Neodyms und Samariums $Cs_3Ln_2(NH_2)_9$. *J. Less-Common Met.* **1980**, *76*, 227–244.

160. Jacobs, H.; Kockelkorn, Uber Kalium- und Rubidiumamidometallate des Europiums, 163. Yttriums und Ytterbiums, $K_3M(NH_2)_6$ und $Rb_3M(NH_2)_6$. *J. Less-Common Met.* **1982**, *164. 85*, 97–110.

161 Linde, G.; Juza, R. Amidometallate von Lanthan und Gadolinium und Umsetzung von Lanthan, Gadolinium und Scandium mit Ammoniak. *Z. Anorg. Allg. Chem.* **1974**, *409*, 191–198.

162. Jacobs, H.; Peters, D.; Hassiepen, K. Caesiumamidometallate des Gadoliniums, Ytterbiums und Yttriums mit perowskitverwandten Atomanordnungen $Cs_3M_2(NH_2)_9$. *J. Less-Common Met.* **1986**, *118*, 31–41.

163. Stuhr, A.; Jacobs, H.; Juza, R. Amide des Yttriums. *Z. Anorg. Allg. Chem.* **1973**, *395*, 291–300.

164. Peters, D.; Jacobs, H. Übergang von dichter Anionenpackung zu perowskitartiger Struktur bei Kalium- und Rubidiumamidyttriat, $KY(NH_2)_4$ und $RbY(NH_2)_4$. *J. Less-Common Met.* **1986**, *119*, 99–113.

165. Hadenfeldt, C.; Jacobs, H. Darstellung, Eigenschaften und Kristallstruktur von $Na_3[Yb(NH_2)_6]$. *170. Z. Anorg. Allg. Chem.* **1972**, *393*, 111–125.

166. Jacobs, H.; Jänichen, K. Darstellung und Kristallstruktur von Tetraamidoaluminaten des Rubidiums und Caesiums, $Rb[Al(NH_2)_4]$ und $Cs[Al(NH_2)_4]$. *J. Less-Common Met.* **1990**, *159*, 315–325.

167. Kreiner, G.; Jacobs, H. Magnetische Struktur von η-Mn_3N_2. *J. Alloys Compd.* **1992**, *183*, 345–362.

168. Richter, T.M.M.; Zhang, S.; Niewa, R. Ammonothermal synthesis of dimorphic $K_2[Zn(NH_2)_4]$. *Z. Kristallogr.* **2013**, *228*, 351–358.

169. Fröhling, B.; Jacobs, H. Positions of the protons in potassium tetraamidozincate, $K_2Zn(NH_2)_4$. *Anorg. Allg. Chem.* **1997**, *623*, 1103–1107.

170. Brec, R.; Novak, A.; Rouxel, J. Etude par spectroscopie infrarouge des amidoaluminates de lithium, sodium et potassium. *Bull. Soc. Chim. Fr.* **1967**, *7*, 2432–2435.

171. Drew, M.; Guémas, L.; Chevalier, P.; Palvadeau, P.; Rouxel, J. Etude structurale de l'amidozincate de rubidium $Rb_2Zn(NH_2)_4$ et de l'amidomanganite de potassium $K_2Mn(NH_2)_4$. *Rev. Chim. Min.* **1975**, *12*, 419–426.

172. Jacobs, H.; Gieger, B.; Hadenfeldt, C. *J. Less-Common Met.* **1979**, *64*, 91–99.

173. Über das System Kalium/Lanthan/Ammoniak. Peters, D.; Jacobs, H. Ammonothermalsynthese von kristallinem Siliciumnitridimid, Si_2N_2NH. *J. Less-Common Met.* **1989**, *146*, 241–249.

174. Harbrecht, B.; Jacobs, H. Hochdrucksynthese von Caesiumamidazid, $Cs_2(NH_2)N_3$ aus Caesiummetall und Ammoniak. *Z. Anorg. Allg. Chem.* **1983**, *500*, 181–187.

175. Höhn, P.; Kniep, R.; Maier, J. $Ba_9N[N_3][TaN_4]_2$ ein Nitridotantalat(V) mit Nitrid- und Azid-Ionen. *Angew. Chem.* **1993**, *105*, 1409–1410.

176. Clarke, S.J.; DiSalvo, F.J. Crystal structure of nonabarium bis(tetranitridoniobate) nitride azide, $Ba_9[NbN_4]_2N[N_3]$. *Z. Kristallogr. NCS* **1997**, *212*, 309–310.

177. Zhang, S.; Zherebtsov, D.; DiSalvo, F.J.; Niewa, R. $Na_5[CN_2]_2[CN]$, $(Li,Na)_5[CN_2]_2[CN]$, and $K_2[CN_2]$: Carbodiimides from high-pressure synthesis. *Z. Anorg. Allg. Chem.* **2012**, *193.* *638*, 2111–2116.

178. Stock, A.; Blix, M. Über das Borimid, $B_2(NH)_3$. *Ber. Dtsch. Chem. Ges.* **1901**, *34*, 3039–3048.

179. Blix, M.; Wirbelauer, W ber das Siliciumsulfochlorid, $SiSCl_2$, Siliciumimid, $Si(NH)_2$, Siliciumstickstoffimid (Silicam), Si_2N_3H und den Siliciumstickstoff, Si_3N_4. *Ber. Dtsch. Chem. Ges.* **1903**, *36*, 4220–4228.

180. Ketchum, D.; Kolis, J. Crystal growth of gallium nitride in supercritical ammonia. *J. Cryst. Growth* **2001**, *222*, 431–434.

181. Jouet, R.J.; Purdy, A.P.; Wells, R.L.; Janik, J.F. Preparation of phase pure cubic gallium nitride, *c*-GaN, by ammonothermal conversion of gallium imide, $\{Ga(NH)_{3/2}\}_n$. *J. Clust. Sci.* **2002**, *13*, 469–486.

182. Zajac, M.; Gosk, J.; Grzanka, E.; Stelmakh, S.; Palczewska, M.; Wysmoek, A.; Korona, K.; Kamiska, M.; Twardowski, A. Ammonothermal synthesis of GaN doped with transition metal ions (Mn, Fe, Cr). *J. Alloys Compd.* **2008**, *456*, 324–338.

183. Jacobs, H.; Rechenbach, D.; Zachwieja, U. Structure determination of γ'-Fe_4N and E-Fe_3N. *J. Alloys Compd.* **1995**, *227*, 10–17.

184. Jacobs, H.; Rechenbach, D.; Zachwieja, D. Untersuchungen zur Struktur und zum Zerfall von Eisennitriden-γ'-Fe_4N und E-Fe_3N. *Härterei Techn. Mitt.* **1995**, *50*, 205–213.

185. Jacobs, H.; Bock, J. Einkristallzüchtung von γ^t-Fe_4N in *209. J. Less-Common Met.* **1987**, *134*, 215–220.

186. überkritischem Ammoniak. Jacobs, H.;Stüve, C. Hochdrucksynthese der η-phase im System Mn-N: Mn_3N_2. *J. Less-Common Met.* **1984**, *96*, 323–329.

187. Leineweber, A.; Jacobs, H.; Hull, S. Ordering of nitrogen in nickel nitride Ni_3N determined by neutron diffraction. *Inorg. Chem.* **2001**, *40*, 5818–5822.

188. Jacobs, H.; Zachwieja, U. Kupferpalladiumnitride, Cu_3Pd_xN mit x = 0, 020 und 0,989, Perowskite mit "bindender $3d^{10}$-$4d^{10}$-Wechselwirkung". *J. Less-Common Met.* **1991**, *170*, 185–190.

189. von Stackelberg, M.; Paulus, R. Untersuchungen über die Kristallstrukturen der Nitride und Phosphide zweiwertiger Metalle. *Z. Physik. Chem. B* **1933**, *22*, 305.

190. Reckeweg, O.; Lind, C.; DiSalvo, F.J. Rietveld refinement of the crystal structure of α-Be_3N_2 and the experimental determination of optical band gaps for Mg_3N_2, Ca_3N_2 and $CaMg_2N_2$. *Z. Naturforsch.* **2003**, *58b*, 159–162.

191. Karau, F.; Schnick, W. Synthese von Cadmiumnitrid Cd_3N_2 durch thermischen Abbau von Cadmiumazid $Cd(N_3)_2$ und Kristallstrukturbestimmung aus Röntgen-Pulverbeugungsdaten. *Z. Anorg. Allg. Chem.* **2007**, *633*, 223–226.

192. Partin, D.E.; Williams, D.J.; O'Keeffe, M. The crystal structures of Mg_3N_2 and Zn_3N_2. *J. Solid State Chem.* **1997**, *132*, 56–59.

193. Reckeweg, O.; DiSalvo, F.J. About binary and ternary alkaline earth metal nitrides. *Z. Anorg. Allg. Chem.* **2001**, *627*, 371–377.

194. Soto, G.; Diaz, J.A.; De la Cruz, W.; Contreras, O.; Moreno, M.; Reyes, A. Epitaxial α-Be_3N_2

thin films grown on Si substrates by reactive laser ablation. *Mater. Sci. Eng., B.* **2002**, *94*, 62–65.

195. García Núñez, C.; Pau, J.; Hernández, M.; Cervera, M.; Ruiz, E.; Piqueras, J. On the zinc nitride properties and the unintentional incorporation of oxygen. *Thin Solid Films* **2012**, *520*, 1924–1929.

196. Soto, G.; Díaz, J.A.; Machorro, R.; Reyes-Serrato, A.; de la Cruz, W. Beryllium nitride thin film grown by reactive laser ablation. *Mater. Lett.* **2002**, *52*, 29–33.

197. Toyoura, K.; Tsujimura, H.; Goto, T.; Hachiya, K.; Hagiwara, R.; Ito, Y. Optical properties of zinc nitride formed by molten salt electrochemical process. *Thin Solid Films* **2005**, *492*, 88–92.

198. Maruska, H.P.; Tietjen, J.J. The preparation and properties of vapor-deposited single-crystalline GaN. *Appl. Phys. Lett.* **1969**, *15*, 327.

199. Perry, P.B.; Rutz, R.F. The optical absorption edge of single-crystal AlN prepared by a close-spaced vapor process. *Appl. Phys. Lett.* **1978**, *33*, 319–321.

200. Wu, J.; Walukiewicz, W.; Shan, W.; Yu, K.; Ager, J.; Haller, E.E.; Lu, H.; Schaff, W.J. Effects of the narrow band gap on the properties of InN. *Phys. Rev. B* **2002**, *66*, 201403.

201. Xie, R.J.; Bert Hintzen, H.T. Optical properties of (oxy)nitride materials: A review. *J. Am. Ceram. Soc.* **2013**, *96*, 665–687.

202. Dwiliński, R.; Doradziński, R.; Garczyński, J.; Sierzputowski, L.; Kucharski, R.; Zajac, M.; Rudziński, M.; Kudrawiec, R.; Serafińczuk, J.; Strupiński, W. Recent achievements in AMMONO-bulk method. *J. Cryst. Growth* **2010**, *312*, 2499–2502.

203. Gogova, D.; Petrov, P.P.; Buegler, M.; Wagner, M.R.; Nenstiel, C.; Callsen, G.; Schmidbauer, M.; Kucharski, R.; Zajac, M.; Dwiliński, R.; *et al.* Structural and optical investigation of non-polar (1–100) GaN grown by the ammonothermal method. *J. Appl. Phys.* **2013**, *113*, 203513.

204. Shubina, T.; Ivanov, S.; Jmerik, V.; Solnyshkov, D.; Vekshin, V.; Kopev, P.; Vasson, A.; Leymarie, J.; Kavokin, A.; Amano, H.; *et al.* Mie resonances, infrared emission, and the band gap of InN. *Phys. Rev. Lett.* **2004**, *92*, 117407.

205. Komissarova, T.A.; Jmerik, V.N.; Ivanov, S.V.; Paturi, P. Detection of metallic in nanoparticles in InGaN alloys. *Appl. Phys. Lett.* **2011**, *99*, 072107.

206. Wu, J.; Walukiewicz, W.; Yu, K.M.; Ager, J.W.; Haller, E.E.; Lu, H.; Schaff, W.J. Small band gap bowing in $In1-_xGa_xN$ alloys. *Appl. Phys. Lett.* **2002**, *80*, 4741.

207. Ranade, M.R.; Tessier, F.; Navrotsky, A.; Marchand, R. Calorimetric determination of the enthalpy of formation of InN and comparison with AlN and GaN. *J. Mater. Res.* **2011**, *16*, 2824–2831.

208. Richter, E.; Stoica, T.; Zeimer, U.; Netzel, C.; Weyers, M.; Tränkle, G. Si doping of GaN in hydride vapor-phase epitaxy. *J. Electron. Mater.* **2013** *42*, 820–825.

209. Steckl, A.J.; Birkhahn, R. Visible emission from Er-doped GaN grown by solid source molecular beam epitaxy. *Appl. Phys. Lett.* **1998**, *73*, 1700–1702.

210. Adekore, B.T.; Callahan, M.J.; Bouthillette, L.; Dalmau, R.; Sitar, Z. Synthesis of erbium-doped gallium nitride crystals by the ammonothermal route. *J. Cryst. Growth* **2007**, *308*, 71–79.

211. Korona, K.P.; Doradziński, R.; Palczewska, M.; Pietras, M.; Kamiska, M.; Kuhl, J. Properties of zinc acceptor and exciton bound to zinc in ammonothermal GaN. *Phys. Status Solidi B* **2003**, *235*, 40–43.

212. Scotti, N.; Kockelmann, W.; Senker, J. Sn_3N_4, a Tin(IV) nitride–syntheses and the first crystal structure determination of a binary tin-nitrogen compound. *Z. Anorg. Allg. Chem.* **1999**, *625*, 1435–1439.

213. Louhadj, A.; Ghezali, M.; Badi, F.; Mehnane, N.; Cherchab, Y.; Amrani, B.; Abid, H.; Sekkal, N. Electronic structure of ScN, YN, LaN and GdN superlattices. *Superlatt. Microstruct.* **2009**, *46*, 435–442.

214. Srivastava, V.; Rajagopalan, M.; Sanyal, S.P. Theoretical investigation on structural, magnetic and electronic properties of ferromagnetic GdN under pressure. *J. Magn. Magn. Mater.* **2009**, *321*, 607–612.

215. Li, D.X.; Haga, Y.; Shida, H.; Suzuki, T.Magnetic properties of ferromagnetic GdN. *Physica B* **1994**, *199/200*, 631–633.

216. Mancera, L.; Rodr, J.A. First principles calculations of the ground state properties and structural phase transformation in YN. *J. Phys.: Condens. Matter* **2003**, *15*, 2625–2633.

217. Niewa, R.Nitridocompounds of manganese: Manganese nitrides and nitridomanganates. *Z. Kristallogr.* **2002**, *217*, 8–23.

218. Takei, W.J.; Heikes, R.R.; Shirane, G. Magnetic structure of Mn_4N-type compounds. *Phys. Rev. B* **1962**, *125*, 1893–1897.

219. Borsa, D.M.; Grachev, S.; Presura, C.; Boerma, D.O. Growth and properties of Cu_3N films and Cu_3N/γ'-Fe_4N bilayers. *Appl. Phys. Lett.* **2002**, *80*, 1823–1825.

220. Sakuma, A. Self-consistent calculations for the electronic structures of iron nitrides, Fe_3N, Fe_4N and $Fe_{16}N_2$. *J. Magn. Magn. Mater.* **1991**, *102*, 127–134.

221. Wang, D.Y. Properties of various sputter-deposited CuN thin films. *J. Vac. Sci. Technol. A* **1998**, *16*, 2084–2092.

222. Desmoulins-Krawiec, S.; Aymonier, C.; Loppinet-Serani, A.; Weill, F.; Gorsse, S.; Etourneau, J.; Cansell, F. Synthesis of nanostructured materials in supercritical ammonia: nitrides, metals and oxides. *J. Mater. Chem.* **2004**, *14*, 228–232.

223. Hasegawa, M.; Yagi, T. Systematic study of formation and crystal structure of 3d-transition metal nitrides synthesized in a supercritical nitrogen fluid under 10 GPa and 1800 K using diamond anvil cell and YAG laser heating. *J. Alloys Compd.* **2005**, *403*, 131–142.

224. Zachwieja, U.; Jacobs, H. Kolumnarstrukturen bei Tri- und Diamminnitraten, $[M(NH_3)_3]NO_3$ und $[M(NH_3)_2]NO_3$ des einwertigen Kupfers und Silbers. *Z. Anorg. Allg. Chem.* **1989**, *571*, 37–50.

225. Hahn, U.; Weber, W. Electronic structure and chemical-bonding mechanism of Cu_3N, Cu_3NPd,and related Cu(I) compounds. *Phys. Rev. B: Condens. Matter* **1996**, *53*, 12684–12693.

226. Leineweber, A.; Niewa, R.; Jacobs, H.; Kockelmann, W. The manganese nitrides η-Mn_3N_2 and Θ-Mn_6N_{5+x}: Nuclear and magnetic structures. *J. Mater. Chem.* **2000**, *10*, 2827–2834.

227. Jacobs,H.; Harbrecht,B.Eine neue Darstellungsmethode für Caesiumhydroxid. *Z. Naturforsch.* **1981**, *36b*, 270–271.

228. Jacobs, H.; Tacke, T.; Kockelkorn, J. Hydroxidmonohydrate des Kaliums und Rubidiums; Verbindungen, deren Atomanordnungen die Schreibweise $K(H_2O)OH$ bzw. $Rb(H_2O)OH$ nahelegen. *Z. Anorg. Allg. Chem.* **1984**, *516*, 67–78.

229. Purdy, A.P. Ammonothermal growth of chalcogenide single crystal materials. US Patent 5,902,396, 11 May 1999.

230. Jacobs, H.; Kirchgässner, R.; Bock, J. Darstellung und Kristallstruktur von Lithiumhydrogensulfid LiHS. *Z. Anorg. Allg. Chem.* **1989**, *569*, 111–116.

231. Jacobs, H.; Erten, C. U¨ ber Kaliumhydrogensulfid, KHS. Z. Anorg. Allg. Chem. 1981, 473, 125–132.

232. Böttcher, P. Beiträge zur Kenntnis der Alkalimetallpolychalkogenide. Habilitation Thesis, RWTH Aachen, Germany, 1980.

233. Böttcher, P. Zur Kenntnis der Verbindung Na_2S_3. *Z. Anorg. Allg. Chem.* **1980**, *467*, 149–157.

234. Böttcher, P. Die Kristallstruktur von K_2S_3 und K_2Se_3. *Z. Anorg. Allg. Chem.* **1977**, *432*, 167–172.

235. Böttcher, P. Preparation and crystal structure of the dialkali metal trichalcogenides Rb_2S_3, Rb_2Se_3 and Cs_2Se_3. *Z. Anorg. Allg. Chem.* **1980**, *461*, 13–21.

236. Böttcher, P. Synthesis and crystal structure of the dirubidiumpentachalcogenides Rb_2S_5 and Rb_2Se_5. *Z. Kristallogr.* **1979**, *150*, 65–73.

237. Böttcher, P.; Kruse, K. Darstellung und Kristallstruktur von Dicaesiumpentasulfid (Cs_2S_5). *J. Less-Common Met.* **1982**, *83*, 115–125.

238. Böttcher, P. Zur Kenntnis von Cs_2Se. *J. Less-Common Met.* **1980**, *76*, 271–277.

239. Böttcher, P. Synthesis and crystal structure of Rb2Te3 and Cs2Te3. J. Less-Common Met. 1980, 70, 263–271.

240. Böttcher, P. Darstellung und Kristallstruktur von Dicaesiumpentatellurid, Cs_2Te_5. *Z. Anorg. Allg. Chem.* **1982**, *491*, 39–46.

3

Direct Energy Supply to the Reaction Mixture during Microwave-Assisted Hydrothermal and Combustion Synthesis of Inorganic Materials

Roberto Rosa *, Chiara Ponzoni and Cristina Leonelli

Dipartimento di Ingegneria "Enzo Ferrari", Università degli Studi di Modena e Reggio Emilia, via Vignolese 905/A, 41125 Modena, Italy; E-Mails: chiara.ponzoni@unimore.it (C.P.); cristina.leonelli@unimore.it (C.L.)

* Author to whom correspondence should be addressed; E-Mail: roberto.rosa@unimore.it

---

Abstract: The use of microwaves to perform inorganic synthesis allows the direct transfer of electromagnetic energy inside the reaction mixture, independently of the temperature manifested therein. The conversion of microwave (MW) radiation into heat is useful in overcoming the activation energy barriers associated with chemical transformations, but the use of microwaves can be further extended to higher temperatures, thus creating unusual high-energy environments. In devising synthetic methodologies to engineered nanomaterials, hydrothermal synthesis and solution combustion synthesis can be used as reference systems to illustrate effects related to microwave irradiation. In the first case, energy is transferred to the entire reaction volume, causing a homogeneous temperature rise within a closed vessel in a few minutes, hence assuring uniform crystal growth at the nanometer scale. In the second case, strong exothermic combustion syntheses can benefit from the application of microwaves to convey energy to the reaction not only during the ignition step, but also while it is occurring and even after its completion. In both approaches, however, the direct interaction of microwaves with the reaction mixture can lead to practically gradient-less heating profiles, on the basis of which the main observed characteristics and properties of the aforementioned reactions and products can be explained.

Keywords: energy transfer; microwaves; hydrothermal synthesis; combustion synthesis; nanomaterials

1. Introduction

For many years, the most widely used method for the preparation of oxides, chalcogenides, silicides and carbides was the direct solid-state reaction between powders, and it probably still represents the most widespread approach for the large-scale production of these important inorganic materials [1,2].

The main limitation of solid-state syntheses, with respect to those exploiting molecular precursors (*i.e.*, those occurring in the liquid or gaseous phases), is the difficulty in bringing the different reactants sufficiently close to each other to allow optimum reduction of diffusion paths. In the solid state, the diffusion limitation is usually compensated by employing high temperatures and long heating periods. These conditions do not permit a subtle control of the reaction stoichiometry nor is it possible to synthesize thermally labile or metastable compounds [2]. The use of microwave (MW) energy, which was initially applied in inorganic chemical synthesis in these same solid-state ceramic routes, allows a significant reduction of the reaction times. This advantage is mainly due to the direct interaction of the electromagnetic field with the inorganic solid reactants, that is the typical feature of volumetric heating. In early studies, when temperature measurement was not properly conducted during microwave irradiation, an additional feature such as lowering of reaction temperatures, with respect to the conventionally performed syntheses [3–9], was reported. However, more accurate experimental set-ups demonstrated that a correct temperature measurement did not confirm this latter observation, while the possibility of obtaining metastable phases was confirmed [5].

With the discovery of the quantum size effect [10,11], solid state synthetic approaches started to be disregarded for the synthesis of advanced inorganic compounds, since engineered nanomaterials rapidly became the main object of study for material scientists and inorganic chemists worldwide [12]. Indeed, the extremely harsh synthetic conditions typical of solid-state synthesis excluded any possibility of a precise control of particle size and shape. An accurate process control, even in non-equilibrium conditions, is fundamental in order to pave the way towards new and exciting advanced applications. Extremely large surface areas and the ability to be functionalized are additional appealing characteristics of engineered nanomaterials. Soft chemistry solution routes [13,14] were consequently performed and investigated in order to explore the different possibilities offered by building desired nanostructures via an almost atom-by-atom assembly (bottom-up approach), rationally designing the final product using reaction mechanism considerations (with similarity to the approach of organic chemists). These synthetic strategies include, but are not limited to, sol-gel [15–19], hydrothermal (more generally solvothermal) [20–24] and the most recently developed solution combustion synthesis [25–31]. By analogy to the evolution seen in solid state ceramic synthesis (as briefly discussed above), and later in organic chemistry [32–37], microwave energy started to be applied in these wet-chemical inorganic syntheses, even if in most of the cases it was considered a mere tool to significantly reduce the reaction times. On the other hand, although much more sporadically than what had occurred for organic syntheses, several additional advantages and characteristics arising from the use of microwave energy were also reported and claimed in the inorganic chemistry field [38]. The most frequently observed feature found when applying microwave heating in a chemical reactor is the resulting uniform particle size distribution; additionally, an effect of the magnetic field on particle morphology has been detected under microwave assisted syntheses of magnetic particles.

The aim of the present work is to focus on these aspects of microwave-assisted inorganic synthesis of engineered nanomaterials, highlighting the main differences that characterize the product obtained with respect to conventional synthesis methods. Results are expected to explain the observed differences on the basis of the microwaves' intrinsic property of being based on a pure energy transfer mechanism rather than heat transfer. Particularly, microwave-assisted hydrothermal synthesis and microwave-assisted combustion synthesis techniques probably represent the most significant examples that can highlight such differences related to the synthetic route. The reason for this is twofold. Firstly, the aqueous medium characterizing both techniques represents one of the most favorable environments for performing microwave-assisted reactions (also approaching a green chemistry perspective [39]). Secondly, both of these syntheses are characterized by synthetic and reactive conditions barely accessible by using conventional heating. Indeed, these adverse conditions can be emphasized by the presence of thick pressurized reaction vessels (utilized in hydrothermal synthesis) and highly exothermic reaction fronts (occurring during combustion syntheses), which in turn make it extremely difficult to transfer heat through the walls of the reactor or to continue to transfer energy after the ignition of the self sustaining reaction.

Even if, nowadays, hydrothermal synthesis has practically reached an optimal, well recognized level of controllability and versatility in the formation of a wide choice of nanostructures [40,41], nevertheless the advantages resulting from the use of microwaves as an innovative energy source need yet to be fully reviewed, especially those which are not related to a mere reduction of reaction time. Moreover, considerations of the scalability of microwave-assisted hydrothermal synthesis, of the possible continuous flow equipment and the resulting process intensification perspectives of this technology will be covered in the next sections. On the other hand, solution combustion synthesis probably represents the most innovative inorganic synthetic strategy, at least for obtaining oxide-based nanostructures. Surely, its intrinsic energy saving characteristics constitute the main attraction of this synthesis approach, and its poor ability to tailor the size and the shape of the target product also represents the main limitation and the main research challenge, to which the application of microwave energy could furnish a partial solution or at least stimulate new findings.

After a brief summary of the fundamentals of microwave energy transfer mechanisms, the next sections will be devoted to an overview of recent advances in microwave-assisted hydrothermal and microwave-assisted solution combustion synthesis for the production of selected engineered nanomaterials, with particular focus on the innovative features characterizing these techniques, which result from both the energy source employed and the synthetic approach used.

2. Microwave Energy Transfer Fundamentals

As is well known, the electromagnetic energy in the microwave band is characterized by frequencies between 300 MHz and 300 GHZ, corresponding to wavelength values of approximately 1 m and 1 mm, respectively, and a photon energy between 1.24×10^{-6} and 1.05×10^{-2} eV, respectively. Microwave heating fundamentally differs from conventional heating mechanisms in that the former is directly generated in the materials by the interaction of the electromagnetic field with electric and magnetic dipoles, which account for the material dielectric, electric and magnetic properties [42].

In the case of aqueous saline solution, the main subject of this review paper, the magnetic field effects on the heating mechanism can be easily neglected. Thus, the interaction of the applied alternating electromagnetic field with the reactive aqueous solution will be uniquely related to its dielectric properties as well as solution conduction.

The ability of a particular dielectric material to polarize in the presence of an applied electric field defines its permittivity (also known as dielectric constant), ε', which can be used to describe the response of the material to the electric field. Polarization phenomena involve dipoles, ions, electrons and interfaces, all of which contribute to the permittivity. In order to account for the alternating nature of the sinusoidal electric field applied (thus for the frequency-dependant nature of the dielectric constant), and for the consequent losses, which are derived from the unavoidable phase shift occurring between the polarization and the electric field, which is responsible for heating, the complex permittivity ε* must be introduced according to the following Equation 1 [43]:

$$\varepsilon^{*} = \varepsilon - j\varepsilon = \varepsilon_0\left(\varepsilon_r - j\varepsilon_r\right) \tag{1}$$

where ε_0 represents the free space permittivity, while the index "r" is referred to values relative to the corresponding empty space values. The imaginary part ε" represents the so-called loss factor, which is related to the ability of the material to store energy.

The ratio between the real and the imaginary parts of the complex permittivity is called loss tangent, tanδ (see Equation 2), and it represents the efficiency of the material to convert absorbed energy into heat (the angle δ is the phase difference between the polarization of the material and the oscillating electric field).

$$\tan\delta = \frac{\varepsilon''}{\varepsilon'} \tag{2}$$

The average power dissipated into the material per unit volume can be represented by the following simplified Equation 3 [43,44]:

$$P_d(x,y,z) = \omega\varepsilon_0\varepsilon''_{eff}\,E^2_{rms} \tag{3}$$

where P_d is the power density in the material at the position (*x*, *y*, *z*) expressed in W/m^3, $\omega = 2\pi f$ (expressed in Hz) with f being the frequency of the incident microwaves and E_{rms} the root mean square of the electric field strength at the position (*x*, *y*, *z*), expressed in V/m.

The relative effective loss factor ε''_{eff} accounts for the polarization losses (dipolar, ionic, electronic and interfacial) combined with the conductivity losses.

At the microwave frequencies, the two predominant loss mechanisms are the dipolar and the ionic ones. It is therefore obvious to what extent the reaction conditions typical of both hydrothermal synthesis and solution combustion synthesis represent the ideal environment for using microwaves as a heating source. Indeed, hydrothermal reaction conditions usually require the presence of inorganic salts (nitrates, chlorides, *etc.*) or other metallic precursors dissolved (or even sometimes suspended) in water combined with a mineralizing agent and an optional surfactant. Solution combustion synthesis conditions, on the other hand, rely on the presence of inorganic nitrate salts and an opportune organic fuel dissolved in water. Thus, in addition to the well-known strong microwave-absorbing ability of water, the addition of ionic species characterizing both these synthetic strategies will lead to higher microwave absorption due to ionic conductivity contributions. A representative example is the increase in the loss tangent of distilled water (at 25 °C and 2.45 GHz) from *ca.* 0.15 to approximately 1.6 with the addition of 0.5 mol/L of sodium chloride (at the same temperature and frequency) [43].

Despite these remarkable characteristics, additional microwave-heating features form the basis of the advantages manifested in the synthesis of several inorganic materials, and they will be discussed separately in the following sub-sections. However, for a more exhaustive discussion of all the parameters/equations defined in this review manuscript, the readers are kindly requested to refer to the literature papers [42–44].

2.1. Volumetric and Selective Nature of MW Heating

On the basis of their interaction with electromagnetic radiation, materials can be classified as reflective, transparent or absorbing ones. In MW-absorbing materials, the microwave power penetration depth (D_p) needs to be evaluated, and for dielectric materials it can be described with the following simplified Equation 4, where λ_0 is the free space wavelength of the incident microwaves.

$$D_p = \frac{\lambda_0 \sqrt{\varepsilon'_r}}{2\pi\varepsilon''_r} \tag{4}$$

D_p represents the depth from the outer surface of the material at which the value of P_d has decreased to 1/e (*i.e.*, ~37%) of its surface value. For several materials and reaction media the value of D_p is significantly higher with respect to the thickness of sample generally used (at least at a laboratory scale), thus the MW power dissipated (and the subsequent heat generation) is reasonably uniform throughout the whole material. This property of microwaves, known as volumetric heating, leads to the inversion of the classical temperature profile and to a significant reduction of the temperature gradient inside the reaction mixture. The latter situation is shown in Figure 1, where a comparison between the temperature profiles in a reaction mixture undergoing both conventional hydrothermal and microwave hydrothermal treatments is depicted.

Figure 1. Example of temperature profiles occurring in a reaction mixture exposed to conventional heating (**left**, e.g., inside a stainless steel autoclave) and to microwave heating (**right**, e.g., inside a microwave transparent polytetrafluoroethylene (PTFE) reaction vessel).

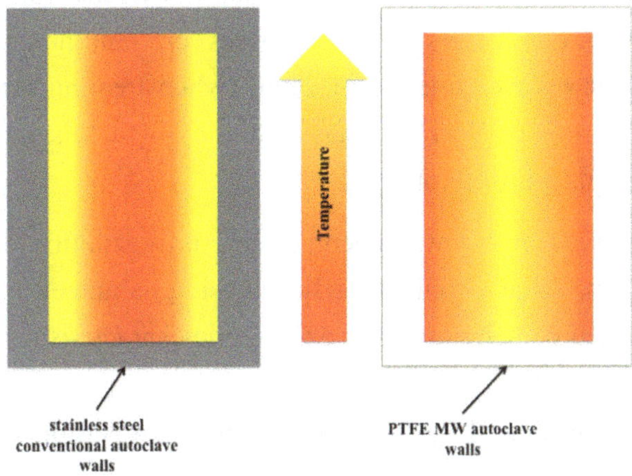

2.2 *Energy Efficiency Issues*

Direct and selective interaction of microwave energy with the reaction mixture, which favors the fast heating of the mixture in solution rather than the reactor's walls or other components, surely represents an intrinsic energy efficiency feature. It is clear that the possibility to transfer energy exactly where needed contributes to increasingly apply microwave-based technologies in a green process intensification perspective, as frequently reported in the literature during recent years [45,46]. Indeed, the obvious enhancement of reaction rates, manifested by most of the inorganic syntheses performed under microwave assistance (as will be detailed in the following section), is probably the main feature which is usually fundamental in defining microwave routes as more energy-efficient.

However, very recently the energy efficiency of microwave-assisted organic synthetic protocols has been critically assessed [47]. Mainly due to the lower electrical energy conversion capability of standard 2.45 GHz magnetrons installed on commercial scientific equipments [48], it has been recommended to perform energy efficiency evaluations on a case by case basis, since the situation has been noted to be significantly different when moving from laboratory-scale single-mode reactors to larger batch reactors like multi-mode ones (for a detailed essay on the different microwave heating cavities and their components, the readers are kindly suggested to refer to [49]).

Concerning the specific field of inorganic synthesis, the same considerations are also valid. However, several strategies exist and are still under systematic investigation, in order to, at least, partially overcome these limitations. In particular, dedicated applicators equipped with proper impedance matching devices can significantly reduce the reflected power and power dissipated in regions different from the desired one. Moreover, the actual tendency of the market to develop new microwave generators, also able to perform a continuous impedance matching by tuning over a particular frequency range, will surely lead to more and more efficient syntheses [50].

The problems encountered with larger batch reactors are related to the penetration depth (D_p, see Equation 4). Indeed, with the increase of the batch scale, D_p could be extremely small leading to an almost heat-unaffected mixture. Understanding the importance of internal agitation in microwave-assisted chemistry [51], the latter circumstance will surely result in more pronounced temperature gradients. A further intriguing and practically un-explored possibility could be the use of microwave frequencies different from that most commonly used worldwide, *i.e.*, 2.45 GHz. Particularly, the use of lower frequencies for a given material or solution will increase its D_p, as is clearly evident from Equation 4. Moreover, continuous flow syntheses are another interesting solution to the limitations of the reduced microwave penetration depth (when treating large batch reactors) [52–55] during the scaling up of microwave-hydrothermal technology. In this latter perspective, the use of multiple MW sources represents a further option.

On the other hand, from Section 4 it will become clear how solution combustion synthesis is not affected by all of these issues, since its energy efficiency is guaranteed by the fact that most of the required energy will be provided by harnessing the exothermic characteristics of the reaction itself.

3. Microwave-Assisted Hydrothermal Synthesis of Engineered Nanomaterials

Among the various wet chemical methods, the hydrothermal approach is the most promising method for the fabrication of nanostructured and microstructured materials in an acceptable crystalline phase as can be seen in Figure 2.

Figure 2. Transmission electron microscopy (TEM) image of ZrO_2 nanopowders doped with 1% Pr obtained via microwave (MW)-assisted hydrothermal synthesis performed at 8 MPa for 30 min, according to the chemical procedure indicated in [56].

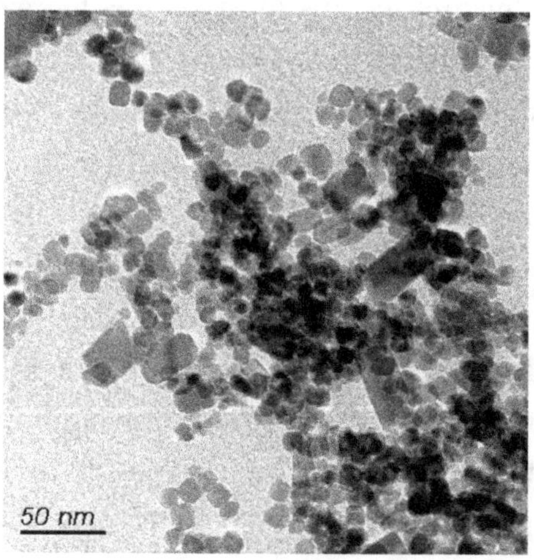

Indeed the ability to precipitate already crystallized powders directly from solution allows controlling the rate and uniformity of nucleation. This results in improved control of size and morphology of crystallites and in significantly reduced aggregation levels [57]. Moreover, crystallization phenomena constitute a self-purifying process, during which the growing crystals/crystallites tend to reject impurities present in the growth environment. The impurities are subsequently removed from the system together with the crystallizing solution, contrary to other synthesis routes, such as high temperature calcination. Thus, with the hydrothermal approach it is possible to synthesize powders in a one-step process, without the contamination typical of a reaction carried out in a closed system. Some key factors such as the reagent concentration, the reaction temperature and the reaction time are crucial for the morphology and crystallite size of the products.

From the pioneering work by Komarneni *et al.* [58], the main advantages of the MW-assisted hydrothermal route over conventional hydrothermal treatments are short reaction time, energy saving, good particle dispersion, high phase purity, high homogeneity in stoichiometry, and small particle size [59,60]. Most of the aforementioned advantages can be easily attributed to the significant reduction of the initial heating period (up to the set temperature), which characterizes dielectric heating over more conventional heat transfer mechanisms, as schematically reported in Figure 3.

Figure 3. Temperature profiles typical of microwave-assisted and conventionally-heated hydrothermal synthesis.

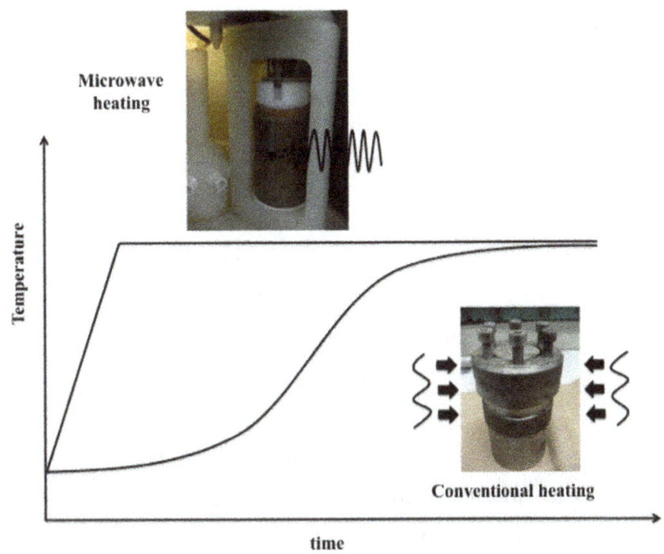

Indeed, the initial part of the heating period characterizing conventional hydrothermal syntheses can last many minutes; thus, it may result in non-uniform temperature profiles inside the autoclave (please refer to Figure 1).

By contrast, the gradient-less volumetric microwaves-based heating (right side of Figure 1) is able to provide the reaction mixture with a significantly uniform distribution of energy, which results in narrower particle size distributions [61–63].

During recent decades, hydrothermal synthetic techniques have been employed for the preparation of a wide variety of advanced engineered nanomaterials, a summary of which is reported in Table 1. Surely, the mostly studied and synthesized compounds are transition metal oxides and multi metal oxide materials [20,59,64,65], although several examples can be found in scientific literature relating to the production of metal sulfides [59,64–68], hydroxyapatite [59,69] and different kinds of carbon-based nanostructures [59,70,71].

Recently, scientific attention has been focused on multiferroic compounds that exhibit the existence of simultaneous ferroelectric and ferromagnetic properties in a certain temperature range, since they have promising applications in spintronics, electromagnetic interference filters, magnetic recording media, sensors, and photovoltaic devices [72–76].

Among the multiferroic compounds, bismuth ferrite $BiFeO_3$ (BFO) has received considerable attention in the past few years because of its characteristics, and a multitude of research work dealing with its conventional hydrothermal synthesis can be found in the scientific literature [77–79]. Most of these studies were devoted to investigating the role of several experimental processing parameters in the synthesis of a pure BFO phase, and reaction times ranging from 6–9 hours were always necessary for that purpose.

Table 1. Some of the different classes of materials synthesized by the hydrothermal technique.

Class of materials	Main examples	Some crystal shapes observed	Ref.
Transition metal oxides	TiO_2, ZnO, ZrO_2, Fe_2O_3, Fe_3O_4	Nanoparticles, nanotubes, nanowires, nanorods, nanocubes, nanoribbons	[20,59,64,65]
Multimetal oxides	$BaTiO_3$, $La_{1-x}Sr_xMnO_3$, $BiFeO_3$	Nanoparticles, polyhedrons, nanoplatelets, nanocubes	[20,64,65]
Metal sulfides	CdS, ZnS, CuS, SnS	Nanoparticles, nanotubes,	[59,64–68]
Biomaterials	Hydroxyapatite	Needle-like (frequently), nanospheres, nanorods, nanowires, whiskers, platelets	[59,69]
Carbon-based nanostructured materials	Carbon nanotubes, nanospheres, nanofibers, graphene-based materials		[59,70,71]

On the other hand, in the early nineties Komarneni *et al.* [80] proposed the microwave hydrothermal synthesis of BFO to obtain highly crystalline agglomerated particles in one heating step. This study demonstrated that the microwave-hydrothermal approach was very efficient in terms of kinetics for the synthesis and processing of both binary and ternary ceramic oxide powders. More recently [81], a comparison of three different methods for preparing BFO polycrystals has been reported: (i) the classic hydrothermal synthesis, (ii) the microwave heating in the solid state and (iii) the microwave-assisted hydrothermal method. The best material in terms of purity and enhanced reproducibility was obtained with the third method, which was also described as the most environmental-friendly, since it employed moderate temperatures (T = 200 °C) and significantly less time (30 min). In the last few years, several experimental studies concerning the microwave-hydrothermal approach have been undertaken in order to obtain single-phase crystalline bismuth ferrite nanocrystals with a high degree of homogeneity and uniformity in particle size at even faster reaction rates and low temperatures.

In a recent study by our group, an easy microwave hydrothermal synthesis route was optimized for the preparation of $BiFeO_3$ pure-phase [82]. In particular, the work focused on the influence of process parameters (*i.e.*, precursor ratio, mineralizer concentration, temperature, time and the use of inorganic chelating agents) on phase formation, particle size distribution and morphology. Single phase BFO was obtained at 180 °C using a concentration of 10 M KOH mineralizer (Figure 4) or at 200 °C using 8M KOH, in a 30 min total reaction time in both cases. Moreover, the presence of Na_2CO_3 acting as an electron donor allowed the reaction to be performed at lower KOH concentrations and temperatures. The particle morphology evolution followed the sequence: spherical, lamellar, lamellar semi-cubic, and cubic. Whereas with Na_2CO_3 addition the evolution was as follows: lamellar semi-cubic, cubic, and prismatic truncated octahedron shape. The main differences of this latter microwave-assisted hydrothermal approach [82] with respect to those mentioned above [80,81] (constituting two of the most significant ones employing MW heating) for the preparation of $BiFeO_3$ nanoparticles and nanostructured particles are summarized in Table 2.

In all of the experiments carried out in [82], microwave heating allowed obtaining a perfect control over the heating rate and reducing the ramp to the reaction temperature to only 5 min (in comparison to the 15 min needed in the example in [81]). This precise and immediate transfer of energy, again, must be emphasized as the key step manifested in the majority of microwave-assisted synthetic inorganic protocols, leading to the observed advantages. Moreover, the temperature adopted in [82] was reduced by approximately 20 °C with respect to the majority of syntheses reported, including those exploiting microwave heating.

Figure 4. Scanning electron microscopy (SEM) images and X-ray diffractometry (XRD) pattern of pure bismuth ferrite ($BiFeO_3$) phase synthesized according to the microwave-assisted hydrothermal procedure optimized in [74], *i.e.*, at 180 °C for 30 min with a KOH concentration of 10 mol/L.

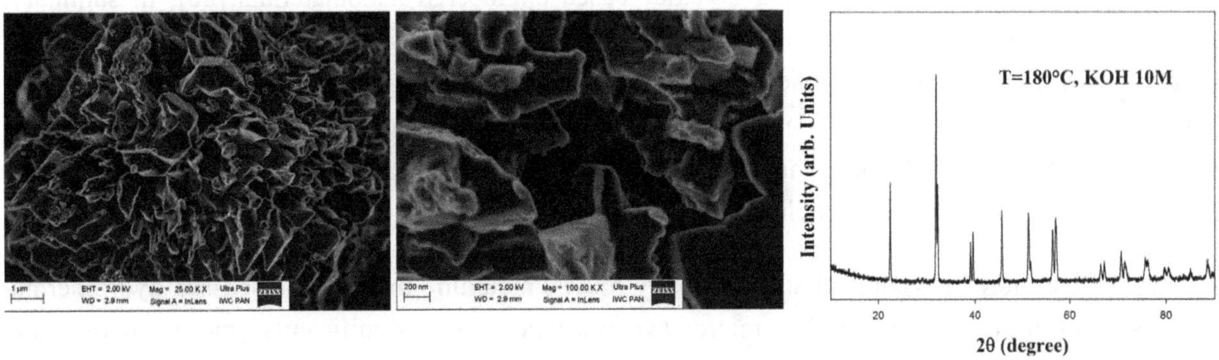

Table 2. Comparison of three different microwave-assisted hydrothermal synthesis strategies proposed for the obtainment of single $BiFeO_3$ phase.

Temperature (°C)	Time (min)	Product	Ref.
ca. 194	120	agglomerated rombohedral $BiFeO_3$ particles	[80]
200	30	polyhedral $BiFeO_3$ nano-structured particles	[81]
180	30	nanocubic $BiFeO_3$ aggregates	[82]

4. Microwave Energy Transfer in the Ignition of Solution Combustion Synthesis

Combustion synthesis (CS) is a relatively new inorganic material manufacturing procedure, originating from the discovery of solid flame phenomena by Merzhanov and co-workers [83,84]. Very quickly it became the most promising method for the synthesis of high temperature ceramics, intermetallics and different kinds of composite materials. The main reason can be found in its intrinsic energy saving characteristics, since most of the energy needed for the synthesis is released by the reaction itself. Indeed, CS exploits exothermic reactions occurring among the reactants; after a proper ignition (*i.e.*, the reaction mixture is heated up to the so-called ignition temperature), the reactants start reacting in a self-sustaining manner, thus requiring no further energy from an external source, since the harnessing of the heat produced by the reaction itself allows completion of the reaction in the whole reactive volume [85]. Typically, two different ignition strategies are distinguishable, leading to the self-propagating high-temperature synthesis (SHS) and to the thermal explosion (TE) modes of combustion synthesis. In the SHS type, the exothermic reaction is ignited at one end of the reactive volume so that the reaction self-propagates to the opposite side of the sample, while in the TE method of combustion the entire reactive volume is heated up homogeneously and uniformly until the reaction takes place almost simultaneously in the entire specimen [85].

Although initially the performance of CS was limited among mixtures of powders, or could benefit from the optional use of a gaseous phase [86], nowadays CS has reached a significant level of versatility and diversity [87]. Particularly, combustion synthesis performed in aqueous solution is a relatively new method for preparation of oxide-based engineered nanomaterials [88]. In summary, solution combustion synthesis exploits exothermic self-sustaining reactions occurring between metal nitrates and different kinds of organic fuels (e.g., glycine, urea, citric acid, polyethylene glycol, *etc.*) in homogeneous solutions. The main steps involved in this synthesis strategy are schematically summarized in Figure 5. Also, in this case the ignition can occur locally (leading to a layer by layer propagation) or volumetrically (leading to a nearly instantaneous reaction).

Most of the more conventional soft-chemistry routes, including the above-discussed hydrothermal synthesis, even if performed by microwave-assistance, are significantly more energy and time consuming.

Moreover—although to a significantly lesser extent with respect to hydrothermal syntheses—with the ever-deepening research findings in the field of CS, a satisfactory control over the crystallite size has been achieved by opportunely modifying the ratio between the metal nitrate and the fuel, or by changing the fuel itself as well as the metal precursor anion [88,89].

Being able to change the particle size distribution by simply changing the ignition strategy from a conventional hot plate to the direct interaction of almost the whole reaction mixture with microwave energy gives more desirable results. Indeed, this latter opportunity has been observed in several works [90–92] accompanied by the typical shorter reaction times, which, as in this case, are all direct consequences of the volumetric nature of microwave heating. Moreover, the fact that microwave ignition of solution combustion synthesis at the laboratory scale is more prone to occur volumetrically (*i.e.*, referred to as Thermal Explosion in traditional solid state performed CS reactions) further assures a more uniform thermal history to the reactive solution. Moreover, the possibility that microwaves continue to furnish energy to the reaction (despite the adverse temperature gradient), even after ignition and/or completion, must be mentioned. This happens based on the dielectric properties of the forming solid products and leads to the possibility of modifying the cooling rates as well.

Figure 5. Schematic representation of the main steps characterizing the solution combustion synthesis approach to produce oxide-based engineered nanomaterials.

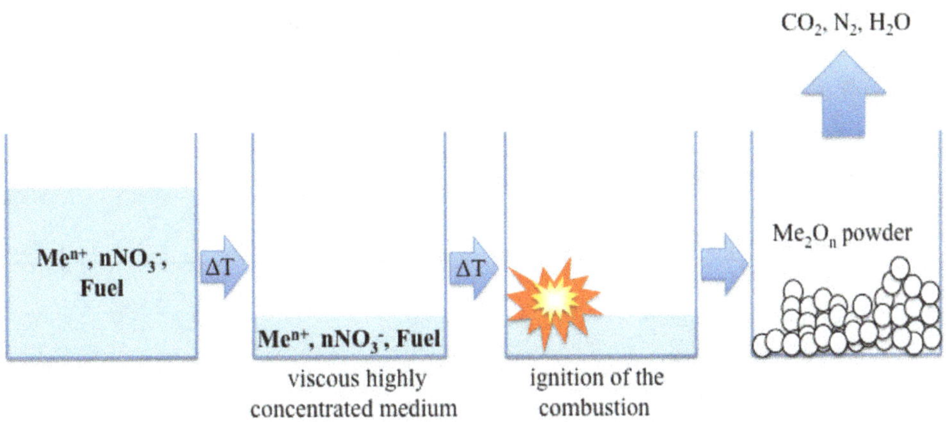

Table 3. Summary of some of the latest modifications of combustion synthesis for the preparation of engineered nanomaterials.

Synthetic approach	Distinctive characteristics	Main classes of materials obtainable	Examples of synthesized compounds	Ref.
Solution combustion synthesis	The solution of the metal precursor(s) nitrate(s) and the fuel is slowly evaporated and then ignited locally or volumetrically. The exothermic reaction occurs between fuel and oxygen-containing species derived from the decomposition of nitrates	- Binary and more complex metal oxide nanopowders - metal nanopowders	- TiO_2, ZnO, $LaFeO_3$, $BiFeO_3$ - Ni, Cu and their alloys	[88,91–93,94]

Table 3. *Cont.*

Impregnated layer and combustion	The reactive solution is impregnated with an inert porous oxide support or with a reactive cellulose paper also acting as a fuel. - a productivity of 0.5-2 kg/h of nanoparticles has been reached by a continuous synthesis approach	- binary and more complex oxides	ZnO, MgO, $Ce_{1-x}Pt_xO_2$, $CuO/ZnO/ZrO_2/Pd$ based catalysts	[95,96]
Carbon combustion synthesis	The exothermic oxidation reaction of carbon to carbon dioxide generates a reaction wave that propagates through the solid reactant mixture. The product of the exothermic reaction is not incorporated into the final product, leading to several advantages (e.g., smaller particles).	- Perovskite oxides	$BaTiO_3$, $SrTiO_3$, $LiNbO_3$, $CoFe_2O_4$	[97,98]

Solution combustion synthesis has recently shown several new trends, which deserve particular attention. Particularly, it has been noted that the nature of the fuel-oxidizer system and its ratio can also cause a radical change in the expected product, leading to the usually expected and desired oxide or eventually to metallic nanoparticles, and thereby suggesting a new synthetic methodology for metal and metallic alloy-based nanomaterials [93]. Moreover, a different combustion synthesis approach, named Carbon Combustion Synthesis, which does not occur properly in solution and which exploits the high exothermicity of the carbon oxidation to CO_2, was recently developed to obtain nanostructured materials [97,98], accompanied by a continuous large-scale nanomaterial preparation procedure based on the principles of solution CS (the so-called impregnated layer and combustion) [95,99]. All of these latter novel approaches are summarized in Table 3 where they are compared to solution combustion synthesis itself in terms of their distinctive characteristics and materials obtainable/obtained. In all likelihood, coupling of these latter novelties with a significantly more energy efficient ignition strategy, like microwaves, is simply a matter of time.

5. Concluding Remarks

This manuscript does not intend to exhaustively cover all the possible applications of microwave energy in the hydrothermal as well as in the solution combustion synthesis of engineered nanomaterials; nevertheless, it does aim to present a wide and descriptive approach of the peculiar characteristics of well conceived modern syntheses using microwave energy as a heating source.

The most recent trends in microwave irradiation of chemical environments are directed toward the utilization of combined techniques [45]. For this reason, we have presented results for microwave-assisted hydrothermal and microwave-assisted solution combustion syntheses. The emerging interest in the production of selected engineered nanosized oxide-based materials pushed us toward this class of inorganic compounds.

Both hydrothermal and solution combustion synthesis approaches significantly benefit from the ability of microwaves to directly generate the necessary heat (for the reaction completion or simply ignition) inside the reactant mixture.

A success story of a well-crystallized single phase complex oxide compound has been described: the microwave-assisted hydrothermal synthesis of perovskite $BiFeO_3$ oxide was performed in significantly shorter reaction times with respect to conventional hydrothermal procedures and with a narrower particles' size distribution.

Similar advantages have also been noted when comparing conventional heating and microwave heating for the ignition of solution combustion synthesis.

In all cases, the ability of microwaves to directly furnish the entire reaction medium with a more uniform energy distribution, with a resulting almost gradient-less temperature profile, has been considered to be the crucial feature marking the originality of these syntheses.

Despite the reported advantages and intrinsic characteristics, the limitations of a microwave energy source have not been ignored in this review paper.

Moreover, considerations of the possible application of dielectric heating to novel aspects of both hydrothermal and solution combustion synthesis approaches or simply considerations of the scaling up of microwave technology have been made.

Acknowledgments

The authors are particularly grateful to Paolo Veronesi, University of Modena and Reggio Emilia for helping with experimental set-up, cavity geometry and electromagnetic field simulations and to Antonino Rizzuti, Polytechnic University of Bari, Italy, for part of the syntheses and characterization work. They express their thanks to Adam Presz for some SEM images, to Statutory Funds of IHHP and to Witold Lojkowski from Institute of High Pressure Physics, Polish Academy of Science, Warsaw, Poland.

The financial support from Project PRIN 2009 "Microwave applicators design: from numerical simulation to materials selection"—Prot. 2009WXXLY2_001 funded by MIUR-Italy was particularly appreciated.

References

1. Rao, K.J.; Vaidhyanathan, B.; Ganguli, M.; Ramakrishnan, P.A. Synthesis of inorganic solids using microwaves. *Chem. Mater.* **1999**, *11*, 882–895.
2. Jansen, M. A concept for synthesis planning in solid-state chemistry. *Angew. Chem. Int. Ed.* **2002**, *41*, 3746–3766.
3. Ramesh, P.D.; Vaidhyanathan, B.; Ganguli, M.; Rao, K.J. Synthesis of β-SiC powder by use of microwave radiation. *J. Mater. Res.* **1994**, *9*, 3025–3027.
4. Ramesh, P.D.; Rao, K.J. Microwave assisted synthesis of aluminum nitride. *Adv. Mater.* **1995**, *7*, 177–179.
5. Rizzuti, A.; Leonelli, C. Crystallization of aragonite particles from solution under microwave irradiation. *Powder Technol.* **2008**, *186*, 255–262.
6. Mastrovito, C.; Lekse, J.W.; Aitken, J.A. Rapid solid-state synthesis of binary group 15 chacogenides using microwave irradiation. *J. Solid State Chem.* **2007**, *180*, 3262–3270.
7. Bhunia, S.; Bose, D.N. Microwave synthesis, single crystal growth and characterization of ZnTe. *J. Crystal Growth* **1998**, *186*, 535–542.
8. Vaidhyanathan, B.; Rao, K.J. Microwave assisted synthesis of technologically important transition metal silicides. *J. Mater. Res.* **1997**, *12*, 3225–3229.
9. Vaidhyanathan, B.; Raizada, P.; Rao, K.J. Microwave assisted fast solid state synthesis of niobates and titanates. *J. Mater. Sci. Lett.* **1997**, *16*, 2022–2025.
10. Wang, Y.; Herron, N. Nanometer-sized semiconductor clusters: materials synthesis, quantum size effects, and photophysical properties. *J. Phys. Chem.* **1991**, *95*, 525–532.
11. Berry, C.R. Structure and optical absorption of AgI microcrystals. *Phys. Rev.* **1967**, *161*, 848–851.
12. Hischier, R.; Walser, T. Life cycle assessment of engineered nanomaterials: state of the art and strategies to overcome existing gaps. *Sci. Total Environ.* **2012**, *425*, 271–282.
13. Gopalakrishnan, J. Chimie douce approaches to the synthesis of metastable oxide materials. *Chem. Mater.* **1995**, *7*, 1265–1275.
14. Hubert-Pfalzgraf, L.G. To what extent can design of molecular precursors control the preparation of high tech oxides? *J. Mater. Chem.* **2004**, *14*, 3113–3123.
15. Roy, R. Ceramics by the solution-sol-gel route. *Science* **1987**, *238*, 1664–1669.
16. Vioux, A. Nonhydrolytic sol-gel routes to oxides. *Chem. Mater.* **1997**, *9*, 2292–2299.
17. Niederberger, M.; Garnweitner, G. Organic reaction pathways in the nonaqueous synthesis of metal oxide nanoparticles. *Chem. Eur. J.* **2006**, *12*, 7282–7302.
18. Giordano, C.; Antonietti, M. Synthesis of crystalline metal nitride and metal carbide nanostructures by sol-gel chemistry. *Nano Today* **2011**, *6*, 366–380.
19. Valizadeh, A.; Mikaeili, H.; Samiei, M.; Farkhani, S.M.; Zarghami, N.; Kouhi, M.; Akbarzadeh, A.; Davaran, S. Quantum dots: Synthesis, bioapplications, and toxicity. *Nanoscale Res. Lett.* **2012**, *7*, 480.
20. Patzke, G.R.; Zhou, Y.; Kontic, R.; Conrad, F. Oxide nanomaterials: Synthetic developments, mechanistic studies, and technological innovations. *Angew. Chem. Int. Ed.* **2011**, *50*, 826–859.
21. Dias, A.; Ciminelli, V.S.T. Electroceramic materials of tailored phase and morphology by hydrothermal technology. *Chem. Mater.* **2003**, *15*, 1344–1352.

22. Hayashi, H.; Hakuta, Y. Hydrothermal synthesis of metal oxide nanoparticles in supercritical water. *Materials* **2010**, *3*, 3794–3817.
23. Demazeau, G. Solvothermal reactions: An original route for the synthesis of novel materials. *J. Mater. Sci.* **2008**, *43*, 2104–2114.
24. Namratha, K.; Byrappa, K. Novel solution routes of synthesis of metal oxide and hybrid metal oxide nanocrystals. *Progr. Crystal Growth Charact. Mater.* **2012**, *58*, 14–42.
25. Kingsley, J.J.; Patil, K.C. A novel combustion process for the synthesis of fine particle alpha-alumina and related oxide materials. *Mater. Lett.* **1988**, *6*, 427–432.
26. Manoharan, S.S.; Patil, K.C. Combustion route to fine particle perovskite oxides. *J. Solid State Chem.* **1993**, *102*, 267–276.
27. Patil, K.C.; Aruna, S.T.; Ekambaram, S. Combustion synthesis. *Curr. Opin. Solid State Mater. Sci.* **1997**, *2*, 158–165.
28. Patil, K.C.; Aruna, S.T.; Mimani, T. Combustion synthesis: an update. *Curr. Opin. Solid State Mater. Sci.* **2002**, *6*, 507–512.
29. Mukasyan, A.S.; Epstein, P.; Dinka, P. Solution combustion synthesis of nanomaterials. *Proc. Comb. Inst.* **2007**, *31*, 1789–1795.
30. Aruna, S.T.; Mukasyan, A.S. Combustion synthesis and nanomaterials. *Curr. Opin. Solid State Mater. Sci.* **2008**, *12*, 44–50.
31. Ruiz-Gomez, M.A.; Gomez-Solis, C.; Zarazua-Morin, M.E.; Torres-Martinez, L.M.; Juarez-Ramirez, I.; Sanchez-Martinez, D.; Figueroa-Torres, M.Z. Innovative solvo-combustion route for the rapid synthesis of MoO_3 and Sm_2O_3 materials. *Ceram. Int.* **2014**, *40*, 1893–1899.
32. Gedye, R.; Smith, F.; Westaway, K.; Ali, H.; Baldisera, L.; Laberge, L.; Rousell, J. The use of microwave ovens for rapid organic synthesis. *Tetrahedron Lett.* **1986**, *27*, 279–282.
33. Giguere, R.J.; Bray, T.L.; Duncan, S.M.; Majetich, G. Application of commercial microwave ovens to organic synthesis. *Tetrahedron Lett.* **1986**, *27*, 4945–4948.
34. Adam, D. Microwave chemistry: out of the kitchen. *Nature* **2003**, *421*, 571–572.
35. Loupy, A. *Microwaves in Organic Synthesis*; Wiley-VCH: Weinheim, Germany, 2002.
36. Kappe, C.O. Controlled microwave heating in modern organic synthesis. *Angew. Chem. Int. Ed.* **2004**, *43*, 6250–6284.
37. Corradi, A.; Leonelli, C.; Rizzuti, A.; Rosa, R.; Veronesi, P.; Grandi, R.; Baldassari, S.; Villa, C. New "green" approaches to the synthesis of pyrazole derivatives. *Molecules* **2007**, *12*, 1482–1495.
38. Leonelli, C.; Lojkowski, W. Main development directions in the application of microwave irradiation to the synthesis of nanopowders. *Chem. Today* **2007**, *25*, 34–38.
39. Anastas, P.T.; Warner, J.C. *Green Chemistry: Theory and Practice*; Oxford University Press: New York, NY, USA, 1998.
40. Huang, J.; Xia, C.; Cao, L.; Zeng, X. Facile microwave hydrothermal synthesis of zinc oxide one-dimensional nanostructure with three-dimensional morphology. *Mater. Sci. Eng. B* **2008**, *150*, 187–193.
41. Shi, W.; Song, S.; Zhang, H. Hydrothermal synthetic strategies of inorganic semiconducting nanostructures. *Chem. Soc. Rev.* **2013**, *42*, 5714–5743.
42. Metaxas, A.C. *Foundations of Electroheat: A Unified Approach*; John Wiley and Sons: Chichester, UK, 1996.

43. Gupta, M.; Eugene, W.W.L. Microwaves-Theory. In *Microwaves and Metals*; John Wiley and Sons: Singapore, 2007 and references therein.
44. Metaxas, A.C.; Meredith, R.J. *Industrial Microwave Heating*; Peter Peregrinus: London, UK, 1983.
45. Leonelli, C.; Mason, T.J. Microwave and ultrasonic processing: Now a realistic option for industry. *Chem. Eng. Process.* **2010**, *49*, 885–900.
46. Rosa, R.; Veronesi, P.; Leonelli, C. A review on combustion synthesis intensification by means of microwave energy. *Chem. Eng. Process.* **2013**, *71*, 2–18.
47. Moseley, J.D.; Kappe, C.O. A critical assessment of the greenness and energy efficiency of microwave-assisted organic synthesis. *Green Chem.* **2011**, *13*, 794–806.
48. Nüchter, M.; Müller, U.; Ondruschka, B.; Tied, A.; Lautenschläger, W. Microwave-assisted chemical reactions. *Chem. Eng. Technol.* **2003**, *26*, 1207–1216.
49. Chan, T.V.C.T.; Reader, H.C. *Understanding Microwave Heating Cavities*; Artech House: Norwood, UK, 2000.
50. For further details see for example: http://www.mksinst.com, www.sairem.com.
51. Ferrero, M.A.; Kremsner, J.M.; Kappe, C.O. Nonthermal microwave effects revisited: on the importance of internal temperature monitoring and agitation in microwave chemistry. *J. Org. Chem.* **2008**, *73*, 36–47.
52. Corradi, A.B.; Bondioli, F.; Ferrari, A.M.; Focher, B.; Leonelli, C. Synthesis of silica nanoparticles in a continuous-flow microwave reactor. *Powder Technol.* **2006**, *167*, 45–48.
53. Katsuki, H.; Furuta, S.; Komarneni, S. Semi-continuous and fast synthesis of nanophase cubic $BaTiO_3$ using a single-mode home-built microwave reactor. *Mater. Lett.* **2012**, *83*, 8–10.
54. Wiles, C.; Watts, P. Continuous flow reactors: A perspective. *Green Chem.* **2012**, *14*, 38–54.
55. Nishioka, M.; Miyakawa, M.; Daino, Y.; Kataoka, H.; Koda, H.; Sato, K.; Suzuki, T.M. Single-mode microwave reactor used for continuous flow reactions under elevated pressure. *Ind. Eng. Chem. Res.* **2013**, *52*, 4683–4687.
56. Bondioli, F.; Ferrari, A.M.; Braccini, S.; Leonelli, C.; Pellacani, G.C.; Opalińska, A.; Chudoba, T.; Grzanka, E.; Palosz, B.; Lojkowski, W. Microwave-hydrothermal synthesis of nanocrystalline Pr-doped zirconia powders at pressures up to 8 MPa, Interfacial Effects and Novel Properties of Nanomaterials. *Solid State Phenom.* **2003**, *94*, 193–196.
57. Riman, R.E. *High Performance Ceramics: Surface Chemistry in Processing Technology*; Pugh, R., Bergstrom, L., Eds.; Marcel-Dekker: New York, NY, USA, 1993.
58. Komarneni, S.; Roy, R.; Li, H.Q. Microwave-hydrothermal synthesis of ceramic powders. *Mater. Res. Bull.* **1992**, *27*, 1393–1405.
59. Byrappa, K.; Adschiri, T. Hydrotermal technology for nanotechnology. *Prog. Cryst. Growth Charact. Mater.* **2007**, *57*, 117–166.
60. Yoshimura, M.; Byrappa, K. Hydrothermal processing of materials: past, present and future. *J. Mater. Sci.* **2008**, *43*, 2085–2103.
61. Van Gerven, T.; Stankiewicz, A. Structure, energy, synergy, time-the fundamentals of process intensification. *Ind. Eng. Chem. Res.* **2009**, *48*, 2465–2474.
62. Schwalbe, T.; Autze, V.; Hohmann, M.; Stirner, W. Novel innovation systems for a cellular approach to continuous process chemistry from discovery to market. *Org. Proc. Res. Devel.* **2004**, *8*, 440–454.

63. Leonelli, C.; Rizzuti, A.; Rosa, R.; Corradi, A.B.; Veronesi, P. Numerical simulation of a microwave reactor used in synthesis of nanoparticles. In Proceedings of IMPI 44th Annual Symposium, Denver, CO, USA, 14–16 July 2010.
64. Baghbanzadeh, M.; Carbone, L.; Cozzoli, P.D.; Kappe, C.O. Microwave-assisted synthesis of colloidal inorganic nanocrystals. *Angew. Chem. Int. Ed.* **2011**, *50*, 11312–11359.
65. Yao, W.T.; Yu, S.H. Recent advances in hydrothermal syntheses of low dimensional nanoarchitectures. *Int. J. Nanotechnol.* **2007**, *4*, 129–162.
66. Yan, S.; Wang, B.; Shi, Y.; Yang, F.; Hu, D.; Xu, X.; Wu, J. Hydrothermal synthesis of CdS nanoparticle/functionalized graphene sheet nanocomposites for visible-light photocatalytic degradation of methyl orange. *Appl. Surf. Sci.* **2013**, *285P*, 840–845.
67. Wan, H.; Ji, X.; Jiang, J.; Yu, J.; Miao, L.; Zhang, L.; Bie, S.; Chen, H.; Ruan, Y. Hydrothermal synthesis of cobalt sulfide nanotubes: The size control and its application in supercapacitors. *J. Power Sources* **2013**, *243*, 396–402.
68. Yan, X.; Michael, E.; Komarneni, S.; Brownson, J.R.; Yan, Z.F. Microwave- and conventional-hydrothermal synthesis of CuS, SnS and ZnS: Optical properties. *Ceram. Int.* **2013**, *39*, 4757–4763.
69. Shojai, M.S.; Khorasani, M.T.; Khoshdargi, E.D.; Jamshidi, A. Synthesis methods for nanosized hydroxyapatite with diverse structures. *Acta Biomat.* **2013**, *9*, 7591–7621.
70. Hu, B.; Wang, K.; Wu, L.; Yu, S.H.; Antonietti, M.; Titirici, M.M. Engineering carbon materials from the hydrothermal carbonization process of biomass. *Adv. Mater.* **2010**, *22*, 813–828.
71. Chen, P.; Yang, J.J.; Li, S.S.; Wang, Z.; Xiao, T.Y.; Qian, Y.H.; Yu, S.H. Hydrothermal synthesis of macroscopic nitrogen-doped graphene hydrogels for ultrafast supercapacitors. *Nano Energy* **2013**, *2*, 249–256.
72. Eerenstein, W.; Mathur, N.D.; Scott, J.F. Multiferroic and magnetoelectric materials, *Nature*, **2006**, *442*, 759–765.
73. Wang, J.; Neaton, J.B.; Zheng, H.; Nagarajan, V.; Ogale, S.B.; Liu, B.; Viehland, D.; Vaidhyanathan, V.; Schlom, D.G.; Waghmare, U.V.; *et al.* Epitaxial $BiFeO_3$ multiferroic thin film heterostructures. *Science* **2003**, *299*, 1719–1722.
74. Hur, N.; Park, S.; Sharma, P.A.; Ahn, J.S.; Guha, S.; Cheong, S.W. Electric polarization reversal and memory in a multiferroic material induced by magnetic fields. *Nature* **2004**, *429*, 392–395.
75. Seidel, J.; Martin, L.W.; He, Q.; Zhan, Q.; Chu, Y.H.; Rother, A.; Hawkridge, M.E.; Maksymovych, P.; Yu, P.; Gajek, M.; *et al.* Conduction at domain walls in oxide multiferroics. *Nat. Mater.* **2009**, *8*, 229–234.
76. Choi, T.; Lee, S.; Choi, Y.J.; Kiryukhin, V.; Cheong, S.W. Switchable ferroelectric diode and photovoltaic effect in $BiFeO_3$. *Science* **2009**, *324*, 63–66.
77. Han, S.H.; Kim, K.S.; Kim, H.-G.; Lee, H.-G.; Kang, H.-W.; Kim, J.S.; Cheon, C.I. Synthesis and characterization of multiferroic $BiFeO_3$ powders fabricated by hydrothermal methods. *Ceram. Int.* **2010**, *36*, 1365–1375.
78. Wang, Y.; Xu, G.; Ren, Z.; Wei, X.; Weng, W.; Du, P.; Shen, G.; Han, G. Mineralizer-assisted hydrothermal synthesis and characterization of $BiFeO_3$ nanoparticles. *J. Am. Ceram. Soc.* **2007**, *90*, 2615–2617.
79. Chen, C.; Cheng, J.; Yu, S.; Che, L.; Meng, Z. Hydrothermal synthesis of perovskite bismuth ferrite crystallites. *J. Cryst. Growth* **2006**, *291*, 135–139.

80. Komarneni, S.; Menon, V.C.; Li, H.Q.; Roy, R.; Ainger, F. Microwave-hydrothermal processing of $BiFeO_3$ and $CsAl_2PO_6$. *J. Am. Ceram. Soc.* **1996**, *79*, 1409–1412.
81. Prado-Gonjal, J.; Villafuerte-Castrejon, M.E.; Fuentes, L.; Moran, E. Microwave-hydrothermal synthesis of the multiferroic $BiFeO_3$. *Mater. Res. Bull.* **2009**, *44*, 1734–1737.
82. Ponzoni, C.; Rosa, R.; Cannio, M.; Buscaglia, V.; Finocchio, E.; Nanni, P.; Leonelli, C. Optimization of BFO microwave-hydrothermal synthesis: Influence of process parameters. *J. Alloys Compds.* **2013**, *558*, 150–159.
83. Merzhanov, A.G.; Shkiro, V.M.; Borovinskaya, I.P. Synthesis of refractory inorganic compounds. US Patent 3726643, April 1973.
84. Merzhanov, A.G.; Shkiro, V.M.; Borovinskaya, I.P. Synthesis of refractory inorganic compounds. *Byull. Izobr.* **1971**, 10.
85. Varma, A.; Lebrat, J.P. Combustion synthesis of advanced materials. *Chem. Eng. Sci.* **1992**, *47*, 2179–2194.
86. Merzhanov, A.G. Reviews: fundamentals, achievements, and perspectives for development of solid-flame combustion. *Russ. Chem. Bull.* **1997**, *46*, 1–27.
87. Morsi, K. The diversity of combustion synthesis processing: A review. *J. Mater. Sci.* **2012**, *47*, 68–92.
88. Rajeshwar, K.; de Tacconi, N.R. Solution combustion synthesis of oxide semiconductors for solar energy conversion and environmental remediation. *Chem. Soc. Rev.* **2009**, *38*, 1984–1998.
89. Nagaveni, K.; Hedge, M.S.; Ravishankar, N.; Subbanna, G.N.; Madras, G. Synthesis and structure of nanocrystalline TiO_2 with lower band gap showing high photocatalytic activity. *Langmuir* **2004**, *20*, 2900–2907.
90. Selvam, N.C.S.; Kumar, R.T.; Kennedy, L.J.; Vijaya, J.J. Comparative study of microwave and conventional methods for the preparation and optical properties of novel MgO-micro and nano-structures. *J. Alloys Compds.* **2011**, *509*, 9809–9815.
91. Nehru, L.C.; Swaminathan, V.; Sanjeeviraja, C. Rapid synthesis of nanocrystalline ZnO by a microwave-assisted combustion method. *Powder Technol.* **2012**, *226*, 29–33.
92. Rosa, R.; Ponzoni, C.; Veronesi, P.; Natali Sora, I.; Felice, V.; Leonelli, C. Solution combustion synthesis of perovskite oxides: Comparison between MWs and conventional ignition. In Proceedings of 14th International Conference on Microwave and High Frequency Heating, Nottingham, UK, 16–19 September 2013. The University of Nottingham: Nottingham, UK, 2013; pp. 250–253.
93. Kumar, A.; Wolf, E.E.; Mukasyan, A.S. Solution combustion synthesis of metal nanopowders: Nickel reaction pathways. *AIChE J.* **2011**, *57*, 2207–2214.
94. Luo, W.; Wang, D.; Peng, X.; Wang, F. Microwave synthesis and phase transitions in nanoscale $BiFeO_3$. *J. Sol-Gel Sci. Technol.* **2009**, *51*, 53–57.
95. Mukasyan, A.S.; Dinka, P. Apparatus and methods for combustion synthesis of nano-powders. US Patent WO2007019332-A1, February 2007.
96. Kumar, A.; Mukasyan, A.S.; Wolf, E.E. Impregnated layer combustion synthesis method for preparation of multicomponent catalysts for the production of hydrogen from oxidative reforming of methanol. *Appl. Catal. A* **2010**, *372*, 175–183.

97. Martirosyan, K.S.; Luss, D. Carbon combustion synthesis of oxides, US Patent 0097419 A1, May 2006.
98. Martirosyan, K.S.; Iliev, M.; Luss, D. Carbon combustion synthesis of nanostructured perovskites. *Int. J. SHS* **2007**, *16*, 36–45.
99. Dinka, P.; Mukasyan, A.S. *In situ* preparation of oxide-based supported catalysts by solution combustion synthesis. *J. Phys. Chem. B* **2005**, *109*, 21627–21633.

4

Synthesis of Diazonium Tetrachloroaurate(III) Precursors for Surface Grafting

Sabine N. Neal [1,†], Samuel A. Orefuwa [1,†], Atiya T. Overton, [1,†], Richard J. Staples [2,†] and Ahmed A. Mohamed [*,1]

1 Department of Chemistry, Delaware State University, 1200 N. DuPont Highway, Dover, Delaware 19901, USA; E-Mails: snneal12@students.desu.edu (S.N.N.); sammybid@yahoo.com (S.A.O.); atiya.overton@gmail.com (A.T.O.)

2 Department of Chemistry, Center for Crystallographic Research, Michigan State University, East Lansing, Michigan 48824, USA; E-Mail: staples@chemistry.msu.edu

† These authors contributed equally to this work.

* Author to whom correspondence should be addressed; E-Mail: amohamed@desu.edu

---

Abstract: The synthesis of diazonium tetrachloroaurate(III) complexes $[R\text{-}4\text{-}C_6H_4N{\equiv}N]AuCl_4$ involves protonation of anilines $CN\text{-}4\text{-}C_6H_4NH_2$, $C_8F_{17}\text{-}4\text{-}C_6H_4NH_2$, and $C_6H_{13}\text{-}4\text{-}C_6H_4NH_2$ with tetrachloroauric acid $H[AuCl_4]\ 3H_2O$ in acetonitrile followed by one-electron oxidation using $[NO]PF_6$. FT-IR shows the diazonium stretching frequency at 2277 cm^{-1} (CN), 2305 cm^{-1} (C_8F_{17}), and 2253 cm^{-1} (C_6H_{13}). Thermogravimetric Analysis (TGA) shows the high stabilities of the electron-withdrawing substituents C_8F_{17} and CN compared with the electron-donating substituent C_6H_{13}. Residual Gas Analysis (RGA) shows the release of molecular nitrogen as the main gas residue among other small molecular weight chlorinated hydrocarbons and chlorobenzene. Temperature-Dependent X-Ray Powder Diffraction (TD-XRD) shows the thermal decomposition in C_6H_{13} diffraction patterns at low temperature of 80 °C which supports the TGA and RGA (TGA-MS) conclusions. X-ray structure shows N≡N bond distance of approximately 1.10 Å and N≡N-C bond angle of 178°.

Keywords: diazonium; tetrachloroaurate(III); X-ray

---

1. Introduction

Modification of surfaces by redox grafting of diazonium salts is a progressing area of materials chemistry [1–12]. Reduction of aryl diazonium salts has been studied on carbon [3,4,6], semiconductors [6], and nanoparticles surfaces [13–19]. Grafted organic layers demonstrated distinctive performance in the formation of corrosion inhibitor film on iron [10,20], modification of graphene [2,11,12] and diamond [5,21], preparation of diazonium-modified enzyme electrodes [22], immobilization of proteins [5,23], grafting polymers [24], and printing gold via soft lithography [25,26]. In addition to surface modification, aryl diazonium salts have been utilized in palladium-catalyzed cross-coupling reactions. In particular, Heck reactions involving aryl diazonium salts have been widely used in the synthesis of natural products. These electrophiles can also be used in Suzuki-Miyaura, carbonylative, Stille, and carbon-heteroatom cross-couplings [27].

Figure 1. Reductive grafting mechanism of diazonium on surfaces.

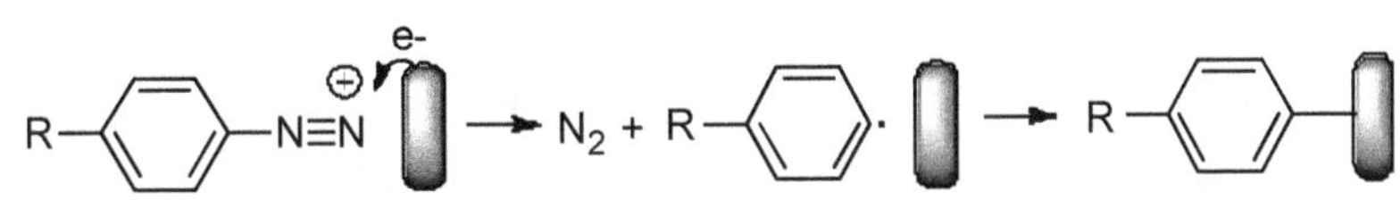

Electrochemical reductive grafting of diazonium has proven to be an effective, versatile, and simple method for producing organic monolayers (Figure 1) [1]. Despite the wide applications of diazonium salts in fundamental and materials chemistry there are many serious drawbacks: their intrinsic instability at room temperature, difficulty to isolate and purify, and their explosive nature [28–30]. In order to avoid some of their limitations diazonium salts must undergo the targeted application in the same medium without isolation. This can be achieved by the *in situ* diazotization of aryl amines followed by reductive grafting [31].

The electron density model described the bonding in diazonium by synergistic N→C σ-bonding and C→N π-back-bonding [32–34]. A stabilizing effect for the C-N bond includes the formation of charge-transfer complexes, metal-based anion, and coordination [35–37]. Based on the relative size of diazonium and the void of the polyether ring, basicity of the oxygen atoms, and steric hindrance in the ring, it is possible to stabilize the diazonium by complexation with macrocyclic polyethers [34,35]. The formation of aryl diazonium ion-crown ether complexes increased the thermal and photochemical stabilities of aryl diazonium ions in solution and in the solid state. Metal-based anion approach was also utilized in order to stabilize the C-N bond [36]. A complex of aryl diazonium with the anion $[ZnCl_4]^{2-}$ was synthesized and X-ray structure was determined [36]. Coordination of the *in situ* generated diazonium demonstrated to be a plausible route for the stabilization of aliphatic diazonium salts [37].

The synthesis of gold complexes of potential applications in materials chemistry is a rising field of interest [38,39]. We describe a facile procedure for the synthesis of stable diazonium tetrachloroaurate(III) complexes. Preliminary results on the synthesis and grafting have been reported in a communication to *Inorganic Chemistry* [39].

2. Results and Discussion

2.1. Synthesis and Characterization

In diazotization reactions, a considerably greater amount of HCl, H_2SO_4, or $HClO_4$ acid should be used even if strong basic amines are used as reagents. The tetrachloroauric(III) acid based procedure can be utilized in preparation of a wide range of diazonium compounds in organic solvents without the need for mineral acids (Scheme 1) [39]. Protonation of CN-4-$C_6H_4NH_2$ with $H[AuCl_4]•3H_2O$ in water resulted in the formation of CN-4-$C_6H_4NH_2AuCl_3$, **1** (Scheme 2). However, the protonation in acetonitrile formed the anilinium tetrachloroaurate(III) complex [CN-4-$C_6H_4NH_3$]$AuCl_4$, **2** (Scheme 3), which was easily oxidized by [NO]PF_6 to form the yellow diazonium tetrachloroaurate(III) complex [CN-4-$C_6H_4N{\equiv}N$]$AuCl_4$, **3**, in a high yield. Similarly, the diazonium tetrachloroaurate(III) complexes [C_8F_{17}-4-$C_6H_4N{\equiv}N$]$AuCl_4$, **4**, and [C_6H_{13}-4-$C_6H_4N{\equiv}N$]$AuCl_4$, **5**, were synthesized in high yields. The diazonium cation was verified in solution by azo coupling with 2-naphthol to give the dark orange color of diazine. The diazonium tetrachloroaurate(III) complexes are soluble in chloroform, acetonitrile, dimethylformamide, and dimethylsulfoxide. The good solubility is an important advantage over diazonium salts of BF_4^-, Cl^-, and PF_6^-. Although caution should be exercised when handling all diazonium salts, there were no indications that gold diazonium compounds are particularly hazardous.

Scheme 1. Synthesis of diazonium tetrachloroaurate(III) complexes.

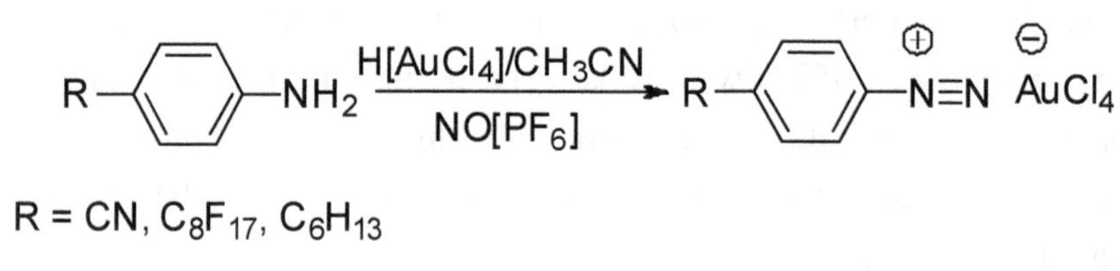

Scheme 2. Protonation of CN-4-$C_6H_4NH_2$ in water formed aniline gold(III) trichloride, **1**.

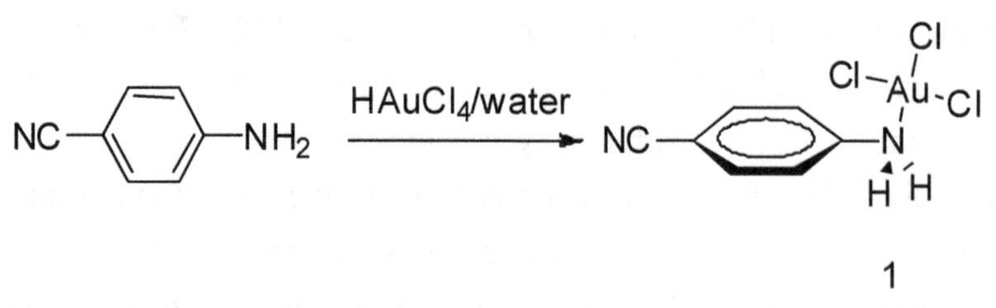

Scheme 3. Protonation of CN-4-$C_6H_4NH_2$ in acetonitrile formed anilinium tetrachloroaurate(III), **2**.

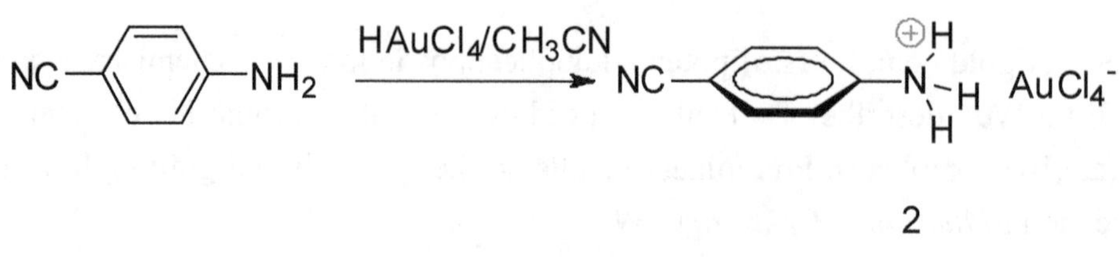

FT-IR spectrum of **3** in the solid state showed the $\nu_{N\equiv N}$ stretching frequency at 2277 cm^{-1} and ^1H NMR spectrum in $CDCl_3$/DMSO-d_6 displayed two doublets assigned to the phenyl protons at δ = 7.71 and 8.33 ppm, remarkably shifted from 6.65 and 7.45 ppm in the 4-aminobenzonitrile ligand (Figure 2). FT-IR spectrum of **4** displayed the $\nu_{N\equiv N}$ stretching frequency at 2305 cm^{-1} and ^1H NMR spectrum in $CDCl_3$/DMSO-d_6 displayed the phenyl protons at δ = 7.67 and 8.59 ppm. FT-IR spectrum of **5** displayed the $\nu_{N\equiv N}$ stretching frequency at 2253 cm^{-1} and ^1H NMR spectrum in $CDCl_3$/DMSO-d_6 displayed the phenyl protons at δ = 7.75 and 8.60 ppm.

Figure 2. FT-IR of diazonium tetrachloroaurate(III) complexes **3** (CN), **4** (C_8F_{17}), and **5** (C_6H_{13}).

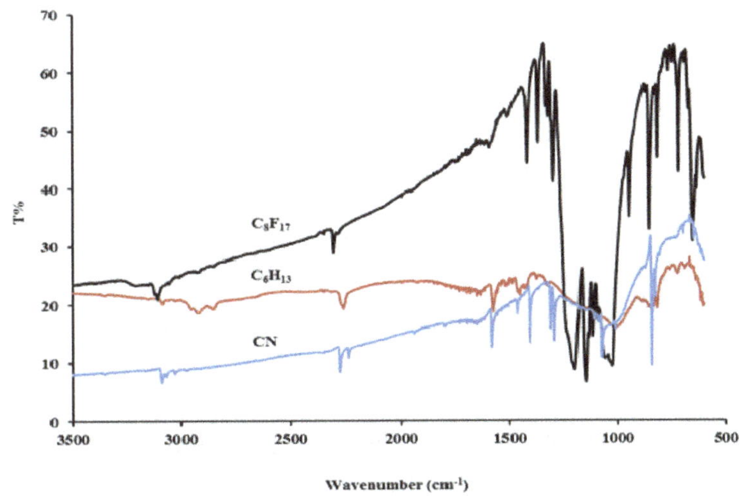

2.2. X-ray Structure

From acetonitrile solvent at room temperature well-shaped yellow crystals of diazonium tetrachloroaurate(III) complexes were obtained. Structures **1–3** have been reported in reference [39]. X-ray structure of **5** is shown in Figure 3 and X-ray data is presented in Table 1. The crystals are quite stable and no sign of decomposition was seen in the X-ray frames. The N≡N distance is typical of diazonium, approximately 1.10 Å. The unit cell shows the arrangement of the tetrachloroaurate anions around the diazonium. Each diazonium group is surrounded by four $[AuCl_4]^-$ normal to the N≡N axis. The N(outer)…Cl distances are 3.217, 3.267, 3.300, and 3.468 Å (Figure 4). The outer nitrogen is connected to four chloride centers in a square-planar arrangement. The close proximity to the N≡N group, Cl…N≡N interaction, and the fact that the chloride is less nucleophilic in the tetrachloroaurate anion than the free chloride are presumably contributing factors for the increased stability of the diazonium salts.

Several diazonium salts have been isolated in the last few decades and their X-ray structures have been reported [40–45]. The structure of benzene diazonium chloride showed N≡N…Cl distance 3.22–3.56 Å. The first example of diazonium stabilization by multiple and close contacts with tosylate anion shows the shortest distances between the diazonium nitrogen and the tosylate oxygen of 2.673–2.770 Å [46]. shorter than the sum of van der Waals radii (2.90 Å) [47]. The ionic structures of metal-based diazonium complexes [R-4-C_6H_4N≡N]X (X = $SbCl_6$, $FeCl_4$, Cu_2Br_2) and [R-4-C_6H_4N≡N]$_2$$ZnCl_4$ show the shortest interionic distances between the diazonium and the metal anion occur between the halide and the outer nitrogen than to the ammonium nitrogen [42,43]. Metal-based diazonium complexes structurally characterized so far indicate that the bond order in diazonium nitrogen-nitrogen is close to a triple bond distance of approximately 1.1 Å.

Table 1. Crystal data, data collection, and structure refinement for $[C_6H_{13}\text{-}4\text{-}C_6H_4N{\equiv}N]AuCl_4$.

Empirical formula	$C_{12}H_{17}AuCl_4N_2$
Formula weight	528.04
Temperature, K	173(2)
Wavelength, Å	0.71073
Crystal system	Monoclinic
Space group	$P2_1/c$
a (Å); (°)	15.5144(3)
b (Å); (°)	38.4270(8) 90.0250(10)
c (Å); (°)	8.7773(2)
Volume, $Å^3$	5232.78(19)
Z	12
Density (cal.), Mg/m^3	2.011
Abs. coefficient, mm^{-1}	21.395
GOF on F^2	1.061
R1, wR2 [$I > 2\sigma(I)$]	0.0568, 0.1530

Figure 3. Structure of $[C_6H_{13}\text{-}4\text{-}C_6H_4N{\equiv}N]AuCl_4$, **5**, at 30% probability. Bond distances (Å): N(1A)-N(2A) 1.086(11) and N(2)-C(1) 1.400(12). Bond angles (°): N(1A)-N(2A)-C(1A) 178.7(9).

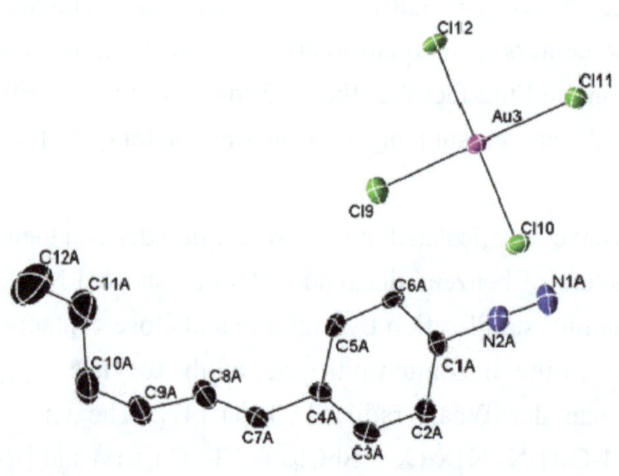

Figure 4. Unit cell of $[C_6H_{13}\text{-}4\text{-}C_6H_4N{\equiv}N]AuCl_4$.

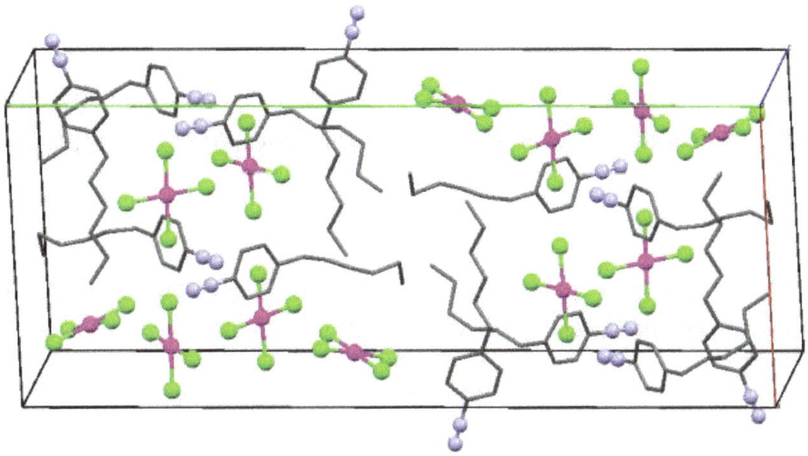

2.3. Thermal Stability and Residual Gas Analysis Studies

The enormous literature on the thermal stability of diazonium salts covers the greatest variety with inorganic stabilizing anions. Study clearly indicates an important influence of the stabilizing anion, substituent, and water in the unit cell. A correlation between the rate of diazonium nitrogen evolution and the electronic configuration and the electronegativity of the metal ion has been studied [48]. It cannot be concluded from the correlation study if the type of metal has a direct effect on diazonium stability due to the failure to stabilize diazonium salts with some of the metal chlorides used.

When examined by Thermal Gravimetric Analysis (TGA) at a heating rate of 2.0 °C/min the diazonium tetrachloroaurate(III) complexes started the decomposition at approximately 100 °C. The TGA curve of **4** displays the loss of nitrogen and chloride in the first stage of 10% which ends at about 100 °C followed by a more abrupt weight loss until 150 °C. From 150 °C to 400 °C the sample losses *ca.* 40% weight (Figure 5). The poor volatility of the fluorinated ligand conceivably is the reason for the sluggish weight loss. The percentage gold residue from **5** is 23% (theoretical 23%). TGA of **3** displays a 40% gold residue (theoretical 42%) and 40% (theoretical 37%) for **5**.

Temperature-Dependent X-Ray Powder Diffraction (TD-XRD) patterns at 25, 50, 100, and 150 °C display peaks associated with the decomposition of **4**, however, the peaks associated with the elemental gold formation are absent within the same temperature range (Figure 6). The diazonium tetrachloroaurate(III) peaks disappeared after heating above 150 °C with the concomitant appearance of a new set of peaks. The cubic gold reflections became more intense within the 250–300 °C range. The TD-XRD associated peaks of **5** (Figure 7) show elemental gold at lower temperature than **4**. The diffraction patters also confirmed the highly crystalline nature of gold residues after complete decomposition of **4** and **5**. Bragg reflections at 2θ 38.1°, 44.5°, 64.8°, and 78.8° could be indexed based on the face-centered cubic gold structure.

The gases released from diazonium tetrachloroaurate(III) pyrolysis were identified by Residual Gas Analysis (RGA) studies. Figure 8 shows the partial pressure of nitrogen for **4** and **5**. In addition to nitrogen, the gases released from each complex in the temperature range 100–450 °C include chlorobenzene, chlorinated small molecular weight hydrocarbons, and chlorine. The fluorous chain and phenyl groups can be clearly seen leaving as PhCl, PhF, and low molecular weight halogenated hydrocarbon fractions CH_2Cl_2, CCl_4, $CHCl_3$, C_2HCl_3, CH_3F, and CF_4 (Table 2). In pyrolysis gas chromatography, the major gases of $[R_2N\text{-}4\text{-}C_6H_4N{\equiv}N]_2ZnCl_4$ decomposition between 160 °C and 250 °C were, in addition to nitrogen, the corresponding chloroaromatics due to the replacement of diazonium by chlorine [49].

Table 2. Partial pressure of gases (Torr) released from Residual Gas Analysis (RGA) decomposition of $[C_8F_{17}\text{-}4\text{-}C_6H_4N{\equiv}N]AuCl_4$ at 450 °C.

Gas	Partial pressure	Gas	Partial pressure
N_2	2.60×10^{-7}	Cl_2	1.08×10^{-11}
CH_3F	9.10×10^{-10}	C_6H_5Cl	6.47×10^{-12}
CH_2Cl_2	5.47×10^{-10}	C_6H_5F	5.21×10^{-12}
$CHCl_3$	2.90×10^{-11}	CCl_4	4.29×10^{-12}
CF_4	1.10×10^{-11}		

Decomposition of the diazonium tetrachloroaurate(III) complexes, ready evolution of nitrogen, under their respective temperatures using melting point apparatus occurred without melting. It appears that for diazonium tetrachloroaurate(III) complexes there are small temperature ranges over which the gases are evolved. When a heat source was applied to **3** the crystals seemed to release gases beginning at about 129 °C and continued until about 134 °C. When a heat source was applied to **5** the gases were released at temperature range of 110-114 °C and in **4** the gases were released at lower range of 96–97 °C.

Figure 5. TGA of $[CN\text{-}4\text{-}C_6H_4N{\equiv}N]AuCl_4$, $[C_8F_{17}\text{-}4\text{-}C_6H_4N{\equiv}N]AuCl_4$, and $[C_6H_{13}\text{-}4\text{-}C_6H_4N{\equiv}N]AuCl_4$.

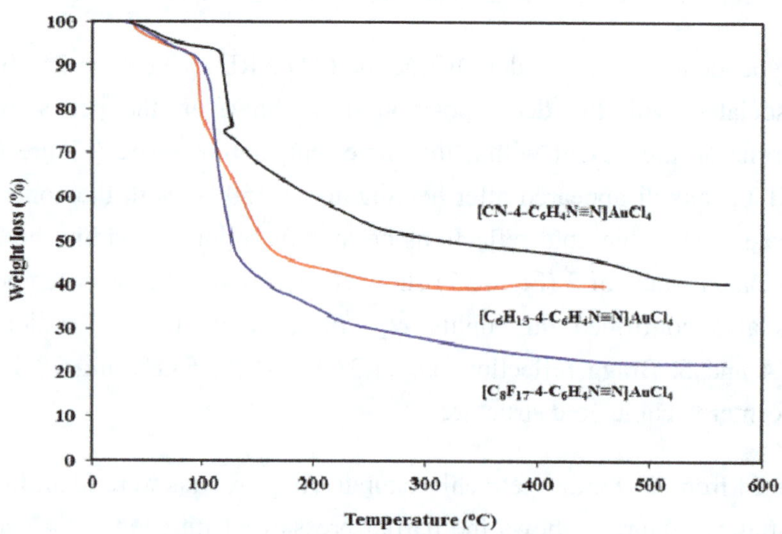

Figure 6. Temperature-Dependent X-ray Powder Diffraction patterns of $[C_8F_{17}\text{-}4\text{-}C_6H_4N{\equiv}N]AuCl_4$.

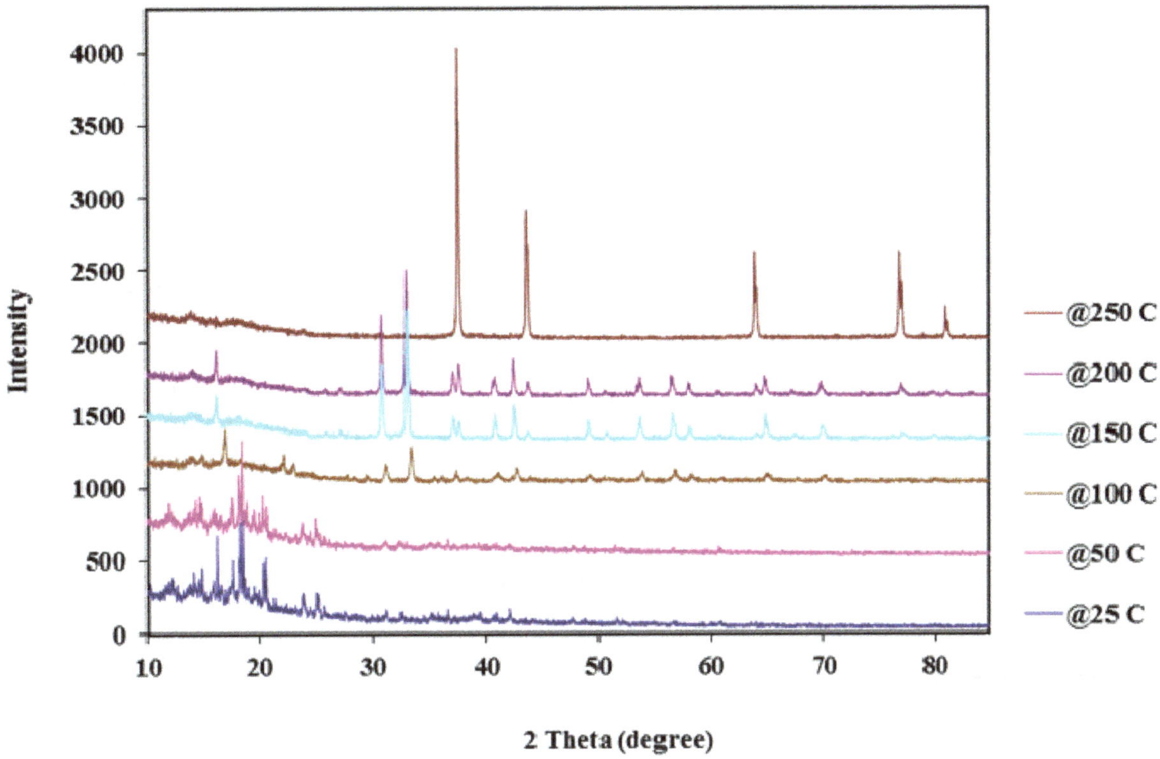

Figure 7. Temperature-Dependent X-ray Powder Diffraction patterns of $[C_6H_{13}\text{-}4\text{-}C_6H_4N{\equiv}N]AuCl_4$.

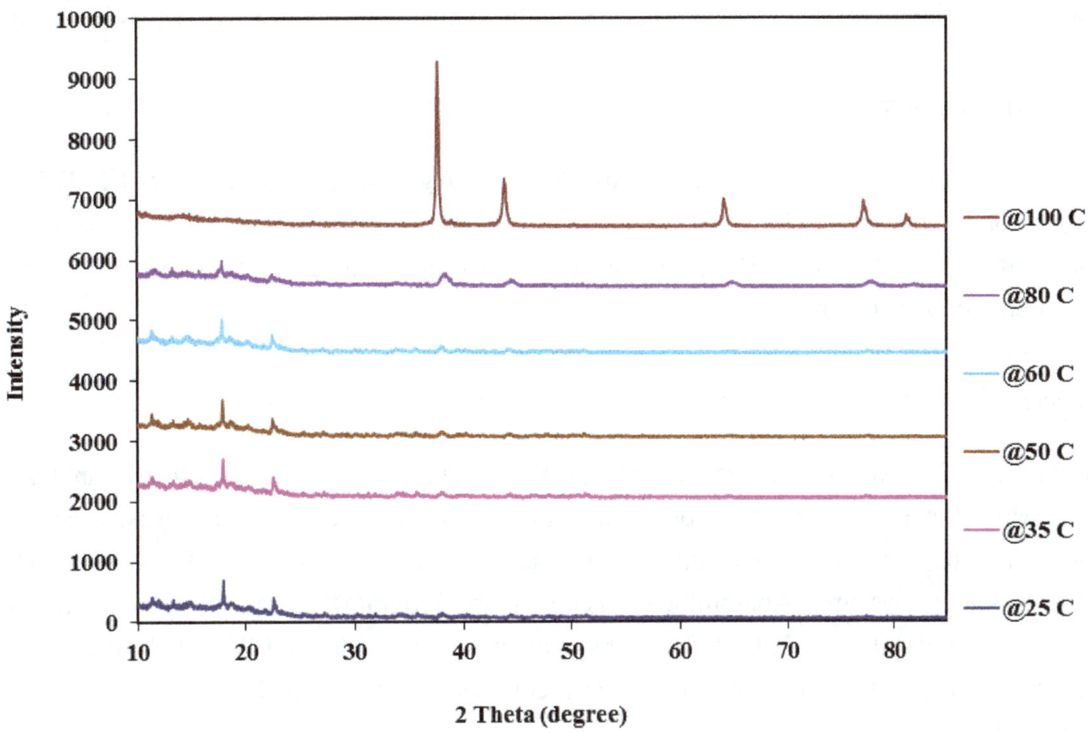

Figure 8. Residual Gas Analysis (RGA) of $[C_8F_{17}\text{-}4\text{-}C_6H_4N{\equiv}N]AuCl_4$ and $[C_6H_{13}\text{-}4\text{-}C_6H_4N{\equiv}N]AuCl_4$.

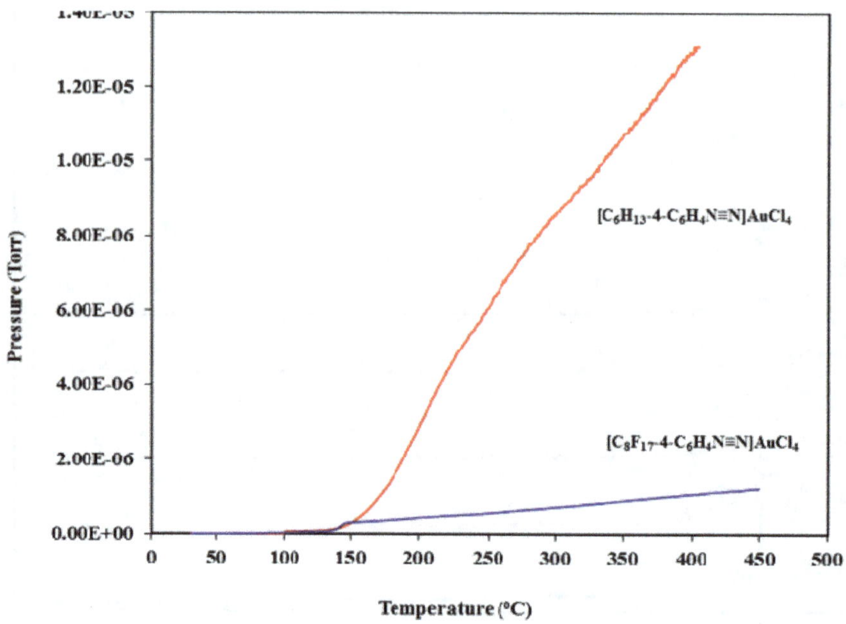

3. Experimental Section

3.1. General Procedures

Unless stated otherwise, all manipulations were carried out in either a N_2-filled Vacuum Atmospheres Co. glove box or on a Schlenk line using N_2. Toluene, acetonitrile, $CN\text{-}4\text{-}C_6H_4NH_2$, $C_8F_{17}\text{-}4\text{-}C_6H_4NH_2$, $C_6H_{13}\text{-}4\text{-}C_6H_4NH_2$, $CDCl_3$, DMSO-d_6, $[NO]PF_6$, and $H[AuCl_4]$ $3H_2O$ were purchased from Sigma-Aldrich. All other commercially available reagents were used as received.

3.2. Physical Measurements

Infrared spectra were recorded in the 4000–400 cm^{-1} range using a Perkin-Elmer FT-IR Spectrum 1000 spectrophotometer. C, H, and N analyses were carried out by Columbia Analytical Services, Tucson, Arizona. ^1H and ^{13}C NMR spectra were recorded on a JEOL 400 MHz. Chemical shifts are reported relative to the chloroform solvent peak.

3.3. Thermogravimetric Analysis and Residual Gas Analysis

Thermal stability was studied using a Perkin-Elmer TG/DTA Thermogravimetric analyzer. The scan rate was set at a 2 °C/min from room temperature to 500 °C. The TGA analysis was carried out in air and in argon atmosphere. Residual gas analyses were carried out using RGA Pro 2000 from Stanford Research Systems. Non-ambient (Temperature Program) X-ray diffraction analysis was carried out on a Panalytical X'Pert Pro X-ray Diffractometer (Model PW3040 Pro) using copper K-alpha radiation (λ = 0.154178 nm). The instrument was operated at 40 KV and 20 mA on an HTK 1200 oven stage.

3.4. X-ray Diffraction

Data for complexes CN-4-$C_6H_4NH_2AuCl_3$, **1**, [CN-4-$C_6H_4NH_3$]$AuCl_4$, **2**, and [CN-4-$C_6H_4N{\equiv}N$]$AuCl_4$, **3**, were collected using a Bruker D8 tricycles diffractometer with APEX II detector CCD based equipped with an LT-2 low-temperature apparatus operating at 110 K. Data for **5** was collected at 213 K with a Siemens SMART CCD diffractometer equipped with an LT-2 low-temperature apparatus. A suitable crystal was chosen and mounted on a glass fiber using grease. Cell parameters were determined and refined using APEX II software on all observed reflections which corrects for Lp and decay [50,51]. Absorption corrections were applied using SADABS supplied by George Sheldrick [52]. The structures are solved by the direct method using the SHELXS-97 program and refined by least squares method on F^2, SHELXL-97, incorporated in SHELXTL-PC V 5.03 [53,54].

3.5. Syntheses

Synthesis of CN-4-$C_6H_4NH_2AuCl_3$, 1: To a 67 mg (0.5 mmol) of CN-4-$C_6H_4NH_2$ suspended in 15 of mL distilled water was added 190 mg (0.5 mmol) H[$AuCl_4$] $3H_2O$ at room temperature. The water insoluble yellow complex was stirred for 2 h. The yellow precipitate was filtered, washed twice with 5 mL of water, and dried under vacuum to give 153 mg. Yield 64.01%. ^{1}H NMR ($CDCl_3$/DMSO-d_6): δ 7.20 (d, 9.7 Hz, 2H), 7.58 (d, 9.7 Hz, 2H). ^{13}C NMR ($CDCl_3$/DMSO-d_6): δ 119.62 (C-1), 125.0 (C-2, C-3), 133.63 (C-4), 143.0 (CN). Elemental analysis calculated for $C_7H_6AuCl_3N_2$: %C = 19.95, %H = 1.43, %N = 6.65. Exp: %C = 19.89, %H = 1.40, %N = 6.75.

Synthesis of [CN-4-$C_6H_4NH_3$]$AuCl_4$, 2: To a 67 mg (0.5 mmol) of CN-4-$C_6H_4NH_2$ dissolved in 10 mL of CH_3CN was added 197 mg (0.5 mmol) H[$AuCl_4$] $3H_2O$ at room temperature. The yellow solution was stirred for 2 h and 30 mL of cold ether was added. The reaction mixture was stirred for additional 2 h and the precipitate was filtered, washed twice with 5 mL cold ether, and dried under vacuum to give 228 mg. Yield 87.69%. ^{1}H NMR ($CDCl_3$/DMSO-d_6): δ 6.70 (d, 9.3 Hz, 2H), 7.17 (d, 9.3 Hz, 2H). ^{13}C NMR ($CDCl_3$/DMSO-d_6): δ 104.5 (C-1), 118.9 (C-2), 133.39 (C-3, C-4), 144.24 (CN). Elemental analysis calculated for $C_7H_7AuCl_4N_2$: %C = 18.36, %H = 1.54, %N = 6.12. Exp: %C = 18.28 , %H = 1.56, %N = 6.09.

Synthesis of [CN-4-$C_6H_4N{\equiv}N$]$AuCl_4$, 3: To a 118 mg (1.0 mmol) of CN-4-$C_6H_4NH_2$ dissolved in 10 mL of CH_3CN was added 391 mg (1.0 mmol) H[$AuCl_4$] $3H_2O$ at 0 °C. A 184 mg (1.0 mmol) [NO]PF_6 was added and stirring continued at 0 °C. The yellow solution was stirred for 2 h and 30 mL of cold ether was added. The reaction mixture was stirred for additional 2 h and the precipitate was filtered, washed twice with 5 mL of cold ether, and dried under vacuum to give 420 mg. Yield 91.90%. ^{1}H NMR ($CDCl_3$/DMSO-d_6): δ 7.71 (d, 9.7 Hz, 2H), 8.33 (d, 9.7 Hz, 2H). ^{13}C NMR ($CDCl_3$/DMSO-d_6): δ 115.4 (C1), 120.3 (C2), 122.4 (C3), 133.3.24 (C4), 134.3 (CN). Elemental analysis calculated for $C_7H_4AuCl_4N_3$: %C = 17.93, %H = 0.86, %N = 8.96. Exp: %C = 17.95, %H = 0.79, %N = 8.84. FT-IR (solid): $\nu_{N\equiv N}$ stretching frequency at 2277 cm^{-1}. UV-Vis (2.13×10^{-4} M in CH_3CN): 318 nm (10,938 $M^{-1} cm^{-1}$).

Synthesis of [C_8F_{17}-4-$C_6H_4N{\equiv}N$]$AuCl_4$, 4: To a 512 mg (1.0 mmol) of C_8F_{17}-4-$C_6H_4NH_2$ dissolved in 20 mL of toluene at 0 °C was added 400 mg (1.0 mmol) H[$AuCl_4$] $3H_2O$ dissolved in 3 mL of acetonitrile. The solution was stirred for 2 h before a 173 mg (1.0 mmol) [NO]PF_6 dissolved in 3 mL of acetonitrile was added. The dark red solution was stirred for an additional 2 h at 0 °C under nitrogen during which canary yellow microcrystals precipitated. The microcrystals were filtered, washed twice with 5 mL of cold toluene, and dried under vacuum to give 835 mg. Yield 96.70%. ^{1}H NMR ($CDCl_3$/DMSO-d_6): δ 7.67 (d, 9.7 Hz, 2H), 8.59 (d, 9.7 Hz, 2H). ^{13}C NMR ($CDCl_3$/DMSO-d_6): δ 120.4 (C-1), 129.5 (C-2), 133.6 (C-3), 138.0 (C-4). Elemental analysis calculated for $C_{14}F_{17}H_4AuCl_4N_2$: %C = 19.51, %H = 0.47, %N = 3.25. Exp: %C = 19.33, %H = 0.66, %N = 3.11. FT-IR (solid): $\nu_{N\equiv N}$ stretching frequency at 2305 cm^{-1}. UV-Vis (1.89×10^{-4} M in CH_3CN): 305 nm (2983 M^{-1} cm^{-1}).

Synthesis of [C_6H_{13}-4-$C_6H_4N\equiv N$]$AuCl_4$, 5: To a 178 mg (1.0 mmol) of C_6H_{13}-4-$C_6H_4NH_2$ dissolved in 20 mL of toluene at 0 °C was added 400 mg (1.0 mmol) H[$AuCl_4$] $3H_2O$ dissolved in 3 mL of acetonitrile. The solution was stirred for 2 h before a 173 mg (1.0 mmol) [NO]PF_6 dissolved in 3 mL of acetonitrile was added. The solution was stirred for an additional 2 h at 0 °C under nitrogen and was kept in the fridge overnight to form yellow microcrystals out of olive green solution. The precipitate was filtered, washed twice with 5 mL of cold toluene, and dried under vacuum to give 425 mg. Yield 80.49%. ^{1}H NMR ($CDCl_3$/DMSO-d_6): δ 7.75 (d, 8.1 Hz, 2H), 8.60 (d, 8.1 Hz, H). Elemental analysis calculated for $C_{12}H_{17}AuCl_4N_2$: %C = 27.31, %H = 3.14, %N = 5.25. Exp: %C = 27.29, %H = 3.24, %N = 5.31. FT-IR (solid): $\nu_{N\equiv N}$ stretching frequency at 2253 cm^{-1}. UV-Vis (5.8×10^{-4} in CH_3CN): 279 nm (14,884 M^{-1} cm^{-1}).

4. Conclusions

We described the synthesis of stable diazonium tetrachloroaurate(III) complexes of potential applications in surface grafting. The structures are similar to inorganic based anions diazonium salts. The synthesized compounds are soluble in several organic solvents and stable at room temperature in the solid state and solution. Research continues on the diazonium tetrachloroaurate(III) complexes to synthesize robust gold-carbon nanoparticles and films.

Acknowledgments

Andrew Goudy of Delaware State University is acknowledged for the TGA, RGA, and TD-XRD measurements.

References

1. Chehimi, M.M. *Aryl Diazonium Salts: New Coupling Agents in Polymer and Surface Science*; Wiley-VCH: Weinheim, Germany, 2012.
2. Sun, Z.; James, D.K.; Tour, J.M. Graphene Chemistry: Synthesis and Manipulation. *J. Phys. Chem. Lett.* **2011**, *2*, 2425–2432.
3. Breton, T.; Bélanger, D. Modification of Carbon Electrode with Aryl Groups having an Aliphatic Amine by Electrochemical Reduction of *in situ* Generated Diazonium Cations. *Langmuir* **2008**, *24*, 8711–8718.

4. Allongue, P.; Delamar, M.; Desbat, B.; Fagebaume, O.; Hitmi, R.; Pinson, J.; Saveant, J.M. Covalent Modification of Carbon Surfaces by Aryl Radicals Generated from the Electrochemical Reduction of Diazonium Salts. *J. Am. Chem. Soc.* **1997**, *119*, 201–207.
5. Salmi, Z.; Lamouri, A.; Decorse, P.; Jouini, M.; Boussadi, A.; Achard, J.; Gicquel, A.; Mahouche-Chergui, S.; Carbonnier, B.; Chehimi, M.M. Grafting Polymer–Protein Bioconjugate to Boron-Doped Diamond using Aryl Diazonium Coupling Agents. *Diamond Rel. Mater.* **2013**, *40*, 60–80.
6. Lehr, J.; Garrett, D.J.; Paulik, M.G.; Flavel, B.S.; Brooksby, P.A.; Williamson, B.E.; Downard, A.J. Patterning of Metal, Carbon, and Semiconductor Substrates with Thin Organic Films by Microcontact Printing with Aryldiazonium Salt Inks. *Anal. Chem.* **2010**, *82*, 7027–7070.
7. Aswal, D.K.; Koiry, S.P.; Jousselme, B.; Gupta, S.K.; Palacin, S.; Yakhmi, J.V. Hybrid Molecule-on-Silicon Nanoelectronics: Electrochemical Processes for Grafting and Printing of Monolayers. *Physica E* **2009**, *41*, 325–344.
8. Gooding, J.J.; Ciampi, S. The Molecular Level Modification of Surfaces: From Self-Assembled Monolayers to Complex Molecular Assemblies. *Chem. Soc. Rev.* **2011**, *40*, 2704–2718.
9. Bélanger, D.; Pinson, J. Electrografting: A Powerful Method for Surface Modification. *Chem. Soc. Rev.* **2011**, *40*, 3995–4048.
10. Combellas, C.; Delamar, M.; Kanoufi, F.; Pinson, J.; Podvorica, F.I. Spontaneous Grafting of Iron Surfaces by Reduction of Aryldiazonium Salts in Acidic or Neutral Aqueous Solution. Application to the Protection of Iron against Corrosion. *Chem. Mater.* **2005**, *17*, 3968–3975.
11. Zhu, Y.; Higginbotham, A.L.; Tour, J.M. Covalent Functionalization of Surfactant-Wrapped Graphene Nanoribbons. *Chem. Mater.* **2009**, *21*, 5284–5291.
12. Lomeda, J.R.; Doyle, C.D.; Kosynkin, D.V.; Hwang, W.-F.; Tour, J.M. Diazonium Functionalization of Surfactant-Wrapped Chemically Converted Graphene Sheets. *J. Am. Chem. Soc.* **2008**, *130*, 16201–16206.
13. Gehan, H.; Fillaud, L.; Felidj, N.; Aubard, J.; Lang, P.; Chehimi, M.M.; Mangeney, C.A. General Approach Combining Diazonium Salts and Click Chemistries for Gold Surface Functionalization by Nanoparticle Assemblies. *Langmuir* **2010**, *26*, 3975–3980.
14. Mirkhalaf, F.; Paprotny, J.; Schiffrin, D.J. Synthesis of Metal Nanoparticles Stabilized by Metal-Carbon Bonds. *J. Am. Chem. Soc.* **2006**, *128*, 7400–7401.
15. Ghosh, D.; Chen, S.W. Palladium Nanoparticles Passivated by Metal–Carbon Covalent Linkages. *J. Mater. Chem.* **2008**, *18*, 755–762.
16. Ratheesh Kumar, V.K.; Gopidas, K.R. Synthesis and Characterization of Gold-Nanoparticle-Cored Dendrimers Stabilized by Metal–Carbon Bonds. *Chem. Asian J.* **2010**, *5*, 887–896.
17. Ratheesh Kumar, V.K.; Gopidas, K.R. Palladium Nanoparticle-Cored G1-Dendrimer Stabilized by Carbon-Pd Bonds: Synthesis, Characterization and use as Chemoselective, Room Temperature Hydrogenation Catalyst. *Tetrahedron Lett.* **2011**, *52*, 3102–3105.
18. Ratheesh Kumar, V.K.; Krishnakumar, S.; Gopidas, K.R. Synthesis, Characterization and Catalytic Applications of Palladium Nanoparticle-Cored Dendrimers Stabilized by Metal–Carbon Bonds. *Eur. J. Org. Chem.* **2012**, 3447–3458.
19. Laurentius, L.; Stoyanov, S.R.; Gusarov, S.; Kovalenko, A.; Du, R.; Lopinski, G.P.; McDermott, M.T. Diazonium-Derived Aryl Films on Gold Nanoparticles: Evidence for a Carbon-Gold Covalent Bond. *ACS Nano* **2011**, *5*, 4219–4227.

20. Adenier, A.; Bernard, M.C.; Chehimi, M.M.; Cabet-Deliry, E.; Desbat, B.; Fagebaume, O.; Pinson, J.; Podvorica, F. Covalent Modification of Iron Surfaces by Electrochemical Reduction of Aryldiazonium Salts. *J. Am. Chem. Soc.* **2001**, *123*, 4541–4549.
21. Nakamura, T.; Suzuki, M.; Ishihara, M.; Ohana, T.; Tanaka, A.; Koga, Y. Photochemical Modification of Diamond Films: Introduction of Perfluorooctyl Functional Groups on Their Surface. *Langmuir* **2004**, *20*, 5846–5849.
22. Harper, J.C.; Polsky, R.; Wheeler, D.R.; Brozik, S.M. Maleimide-Activated Aryl Diazonium Salts for Electrode Surface Functionalization with Biological and Redox Active Molecules. *Langmuir* **2008**, *24*, 2206–2211.
23. Corgier, B.P.; Marquette, C.A.; Blum, L.J. Diazonium-Protein Adducts for Graphite Electrode Microarrays Modification: Direct and Addressed Electrochemical Immobilization. *J. Am. Chem. Soc.* **2005**, *127*, 18328–18332.
24. Mévellec, V.; Roussel, S.; Tessier, L.; Chancolon, J.; Mayne-L'Hermite, M.; Deniau, G.; Viel, P.; Palacin, S. Grafting Polymers on Surfaces: A New Powerful and Versatile Diazonium Salt-Based One-Step Process in Aqueous Media. *Chem. Mater.* **2007**, *19*, 6323–6330.
25. Garrett, D.J.; Lehr, J.; Miskelly, G.M.; Downard, A.J. Microcontact Printing using the Spontaneous Reduction of Aryldiazonium Salts. *J. Am. Chem. Soc.* **2007**, *129*, 15456–15457.
26. Hossain, M.Z.; Walsh, M.A.; Hersam, M.C. Scanning Tunneling Microscopy, Spectroscopy, and Nanolithography of Epitaxial Graphene Chemically Modified with Aryl Moieties. *J. Am. Chem. Soc.* **2010**, *132*, 15399–15403.
27. Roglans, A.; Pla-Quintana, A.; Moreno-Manas, M. Diazonium Salts as Substrates in Palladium-Catalyzed Cross-Coupling Reactions. *Chem. Rev.* **2006**, *106*, 4622–4643.
28. Patai, S. The Chemistry of Diazonium and Diazo Groups. *The Chemistry of Functional Groups*; John Wiley & Sons Inc: New York, NY, USA, 1978; Volume 2.
29. Saunders, K.H. *The Aromatic Diazo-Compounds and Their Technical Applications*; Edward Arnold & Co.: London, UK, 1949.
30. Zollinger, H. Diazotization of Amines and Dediazoniation of Diazonium Ions. In *The Chemistry of Amino, Nitroso, Nitro and Related Groups*; Patai, S., Ed.; Wiley & Sons: New York, NY, USA, 1996; pp. 636–637.
31. Chamoulaud, G.; Bélanger, D. Spontaneous Derivatization of a Copper Electrode with *in situ* Generated Diazonium Cations in Aprotic and Aqueous Media. *J. Phys. Chem. C* **2007**, *111*, 7501–7507.
32. Glaser, R.; Horan, C.J. Benzenediazonium Ion. Generality, Consistency, and Preferability of the Electron Density Based Dative Bonding Model. *J. Org. Chem.* **1995**, *60*, 7518–7528.
33. Glaser, R.; Horan, C.J.; Lewis, M.; Zollinger, H. σ-Dative and π-Backdative Phenyl Cation–Dinitrogen Interactions and Opposing Sign Reaction Constants in Dual Substituent Parameter Relations. *J. Org. Chem.* **1999**, *64*, 902–913.
34. Gokel, G.W.; Cram, D.J. Molecular Complexation of Arenediazonium and Benzoyl Cations by Crown Ethers. *J. Chem. Soc. Chem. Commun.* **1973**, 481–482.
35. Kuokkanen, T.; Haataja, A. Effect of Solvent on the Complexation and Thermal Stability of Benzenediazonium Tetrafluoroborate in the Presence of Crown Ethers. *Acta Chem. Scand.* **1993**, *47*, 872–876.

36. Morosin, B.; Lingafelter, E.C. The Crystal Structure of Tetramethylammonium Tetrachlorozincate and Tetrachlorocobaltate. *Acta Crystallogr.* **1959**, *12*, 611–612.
37. Doctorovich, F.; Escola, N.; Trápani, C.; Estrin, D.A.; Lebrero, M.C.G.; Turjanski, A.G. Stabilization of Aliphatic and Aromatic Diazonium Ions by Coordination: An Experimental and Theoretical Study. *Organometallics* **2000**, *19*, 3810–3817.
38. Mohamed, A.A. Advances in the Coordination Chemistry of Nitrogen Ligand Complexes of Coinage Metals. *Coord. Chem. Rev.* **2010**, *254*, 1918–1947.
39. Overton, A.T.; Mohamed, A.A. Gold Diazonium Complexes for Electrochemical Reductive Grafting. *Inorg. Chem.* **2012**, *51*, 5500–5502.
40. Gougoutas, J.Z.; Johnson, J. Structure and Solid-State Chemistry of 3-Carboxy-2-Naphthalenediazonium Bromide. *J. Am. Chem. Soc.* **1978**, *100*, 5816–5820.
41. Greenberg, B.; Okaya, Y. Crystal and Molecular Structure of 2-Diazonium-4-Phenolsulfonate Monohydrate, $C_6H_3N_2^+.SO_3^-.OH.H_2O$. *Acta Cryst. B* **1969**, *25*, 2101–2108.
42. Mostad, A.; Rømming, C. The Crystal Structure of p-Benzenebisdiazonium Tetrachlorozincate. *Acta Chem. Scand.* **1968**, *22*, 1259–1266.
43. Romming, C. The Structure of Benzene Diazonium Chloride. *Acta Chem. Scand.* **1963**, *17*, 1444–1454.
44. Romming, C. The Crystal Structure of p-Benzenediazonium Sulphonate. *Acta Chem. Scand.* **1972**, *26*, 523–533.
45. Romming, C.; Tjornhom, T. The Crystal Structure of the 1:1 Complex Benzenediazonium Chloride-Acetic Acid. *Acta Chem. Scand.* **1968**, *22*, 2934–2942.
46. Filimonov, V.D.; Trusova, M.; Postnikov, P.; Krasnokutskaya, E.A.; Lee, Y.M.; Hwang, H.Y.; Kim, H.; Chi, K.-W. Unusually Stable, Versatile, and Pure Arenediazonium Tosylates: Their Preparation, Structures, and Synthetic Applicability. *Org. Lett.* **2008**, *10*, 3961–3964.
47. Pauling, L. *The Nature of the Chemical Bond*, 3rd ed.; Cornell University Press: Ithaca, NY, USA, 1960.
48. Gremillion, A.F.; Jonassen, H.B.; O'Connor, R.J. The Thermal Stabilities and Infrared Spectra of some Metal Salt Stabilized Diazonium Salts. *J. Am. Chem. Soc.* **1959**, *81*, 6134–6138.
49. Savitsky, A.; Siggia, S. Analysis of Primary Aromatic Amines and Nitrite by Diazotization and Pyrolysis Gas Chromatography. *Anal. Chem.* **1974**, *46*, 149–152.
50. *SMART*, Version 4.043; Software for the CCD Detector System; Bruker Analytical X-ray Systems: Madison, WI, USA, 1995.
51. SAINT, Version 4.035; Software for the CCD Detector System; Bruker Analytical X-ray Systems: Madison, WI, USA, 1995.
52. Blessing, R.H. SADABS. Program for absorption corrections using Siemens CCD based on the method of Robert Blessing. *Acta Cryst. A* **1995**, *51*, 33.
53. Scheldrick, G.M. *SHELXS-97*; Program for the Solution of Crystal Structure; University of Göttingen, Göttingen, Germany, 1997.
54. *SHELXTL 5.03*; Program Library for Structure Solution and Molecular Graphics; Bruker Analytical X-ray Systems: Madison, WI, USA, 1995.

5

Long Alkyl Chain Organophosphorus Coupling Agents for *in Situ* Surface Functionalization by Reactive Milling

Annika Betke and Guido Kickelbick *

Inorganic Chemistry, Saarland University, Am Markt Zeile 3, Saarbrücken 66125, Germany;
E-Mail: annika.betke@uni-saarland.de

* Author to whom correspondence should be addressed; E-Mail: guido.kickelbick@uni-saarland.de

---

Abstract: Innovative synthetic approaches should be simple and environmentally friendly. Here, we present the surface modification of inorganic submicrometer particles with long alkyl chain organophosphorus coupling agents without the need of a solvent, which makes the technique environmentally friendly. In addition, it is of great benefit to realize two goals in one step: size reduction and, simultaneously, surface functionalization. A top-down approach for the synthesis of metal oxide particles with *in situ* surface functionalization is used to modify titania with long alkyl chain organophosphorus coupling agents. A high energy planetary ball mill was used to perform reactive milling using titania as inorganic pigment and long alkyl chain organophosphorus coupling agents like dodecyl and octadecyl phosphonic acid. The final products were characterized by IR, NMR and X-ray fluorescence spectroscopy, thermal and elemental analysis as well as by X-ray powder diffraction and scanning electron microscopy. The process entailed a tribochemical phase transformation from the starting material anatase to a high-pressure modification of titania and the thermodynamically more stable rutile depending on the process parameters. Furthermore, the particles show sizes between 100 nm and 300 nm and a degree of surface coverage up to 0.8 mmol phosphonate per gram.

Keywords: reactive milling; titania particles; long alkyl chain organophosphorus coupling agents; tribochemical phase transition

---

1. Introduction

Surface modified particles are important building blocks for various applications in optics [1], electronics [2], material science [3] and in the biomedical field [4]. For this reason, there is increasing interest in economic synthetic strategies. In this context, there has been significant interest in mechanochemical approaches [5,6]. Reactive milling turned out to be a promising technique because it is a solvent-free and environmentally friendly alternative to conventional synthetic strategies [7,8]. Reactive milling has proven its suitability for the synthesis of various materials, for example calcium hydride by reactive milling of calcium and phenylphosphonic acid (PPA) under an inert atmosphere [9] or $ZrTiO_4$ [10] to name only two examples out of many others. Recently, we investigated the suitability of reactive milling for the preparation of surface functionalized titania nanoparticles. We were able to show that an *in situ* surface functionalization using PPA as coupling agent is possible and that the organophosphorus surface functionalization can stabilize crystallographic high-pressure phases [11,12]. Often the surface functionalization of particles with organic coupling agents involves the use of organic solvents; our technique allows a solvent free, ecologically more friendly approach. In case of PPA, the surface functionalization of titania can also be obtained in water as solvent [13]. This is not the case for long alkyl chain phosphonic acids. Here, definitely an organic solvent is necessary for the complete dissolution of the respective coupling agent [14,15]. This was the reason for us to expand our previous studies to this type of surface modifier. Dodecylphosphonic acid (DDPA) was used to perform a systematic study on the behavior at different process parameters. Afterwards, octadecylphosphonic acid (ODPA) was used to demonstrate that the approach is assignable on other long alkyl chain organophosphorus species.

2. Results and Discussion

Reactive milling was carried out using titania as inorganic pigment and long alkyl chain organophosphorus coupling agents. The experiments were performed in a high-energy planetary ball mill using a ZrO_2-ceramic grinding bowl and varying the reaction conditions such as milling time and rotational speed (revolutions per minute, rpm). Titania has several crystallographic phases that can be converted into each other depending on the energy input into the system. Consequently, the ratio between the different crystallographic phases is an indicator for the energy applied to the samples. After the milling process, the obtained material was thoroughly washed and dried. The washing solution was analyzed to demonstrate that the coupling agent is not degraded due to the high-energy impact during the milling process. ^{1}H, ^{13}C and ^{31}P NMR spectra confirm that DDPA was the only species present in the washing solution (see Supplementary Information Figures S1–S6).

The final products were analyzed applying various techniques. A systematic study on the influence of different process parameters was performed for the DDPA functionalized particles. FTIR analyses show that the milling process results in a bonding between the coupling agent and the titania phase. The starting material reveals no absorption in the typical regions for organic molecules. In contrast, the samples after the milling process display bands that are characteristic for C–H vibrations (2900 cm^{-1} and 1450 cm^{-1}) as well as P–O oscillations (1000 cm^{-1}) (Figure 1 and Supplementary Information Figures S7 and S8). The broad band at approximately 1000 cm^{-1} is characteristic for surface bonded phosphonates; the width of this band is due to the different bonding types (e.g., bi- and tridentate) of the phosphonates on the titania surface [16].

Figure 1. FTIR-spectra: (**a**) starting material titania; (**b**) coupling agent dodecylphosphonic acid and samples after the milling process: (**c**) 300 rpm/12 h; (**d**) 300 rpm/24 h; (**e**) 300 rpm/36 h; (**f**) 300 rpm/48 h; surface modification has taken place after the milling process, which is indicated by the characteristic bands for the C–H oscillation (1450 cm^{-1} and 2900 cm^{-1}) and the wide band at 1000 cm^{-1} (P–O region).

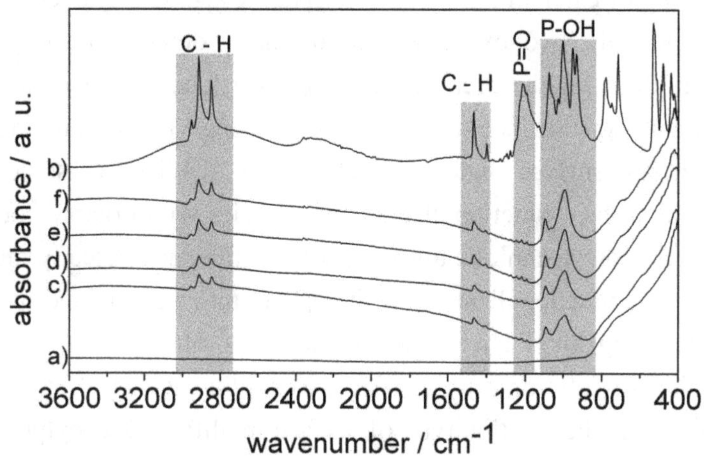

To determine the amount of surface coverage, thermal gravimetric (TGA) and elemental (CHN) analyses were performed and the results were correlated with the process parameters rotational speed and milling time (Figure 2). It turned out that the degree of surface coverage increases with increasing milling time as well as increasing milling speed. This correlation can have two reasons. On the one hand, a better binding of the organic groups to the titana surface can have its origin in the increased energy input. On the other hand, the surface area of the particles is increased which could lead to a higher portion of coupling agent bound to the surface. The results show that the correlation between the surface coverage and the milling time or speed is almost linear. CHN and TGA results show similar development, the slight discrepancies are due to measuring inaccuracies of both methods.

The binding modes of the DDPA to the titania surface were studied by the use of solid-sate NMR spectroscopy. The results reveal that DDPA was bonded to the titania surface. Pure DDPA yields in a solid-state ^{31}P NMR signal at $\delta = 31.9$ ppm [17], whereas the samples after the milling process show a series of partially broad signals in the range of $\delta = 10$–35 ppm. Furthermore, a sharp peak is observed at $\delta = 11$ ppm (Figure 3 and Supplementary Information Figures S9–S12). This is in good agreement with the variety of possible binding modes of the phosphonate to the metal oxide surface. In literature, for tridentate coordination modes, sharp peaks with chemical shifts of around 20 ppm to lower fields with respect to the free phosphonic acid are reported [18]. For bi- and monodentate binding modes the downfield shift is less pronounced and the corresponding peaks are significantly broader [19]. Consequently, the peaks in the range of 10–35 ppm correspond to the bi- and monodentate bonded phosphonates and the sharp signal at around 11 ppm can be assigned to the tridentate coordinated species for the samples obtained by reactive milling of titania with DDPA. Theoretically, the signal for tridentate phosphonate species at around 11 ppm could be attributed to layered bulk titanium alkylphosphonates [19]. However, these species show a significantly increased decomposition temperature (~750 °C) compared to alkyl phosphonate modified titania nanoparticles (~550 °C) represented by an additional exothermic step in TG measurements. The TG measurements of the samples obtained by reactive milling of titania with DDPA clearly show only one decomposition step for the degradation of phosphonate species at around 550 °C (Figure 4). Therefore, the presence of layered titanium phosphonate compounds can be excluded [19]. The solid-state ^{13}C NMR spectra of all samples show signals in the alkyl chain region at δ = 5–30 ppm which reveals that the alkyl chain is still present after the milling process (Figure 3 and Supplementary Information Figures S9–S12).

Figure 2. Surface modification in dependence of (**a**) the rpm or (**b**) the milling time, results calculated from TG analyses (straight line) and from CHN analyses (dashed line).

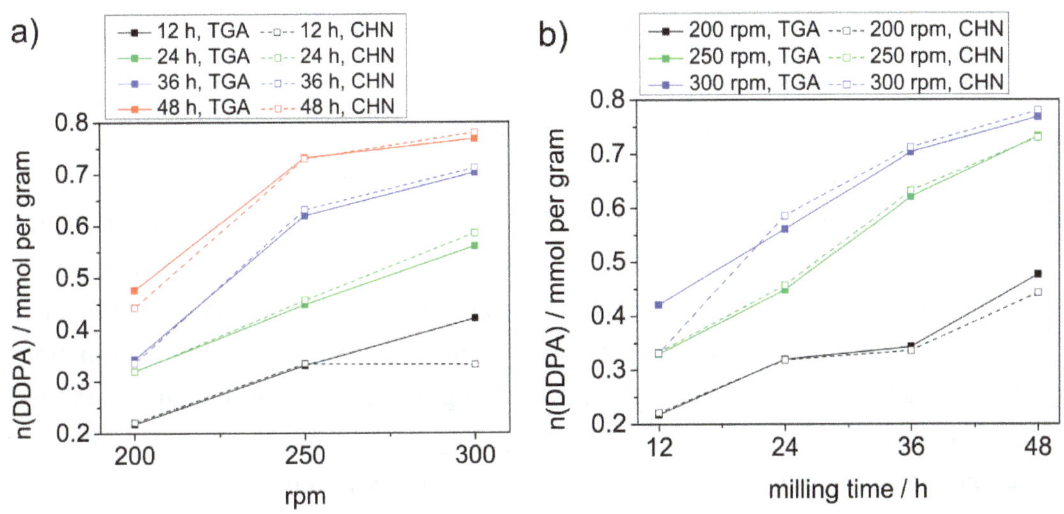

Figure 3. CP-MAS ^{31}P (**left**) and ^{13}C (**right**) spectra of titania milled with dodecylphosphonic acid at 200 rpm for 48 h.

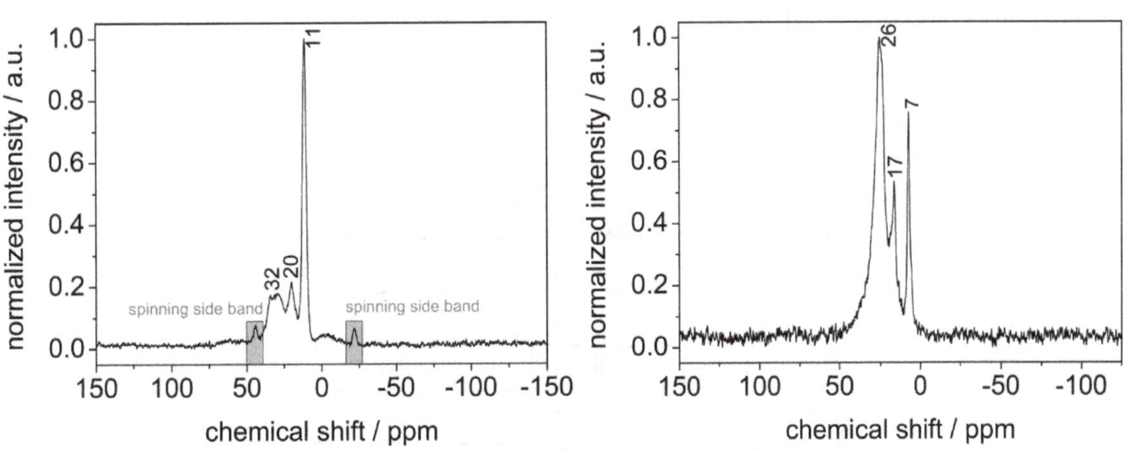

Figure 4. Thermal gravimetric analysis (TGA) of a sample obtained by reactive milling of titania with DDPA. All samples reveal the same TGA profile and differ only in the overall mass loss. The following figure shows the sample which was exposed to the highest energy impact (300 rpm, 48 h).

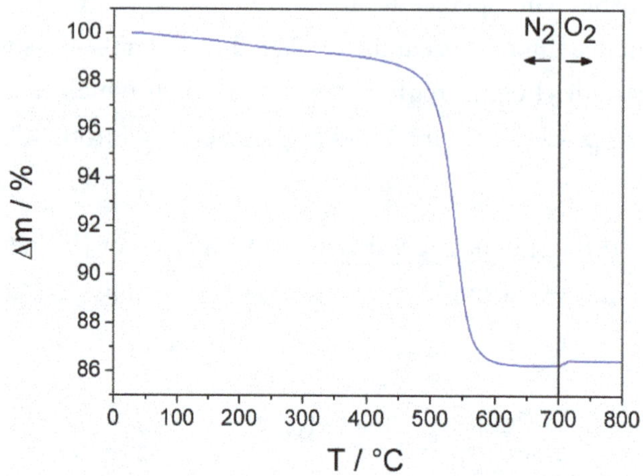

X-ray powder diffraction was used to study the phase composition of the final products. The results show that a tribochemical phase transition from the starting material anatase to the thermodynamic more stable rutile as well as a high-pressure modification of titania occurs. It turned out that the degree of the phase transition is dependent on the process parameters. The reflections belonging to the rutile and the high-pressure phase (2θ = 27° or 31°, 42° and 66°, respectively) become more intensive when the milling time or the rotational speed are enhanced (Figure 5 and Supplementary Information Figures S13 and S14). The difference between the rutile and the high-pressure modification present itself in the orientation of the vertex-connected TiO_6-octahedra along the *c*-axis. In case of rutile, they form straight lines and in case of high-pressure titania they build zigzag chains.

Figure 5. XRD patterns of (**a**) the starting material anatase (*) and some samples after milling with dodecylphosphonic acid (DDPA): (**b**) 300 rpm/12 h; (**c**) 300 rpm/24 h; (**d**) 300 rpm/36 h; (**e**) 300 rpm/48 h. The selection indicates the main differences in the reflections: The presence of high-pressure TiO_2 (#) (2θ = 31°, 42° and 66°) and rutile (°) (2θ = 27°) is indicated by additional reflections.

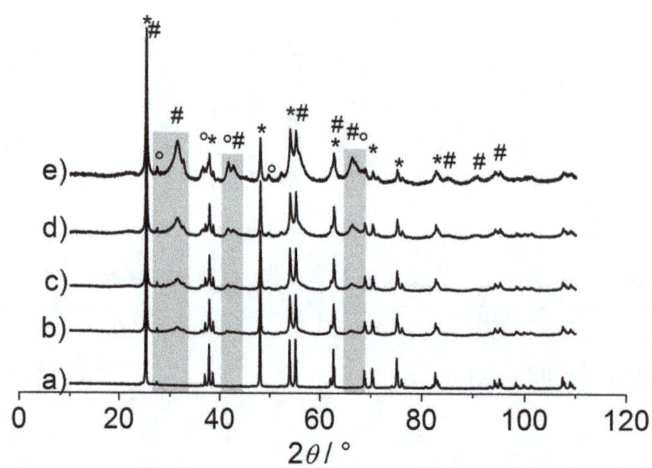

Additional to the qualitatively visible change in the reflection intensity a quantitative Rietveld phase analysis was performed. The development of the background gives no evidence for the presence of an amorphous phase. Nevertheless, the presence of such a phase cannot be excluded but if it should be present its proportion is estimated to be less than 5%. The summarized results are presented in Figure 6 and some results are listed in Table 1 (see also Supplementary Information Table S1). The Rietveld analysis confirms that the fraction of rutile and high-pressure TiO_2 increases with longer milling time and higher milling speed.

Table 1. Titania milled with dodecylphosphonic acid (DDPA), phase composition of some selected samples.

Process parameter	Anatase/wt%	Rutile/wt%	hp-TiO_2/wt%
0 h/0 rpm	98.2 ± 0.1	1.8 ± 0.1	0
12 h/200 rpm	91.9 ± 0.2	1.8 ± 0.1	6.3 ± 0.2
48 h/200 rpm	77.8 ± 0.3	2.6 ± 0.1	19.6 ± 0.3
12 h/300 rpm	66.5 ± 0.2	2.9 ± 0.2	30.6 ± 0.2
48 h/300 rpm	24.9 ± 0.2	3.5 ± 0.2	71.6 ± 0.2

Figure 6. Titania milled with dodecylphosphonic acid (DDPA): Phase composition of the samples after the milling process determined by X-ray powder diffraction and Rietveld analysis. Phase composition occurs from the starting material anatase to rutile and high-pressure (hp) TiO_2.

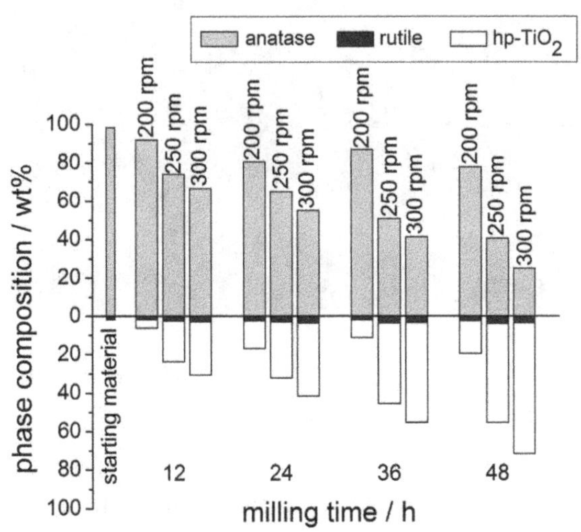

Besides the surface functionalization and phase transition, further important characteristics of the samples are the crystallite as well as the particle size. The crystallite sizes were derived from XRD pattern applying the Pawley method. The starting material was granular crystalline anatase, which exhibits crystallite sizes above 300 nm. The high-pressure TiO_2 formed during the milling process has crystallite sizes around 6 nm but the residual anatase phase remains in the granular crystalline dimension. Compared to the starting material there is only a reduction of about 100 nm. For the rutile phase no reliable refinement of crystallite size could be performed due to the low percentage of this phase in the mixture. These results show that the milling time and the rotational speed primarily have influence on the phase transition but not on the particle size. Regardless of how the process parameters were selected, the resulting high-pressure TiO_2 is nano crystalline and the residual anatase remains as granular crystalline. Furthermore, the morphology of the obtained samples was analyzed applying SEM measurements (Figure 7). The SEM images display that the particles are not uniformly shaped and agglomerated with particle sizes in a range of 100–300 nm. It is evident that the particle size decreases with increasing milling time and rotational speed.

In order to determine the particle size, dynamic light scattering was performed. Various organic solvents such as ethanol, hexane, ethyl acetate, toluene or tetrahydrofuran were tested but, even with the use of ultrasound, it was not possible to obtain a monodisperse suspension. Consequently, in regards to the particle size, no reliable results could be obtained. This effect is characteristic for long alkyl chain functionalized particles because of the zipper effect which indicates that the surface modifying compoundsmolecules get caught to each other due to Van-der-Waals interactions forming very stable agglomerates [20].

Figure 7. SEM images of two selected samples: titania milled with dodecylphosphonic acid (DDPA) at 200 rpm for 12 h (left) and at 300 rpm for 48 h (right). The size of the particles decreases with increasing energy input, but both samples are agglomerated and not uniformly shaped.

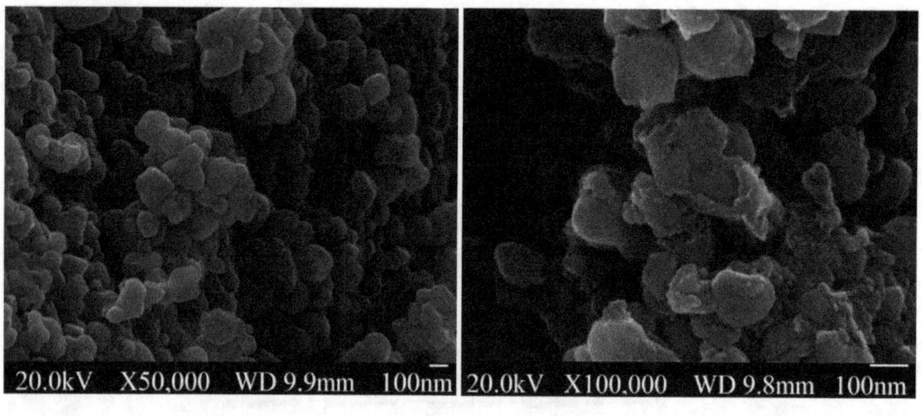

Due to abrasion of the grinding bowl, a contamination of the samples cannot be avoided but it is expected to be insignificantly small. X-ray fluorescence spectroscopy was used to analyze these impurities (Figure 8). The results confirm that the contamination of the samples due to abrasion of the grinding bowl is negligible. The contamination increases with increasing milling time and rotational speed.

The samples after the milling process show a grey discoloration which becomes more intensive with increasing milling time and speed (Figure 9). Based on the fact that starting materials, TiO_2 and DDPA, as well as the grinding bowl material are white, this discoloration must have another origin than the impurities due to the abrasion of the grinding bowl. Electron paramagnetic resonance was applied to clarify the reason for the mentioned discoloration (see Supplementary Information Figure S15). The results indicate the presence of Ti^{3+} ions which are known to entail a darkening of the material [21]. An explanation for the appearance of a subvalent Ti species is the fact that the DDPA acts not only as a coupling but also as a reducing agent under the harsh conditions present during the milling process. Most likely, the organic moiety is oxidized to phenolic/chinoide species or split off from the coupling agent molecule, respectively. However, as the concentration of Ti^{3+} is low, the amount of possible oxidation products for DDPA in the washing solution is clearly below the detection limit of NMR spectroscopy. Therefore, an identification of these side products was not successful.

Figure 8. Impurities introduced by the grinding bowl according to (**a**) rotational speed (rpm) or (**b**) milling time determined by X-ray fluorescence spectroscopy, titania milled with dodecylphosphonic acid.

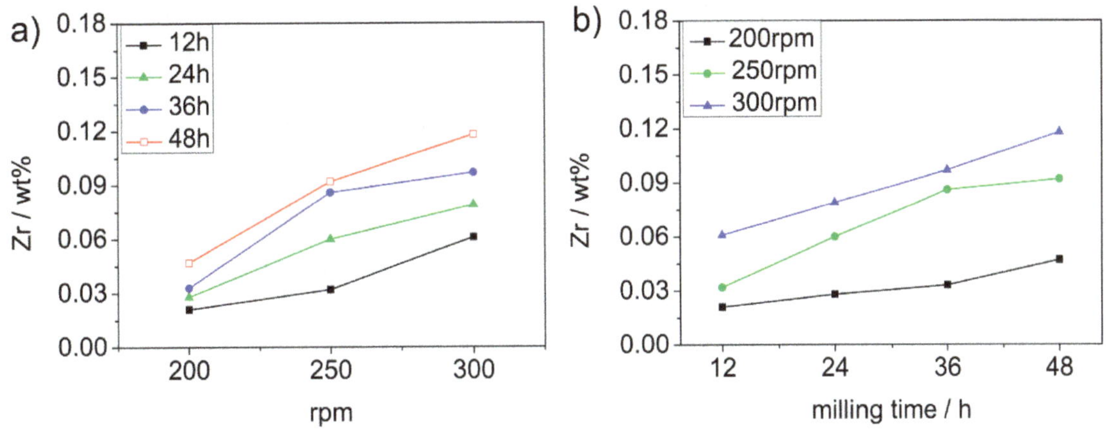

Figure 9. Images of some selected samples: (**a**) TiO_2 (**b**) 200 rpm/12 h (**c**) 200 rpm/48 h (**d**) 300 rpm/12 h (**e**) 300 rpm/48 h. The grey discoloration increases with increasing energy input.

(**a**) (**b**) (**c**) (**d**) (**e**)

After the successful studies on DDPA as long alkyl chain phosphonate coupling agent, again ODPA was used to demonstrate that this approach is assignable on other long alkyl chain organophosphorus species. The sample was prepared in the same way using those process parameters which end up in the highest surface functionalization (300 rpm and 48 h). FTIR analysis showed that a surface functionalization has taken place in case of using ODPA as well (Figure 10). The degree of surface functionalization was determined to be 0.6 mmol/g by the use of TGA as well as CHN analysis. The binding modes of the ODPA on the titania surface were studied using solid state ^{31}P-NMR spectroscopy (Figure 11). Similar to the results of the studies with DDPA, the spectrum show a series of partially broad signals in the range of δ = 10–30 ppm and a sharp peak at δ = 6 ppm. Pure ODPA shows a solid-state ^{31}P NMR signal at δ = 30 ppm [22]. The ^{31}P NMR is in good agreement with the variety of possible binding modes of the phosphonate to the metal oxide surface (tridentate coordination modes shift around 20 ppm to lower fields with respect to the free phosphonic acid; for bi- and monodentate binding modes a downfield shift of 5–15 ppm compared to the free acid is reported) [18]. Consequently, the peaks in the range of 10–30 ppm correspond to the bi- and monodentate bonded phosphonates, and the sharp signal at around 6 ppm can be assigned to the tridentate coordinated species. These shifts confirm the covalent bonding of the ODPA on the titania surface and demonstrate the different binding modes. Analogous to DDPA functionalized titania, the presence of layered alkyl phosphonates can be excluded by TG measurements of the ODPA functionalized sample (Figure 12).

Figure 10. FTIR-spectra of the starting material, the coupling agent OPDA, and the sample after the milling process.

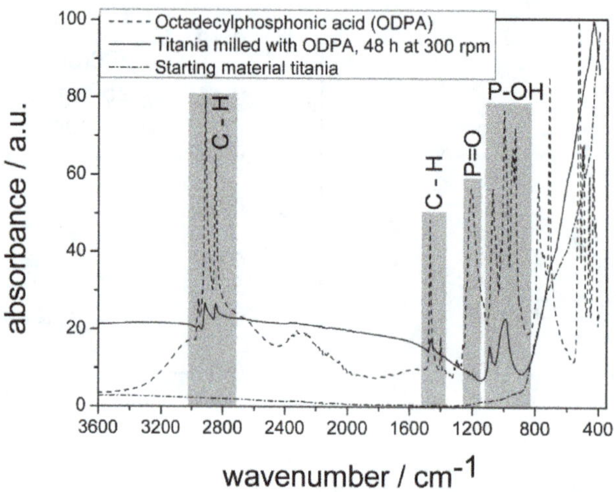

Figure 11. CP-MAS ^{31}P NMR spectrum of titania milled with octadecylphosphonic acid.

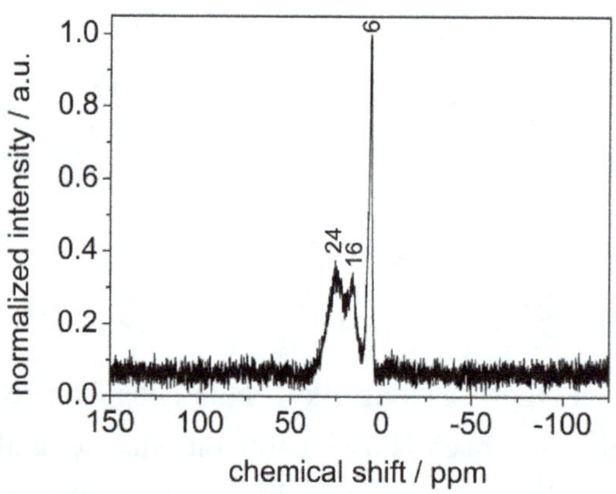

Figure 12. TGA of the sample obtained by reactive milling of titania with ODPA.

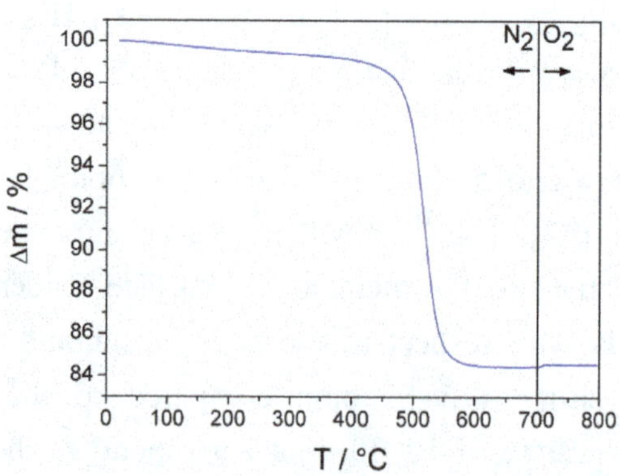

The sample with ODPA as coupling agent was analyzed using X-ray diffraction as well determining the phase composition or rather to describe the tribochemical phase transformation. The sample was composed of 29.1% ± 0.2% anatase, 2.1% ± 0.2% rutile and 68.8% ± 0.3% high-pressure TiO_2. Comparing these results with the findings of DDPA as coupling agent, it emphasizes that, in case of DDPA, a slightly higher degree of functionalization can be reached. In contrast, the degree of the tribochemical phase transformation is almost similar. Just as in the case of DDPA, the particles functionalized with ODPA are not uniformly shaped and agglomerated which was determined applying SEM techniques (Figure 13).

Figure 13. SEM image of titania functionalized with ODPA.

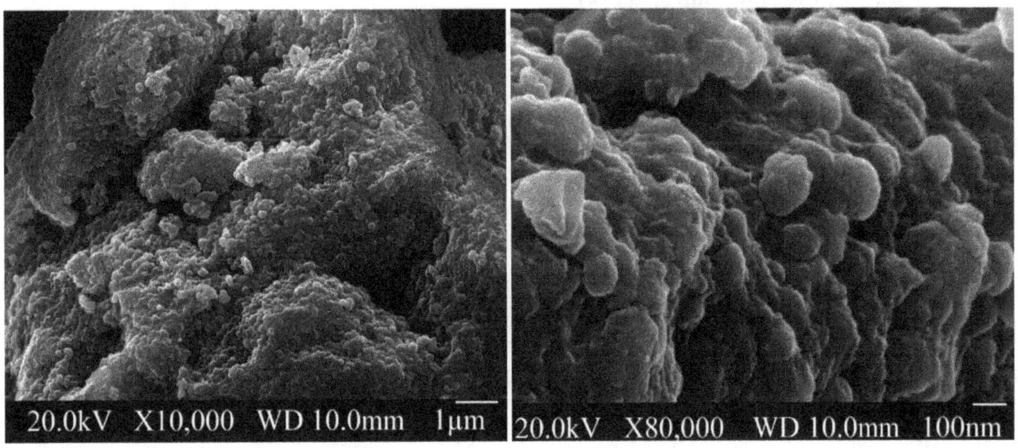

3. Experimental Section

3.1. Materials

Titanium(IV)-oxide powder anatase (99.8%) was purchased from Sigma Aldrich (Steinheim, Germany) and has been used without further purification. Dodecylphosphonic and octadecylphosphonic acid were synthesized in the working group.

3.2. Instruments and Characterization

FTIR measurements were performed under ambient air (40 scans at a resolution of 4 cm^{-1}) in attenuated total reflectance (ATR) mode on a Bruker (Billerica, MA, USA) Vertex 70 spectrometer. X-ray powder diffraction was carried out on two different diffractometers, a Panalytical (Almelo, Netherlands) X'Pert and a Bruker (Billerica, MA, USA) D8 Advance-system, and Bragg-Brentano geometry and CuKα radiation was used in both cases. The quantitative analysis was carried out by the Rietveld-method using the program TOPAS [23] and crystallographic data for the modifications of titania (anatase [24], rutile [25], high-pressure TiO_2 [26]). The same program package was used for the determination of the crystallite sizes applying the Pawley-method. X-ray fluorescence spectroscopy measurements were performed on an EDAX (Wiesbaden, Germany) Eagle II instrument. Thermogravimetric analysis (TGA) was carried out on a Netzsch (Selb, Germany) Iris TG 209. The sample was placed in an alumina crucible which was then heated from room temperature to 700 °C under nitrogen atmosphere followed by heating to 800 °C under oxygen atmosphere with a rate of

20 K min^{-1}. The aim of this procedure is the removal of organic residues which were not completely decomposed during the heating step up to 700 °C. The mass loss during the thermal analysis was used to calculate the amount of coupling agent on the surface of the titania particles. Here, one has to consider that the residue is not only titania, because the phosphonate groups are oxidized to phosphate which remains on the surface of the particles. Equation 1 gives the molar amount of surface-modifying agent per gram of particles.

$$c_{CA} = \frac{\Delta m \cdot 10}{M_{\Delta m}} \quad (1)$$

c_{CA}: molar concentration of surface-modifying agent per gram (mmol g^{-1})
Δm: mass loss during thermal analysis (%)
$M_{\Delta m}$: molar mass of the molecule which leaves during thermal analysis (g mol^{-1})

Elemental analysis was carried out on a Leco (St. Joseph, MI, USA) CHN-900 analyzer. The amount of coupling agent on the surface of the titania particles was calculated by using the percentage of carbon from elemental analysis. The molar amount of coupling agent per gram of particles can be calculated using Equation (2).

$$c_C = \frac{m_C \cdot 10}{M_C \cdot N_C} \quad (2)$$

c_{CA}: molar concentration of coupling agent per gram (mmol g^{-1})
m_C: mass of carbon from elemental analysis (%)
M_C: molar mass of carbon (g mol^{-1})
N_C: number of carbon atoms per coupling agent

Liquid state NMR spectra were recorded on a Bruker (Billerica, MA, USA) Avance 300 spectrometer operating at 300.13 MHz for ^{1}H, 75.47 MHz for ^{13}C and at 121.49 MHz for ^{31}P. Solid state NMR spectroscopy was carried out on a Bruker (Billerica, MA, USA) Avance DPX 300 instrument equipped with a 4 mm broad band cross-polarization Magic Angle Spinning probe head operating at 75.40 MHz for ^{13}C and 121.39 MHz for ^{31}P or on a Bruker (Billerica, MA, USA) DSX Avance NMR spectrometer (125.78 MHz for ^{13}C and 202.48 MHz for ^{31}P). Scanning electron microscopy (SEM) images were recorded on a JEOL (Tokyo, Japan) SEM-7000 microscope. The SEM samples were prepared by placing some grains on a specimen stub with attached carbon adhesive foil followed by deposition of a gold layer.

3.3. Synthesis

The experiments were performed in a high energy planetary ball mill Retsch (Haan, Germany) PM 100. The material of the grinding bowl was a ZrO_2-ceramic. We used a 50 mL grinding bowl and 200 corresponding milling balls with a diameter of 5 mm. Systematic studies on the effect of the milling time and the rpm on the result of the milling process were carried out. The milling time was varied between 12 h and 48 h and the rpm had values of 200 rpm, 250 rpm and 300 rpm.

In each case, 4 g of titania (anatase) and 1 g of the respective coupling agent (dodecylphosphonic and octadecylphosphonic acid) were placed in the grinding bowl and milled at specific parameters.

After the milling process, the product was washed with ethanol and water. Finally, the functionalized particles were separated by centrifugation (13,000 rpm) and dried at 100 °C.

4. Conclusions

Reactive milling can be used for the synthesis of metal oxide particles with *in situ* surface functionalization. It was shown that this technique can be successfully conducted using long alkyl chain organophosphorus coupling agents. DDPA was used to conduct systematic studies on the influence of the process parameters milling time and rotational speed. The results reveal that increasing milling time as well as increasing rotational speed results in an increasing phase transition of the starting material anatase to rutile and high-pressure titania. Furthermore, an increase of the mentioned process parameters results in an increase of surface functionalization. Afterwards, ODPA was used to successfully demonstrate that the approach is assignable on other long alkyl chain organophosphorus species. The surface functionalization of inorganic particles with long alkyl chain coupling agents is useful for further applications. For instance, such surface functionalized particles can be incorporated into a polymer matrix to obtain a composite material with improved properties. Furthermore, the approach could be conducted applying other organic coupling agents, for example antenna molecules for photo catalysis.

Acknowledgments

The authors thank Susanne Harling (Saarland University) for the elemental analysis and Matthias Gasthauer (Saarland University) for the synthesis of the coupling agents. Furthermore, we thank Michael Puchberger (Vienna University of Technology) and Dirk Schaffner (University of Kaiserslautern) for the solid state NMR spectroscopy and Martin Hartmann (Friedrich-Alexander University Erlangen-Nürnberg) for the ESR measurements.

Author Contributions

Annika Betke carried out the experiments and characterizations except elemental analysis, solid state NMR and ESR (see Acknowledgments). Guido Kickelbick provided the resources and infrastructure that allowed the development of this work. He also provided mentoring with data analysis and sample characterization. All co-authors compiled and discussed the manuscript.

References

1. Hameed, S.; Predeep, P.; Baiju, M.R. Polymer Light Emmiting Diodes—A Review on Materials and Techniques. *Rev. Adv. Mater. Sci.* **2010**, *26*, 30–42.
2. Chen, W.; Qiu, Y.; Yang, S. Branched ZnO nanostructures as builing blocks of photoelectrodes for efficient solar energy conversion. *Phys. Chem. Chem. Phys.* **2012**, *14*, 10872–10881.

3. Leung, K.C.-F.; Xuan, S.; Zhu, X.; Wang, D.; Chak, C.-P.; Lee, S.-F.; Ho, W.K.-W.; Chung, B.C.-T. Gold and iron oxide hybride nanocomposite materials. *Chem. Soc. Rev.* **2012**, *41*, 1911–1928.
4. Mieszawska, A.J.; Mulder, W.J.M.; Fayad, Z.A.; Cormode, D.P. Multifunctional Gold Nanoparticles for Diagnosis and Therapy of Disease. *Mol. Pharm.* **2013**, *10*, 831–847.
5. Fernández-Bertran, J.F. Mechanochemistry: An overview. *Pure Appl. Chem.* **1999**, *71*, 581–586.
6. Garay, A.L.; Pichon, A.; James, S.L. Solvent-free synthesis of metal complexes. *Chem. Soc. Rev.* **2007**, *36*, 846–855.
7. Kaupp, G. Waste-free large-scale synthesis without auxiliaries for sustainable production omitting purifying workup. *CrystEngComm* **2006**, *8*, 794–804.
8. James, S.L.; Adams, C.J.; Bolm, C.; Braga, D.; Collier, P.; Friscic, T.; Grepioni, F.; Harris, K.D.M.; Hyett, G.; Jones, W.; *et al.* Mechanochemistry: Opportunities for new and cleaner synthesis. *Chem. Soc. Rev.* **2012**, *41*, 413–447.
9. Ney, C.; Kohlmann, H.; Kickelbick, G. Metal hydride synthesis through reactive milling of metals with solid acids in a planetary ball mill. *Int. J. Hydrogen Energy* **2011**, *36*, 9086–9090.
10. Fuentes, A.F.; Takacs, L. Preparation of multicomponent oxides by mechanochemical methods. *J. Mater. Sci.* **2013**, *48*, 598–611.
11. Fischer, A.; Ney, C.; Kickelbick, G. Synthesis of Surface-Functionalized Titania Particles with Organophosphorus Coupling Agents by Reactive Milling. *Eur. J. Inorg. Chem.* **2013**, *33*, 5701–5707.
12. Betke, A.; Kickelbick, G. Important reaction parameters in the synthesis of phenylphosphonic acid functionalized titania particles by reactive milling. *New J. Chem.* **2014**, *38*, 1264–1270.
13. Guerrero, G.; Mutin, P.H.; Vioux, A. Anchoring of Phosphonate and Phosphinate Coupling Molecules on Titania Particles. *Chem. Mater.* **2001**, *13*, 4367–4373.
14. Convertino, A.; Leo, G.; Tamborra, M.; Sciancalepore, C.; Striccoli, M.; Curri, M.L.; Agostiano, A. TiO_2 colloidal nanocrystals functionalization on PMMA: A tailoring of optical properties and chemical adsorption. *Sens. Actuators B* **2007**, *126*, 138–143.
15. Cozzoli, P.D.; Kornowski, A.; Weller, H. Low-Temperature Synthesis of Soluble and Processable Organic-Capped Anatase TiO_2 Nanorods. *J. Am. Chem. Soc.* **2003**, *125*, 14539–14548.
16. El Malti, W.; Laurencin, D.; Guerrero, G.; Smith, M.E.; Mutin, P.H. Surface modification of calcium carbonate with phosphonic acids. *J. Mater. Chem.* **2012**, *22*, 1212–1218.
17. Brodard-Severac, F.; Guerrero, G.; Maquet, J.; Florian, P.; Gervais, C.; Mutin, P.H. High-Field ^{17}O MAS NMR Investigation of Phosphonic Acid Monolayers on Titania. *Chem. Mater.* **2008**, *20*, 5191–5196.
18. Pica, M.; Donnadio, A.; Troni, E.; Capitani, D.; Casciola, M. Looking for New Hybrid Polymer Fillers: Synthesis of Nanosized α-Type Zr(IV) Organophosphonates through an Unconventional Topotactic Anion Exchange Reaction. *Inorg. Chem.* **2013**, *52*, 7680–7687.
19. Souma, H.; Chiba, R.; Hayashi, S. Solid-State NMR Study of Titanium Dioxide Nanoparticles Surface-Modified by Alkylphosphonic Acids. *Bull. Chem. Soc. Jpn.* **2011**, *84*, 1267–1275.
20. Bedia, A.; Cuccia, L.; Demers, L.; Morin, F.; Lennox, R.B. Structure and Dynamics in Alkanethiolate Monolayers Self-Assembled on Gold Nanoparticles: A DSC, FT-IR, and Deuterium NMR Study. *J. Am. Chem. Soc.* **1997**, *119*, 2682–2692.

21. Xiong, L.-B.; Li, J.-L.; Yang, B.; Yu, Y. Ti^{3+} in the Surface of Titanium Dioxide: Generation, Properties and Photocatalytic Application. *J. Nanomater.* **2012**, *2012*, 13, doi:org/10.1155/2012/831524.
22. Tienes, B.M.; Perkins, R.J.; Shoemaker, R.K.; Dukovic, G. Layered Phosphonates in Colloidal Synthesis of Anisotropic ZnO Nanocrystals. *Chem. Mater.* **2013**, *25*, 4321–4329.
23. *Topas*, V4.2; General Profile and Structure Analysis Software for Powder Diffraction Data, User Manual; Bruker AXS: Karlsruhe, Germany, 2008.
24. Kim, D.-W.; Enomoto, N.; Nakagawa, Z.; Kawamura, K. Molecular Dynamic Simulaion in Titanium Dioxide Polymorphs: Rutile, Brookite, and Anatase. *J. Am. Ceram. Soc.* **1996**, *4*, 1095–1099.
25. Hill, R.J.; Madsen, C. Rietveld analysis using para-focusing and Debye-Scherrer geometry data collected with a Bragg-Brentano diffractometer. Z. *Kristallogr.* **1991**, *196*, 73–92.
26. Fulatov, S.K.; Bendeliani, N.A.; Albert, B.; Kopf, J.; Dyuzheva, T.; Lityagina, L.M. Crystalline structure of the TiO_2 II high-pressure phase at 293, 223, and 133 K according to single-crystal x-ray diffraction data. *Dokl. Phys.* **2007**, *52*, 195–199.

6

Facile and Selective Synthetic Approach for Ruthenium Complexes Utilizing a Molecular Sieve Effect in the Supporting Ligand

Dai Oyama [1,*], Ayumi Fukuda [1,†], Takashi Yamanaka [1,†] and Tsugiko Takase [2]

1 Department of Industrial Systems Engineering, Cluster of Science and Engineering, Fukushima University, 1 Kanayagawa, Fukushima 960-1296, Japan; E-Mails: daioyama@yahoo.co.jp (A.F.); s1070050@yahoo.co.jp (T.Y.)

2 Center for Practical and Project-Based Learning, Cluster of Science and Engineering, Fukushima University, 1 Kanayagawa, Fukushima 960-1296, Japan; E-Mail: ttakase@sss.fukushima-u.ac.jp

† These authors contributed equally to this work.

* Author to whom correspondence should be addressed; E-Mail: daio@sss.fukushima-u.ac.jp

---

Abstract: It is extremely important for synthetic chemists to control the structure of new compounds. We have constructed ruthenium-based mononuclear complexes with the tridentate 2,6-di(1,8-naphthyridin-2-yl)pyridine (dnp) ligand to investigate a new synthetic approach using a specific coordination space. The synthesis of a family of new ruthenium complexes containing both the dnp and triphenylphosphine (PPh_3) ligands, $[Ru(dnp)(PPh_3)(X)(L)]^{n+}$ (X = PPh_3, NO_2^-, Cl^-, Br^-; L = OH_2, CH_3CN, C_6H_5CN, SCN^-), has been described. All complexes have been spectroscopically characterized in solution, and the nitrile complexes have also been characterized in the solid state through single-crystal X-ray diffraction analysis. Dnp in the present complex system behaves like a "molecular sieve" in ligand replacement reactions. Both experimental data and density functional theory (DFT) calculations suggest that dnp plays a crucial role in the selectivity observed in this study. The results provide useful information toward elucidating this facile and selective synthetic approach to new transition metal complexes.

Keywords: synthesis; ruthenium complex; polypyridyl ligand; crystal structure; DFT calculation

1. Introduction

A more accurate regulation of the structure leads to the construction of appropriate reaction systems. For example, some isomeric pairs of $[Ru(tpy)(L)Cl]^{n+}$ type complexes (tpy = 2,2':6',2''-terpyridine, L = asymmetrical pyridyl-based bidentate ligands; quinaldic acid (qu) and 2-(1-methyl-3-pyrazolyl)pyridine (pypz-Me)) have been prepared and structurally characterized as precatalysts to investigate the effect of isomeric structural features on the catalytic epoxidation process [1,2]. In the complex $[Ru(tpy)(qu)(OH_2)]^+$, the *cis* isomer has been established to be an excellent catalyst for the chemoselective epoxidation through limited differences in electronic structural features exist between the isomeric pair [1]. In the complex $[Ru(tpy)(pypz\text{-}Me)(OH_2)]^{2+}$, on the other hand, the *trans* isomer exhibits better activity because it contains a pyridine C–H bond nearly parallel to the Ru–O bond, whereas for the *cis* isomer this position is occupied by a $C–CH_3$ group and thus exerts a much stronger steric effect [2]. These examples indicate that it is important to clearly distinguish the molecular structures of the compounds. Therefore, much attention has been paid to study both facile and selective synthesis for complex compounds.

In an effort to discover new molecular systems, we have recently reported ruthenium-based mononuclear complexes with the pyridyl-bridged di(1,8-naphthyridin-2-yl) (dnp) ligand [3]. The tridentate dnp has two noncoordinating nitrogen atoms and is able to form intramolecular hydrogen bonds between the coordinated water and dnp. In addition, dnp has a narrower coordination space on the equatorial site than the parent ligand, tpy (Chart 1). Therefore, size complementarity of the entering molecule with the coordination space of dnp plays an important role and is apparent in size-selective recognition of entering molecules such as nitriles and halides. We report that dnp in the present complex system behaves like a "molecular sieve" in ligand replacements and it leads to a facile and selective synthesis of a new family of ruthenium complexes.

Chart 1. Chemical structures of tpy (**left**) and dnp (**right**).

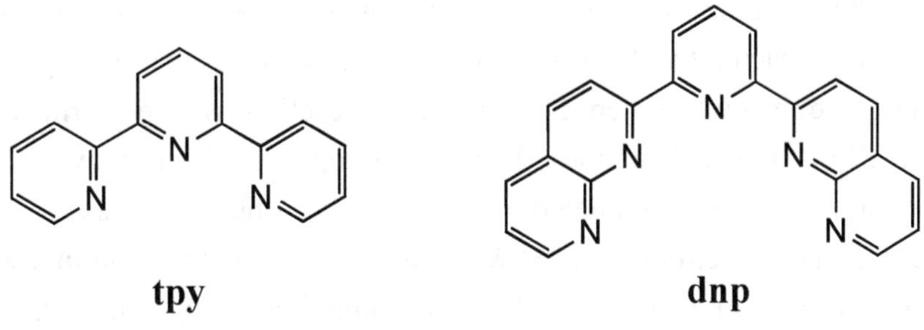

2. Results and Discussion

2.1. Synthesis of Ruthenium Complexes Utilizing the Molecular Sieve Effect of the dnp Ligand

2.1.1. Synthesis of the Precursor

Dnp was prepared by the Friedländer condensation of two equivalents of 2-aminonicotinaldehyde with 2,6-diacetylpyridine according to procedures that have been reported earlier [4,5]. Precursor **1** was synthesized by treating $[RuCl_2(CH_3CN)_2(PPh_3)_2]$ [6] with dnp in an ethanolic solution (Scheme 1). The analogous complexes containing tpy initially formed chlorido complexes, which can only be hydrated with the assistance of the silver ion to cleave the Ru–Cl bond [7,8]. However, the present system spontaneously hydrates under the reaction conditions. Dnp has two additional fused pyridine rings attached to two side-coordinating pyridines of tpy. These rings presumably promote hydration both by electronic repulsion of the chloride leaving group and stabilization of the entering water through hydrogen bonds.

Scheme 1. Preparation of the precursor.

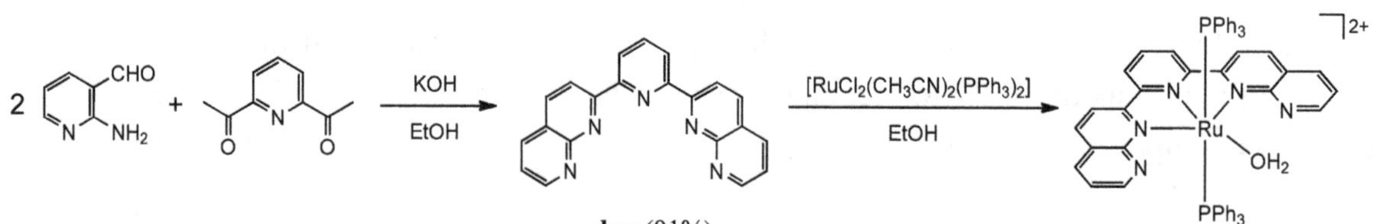

dnp (91%)

1 (red, 45%)

1 was characterized by both ^1H and $^{31}P\{^1H\}$ nuclear magnetic resonance (NMR) spectroscopy in acetone-d_6, and the resulting spectra showed well-resolved signals. A signal observed at 8.23 ppm disappeared on the addition of D_2O, which could be attributed to the protons of the aqua ligand [9]. The $^{31}P\{^1H\}$ NMR spectrum of **1** showed a singlet at 27.29 ppm, which indicates two triphenylphosphine (PPh_3) ligands situated *trans* to each other. The crystal structure of **1** by single-crystal X-ray analysis also supports this result in solution [3].

2.1.2. Synthesis of Complexes

In dnp, the coordination space of the equatorial position is narrower than that of the similar tpy. By using this steric feature, it is possible to impart selectivity when replacing the labile aqua ligand with another one. The aqua ligand of **1** was replaced with the rodlike molecule CH_3CN at room temperature to give the corresponding **2** (acetonitrile complex) (Figure 1a). In addition, **3** (benzonitrile (PhCN)-bound complex) formed in the reaction with PhCN, though PhCN contains a bulky phenyl group. It was confirmed by both X-ray structural and various spectroscopic analyses that the aqua ligand of **1** can be replaced by nitriles (see below). When a substitution reaction was performed using the linear thiocyanate anion (SCN^-) to investigate the effect of electronic repulsion between dnp and incoming ligands, corresponding **4** (SCN-coordinated complex) was obtained. Based on ^1H NMR and Electrospray ionization mass spectrometry (ESI-MS) analyses, the SCN ligand in **4** was observed to be coplanar with dnp. It is well-known that SCN^- is an ambidentate ligand. Although we could not identify the coordination atom from the infrared spectrum data ($\nu_{N\equiv C}$ = 2134 cm^{-1}) of **4**, SCN^- would likely coordinate with its smaller (relative to sulfur) nitrogen atom in **4** because SCN^- was coordinated selectively to the ruthenium center at the nitrogen atom in a similar complex containing dnp [10].

On the other hand, the aqua ligand of **1** did not exchange with the nitrite ion (NO_2^-), which is another ambidentate ligand; instead, the axial ligand (PPh_3) underwent substitution to produce **5** (Figure 1b). Two characteristic bands in the fingerprint region (1438 and 1304 cm^{-1}) from the IR (infrared) spectrum of **5** suggest that **5** is a nitrogen atom-coordinated nitro species [11,12]. In addition to NO_2^-, **1** reacted with halide ions Cl^- and Br^- to give the axial ligand-substituted **6** (chlorido complex) and **7** (bromido complex), respectively. It was confirmed by both ^1H NMR and mass spectral analyses that complexes **5**–**7** each have one PPh_3 ligand and retain the aqua ligand.

Figure 1. Synthetic routes for complexes **2**–**8**. Eq-subst. and ax-subst. denote equatorial position substitution and axial position substitution, respectively.

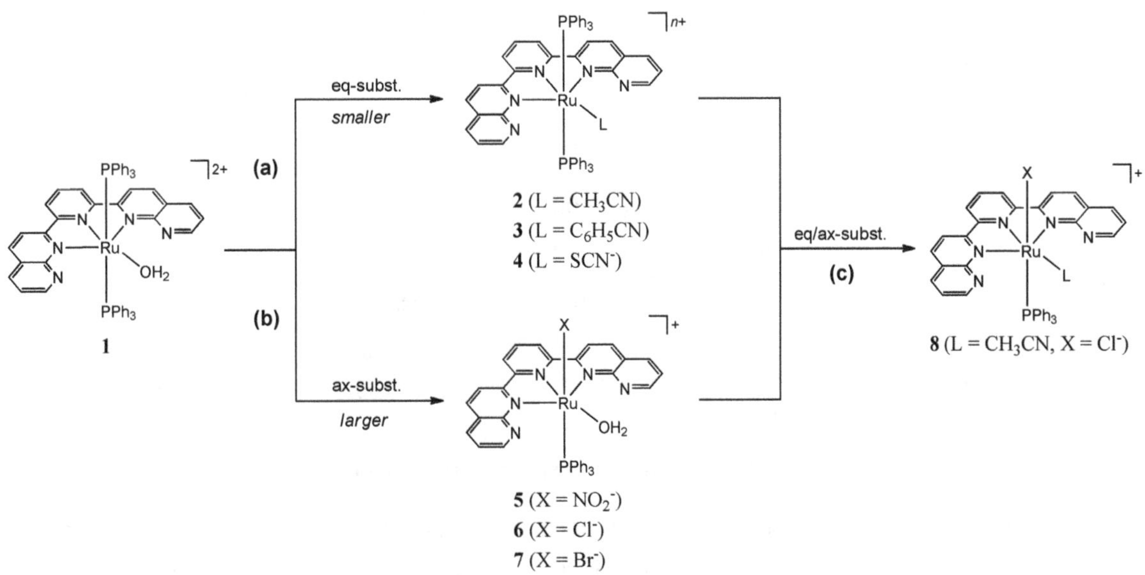

Reaction of the acetonitrile-substituted complex with chloride (*i.e.*, **2** and Cl^-) and the chlorido complex with acetonitrile (*i.e.*, **6** and CH_3CN) both lead to **8** (double-substituted complex) (Figure 1c). To summarize, the synthetic reactions using **1** are classified into two groups based on the combination of size selectivity of dnp and complementarity of size and shape of the incorporated molecule. The ligand having both linear molecular shape and smaller coordination atoms (the second period elements) incorporates into the coordination space of dnp to replace the coordinated water molecule (Figure 1a). On the other hand, nonlinear molecules or the third or later period elements are unable to approach the coordination space of dnp. As a result, an axial ligand, rather than the aqua ligand, is involved in the substitution reaction (Figure 1b). Thus, the supporting dnp acts as a so-called "molecular sieve."

2.2. Spectral and Structural Features of the Complexes

2.2.1. Electronic Absorption Spectra

The electronic absorption spectral data for the series complexes are summarized in Table 1. The complexes exhibit intense high-energy *d*→π* metal-to-ligand charge transfer (MLCT) and π→π* intraligand transitions in the UV region. The weak UV absorptions observed in the 310–360 nm range have been assigned to unresolved ligand-field *d*→*d* transitions on the ruthenium(II) center. In addition, the stronger visible transitions in the 370–600 nm range have been assigned to low-energy MLCT transitions [13]. Dnp may be considered as a dipyrido-fused analogue of tpy, which is both more delocalized and more electronegative [14]. As a result, dnp is a much better charge acceptor. This difference is evidenced by the absorption maxima of **8** and the corresponding terpyridine analogue ($[Ru(tpy)(PPh_3)(CH_3CN)Cl]^+$), where the latter (tpy-complex) MLCT band is at 466 nm [15], and **8** absorbs at 563 nm. It is worth noting that the more planar system gives rise to a better delocalization and consequently exhibits lower energy absorption.

Table 1. Electronic absorption data for complexes **1–8** [a].

Complex	λ_{max}/nm (ε/M^{-1} cm^{-1})			Solvent
1	523 (3500) 327 (22400)	373 (16700)	355 (15800)	acetone
2	477 (3100) 319 (24300)	374 (20400) 274 (42000)	356 (16300)	methanol
3	492 (4300) 312 (28700)	453 (4600)	372 (18600)	methanol
4	547 (3400) 279 (28800)	373 (12800)	321 (19100)	acetonitrile
5	551 (4100) 327 (18800)	370 (15900)	343 (18800)	acetone
6	580 (3700)	352 (19900)	330 (27400)	acetone
7	590 (4600)	350 (25500)	327 (21800)	acetone
8	563 (3400) 320 (25300)	374 (15800) 240 (37200)	353 (18600)	methanol

[a] $c = 1.0 \times 10^{-4}$ M.

Figure 2 shows the visible absorption spectra of $[Ru(dnp)(PPh_3)(OH_2)X]^{n+}$-type aqua complexes (X = PPh_3: **1**, NO_2: **5**, Cl: **6**, and Br: **7**) in acetone solution. The MLCT bands of anion-bound complexes (**5**, **6**, and **7**) are red-shifted with regard to the corresponding **1** (phosphine complex) because of the relative destabilization of dπ(Ru) levels caused by the anionic ligands. The relative shifts in the MLCT transitions to lower energies (X = Br^- < Cl^- < NO_2^- < PPh_3) follow the spectrochemical series. Ligand variation has altered the MLCT band energies, suggesting that substitution of these ligands moves the dπ orbital energies of the ruthenium(II) metal center.

Figure 2. Visible absorption spectra of aqua complexes **1** and **5**–**7** (1.0×10^{-4} M in acetone).

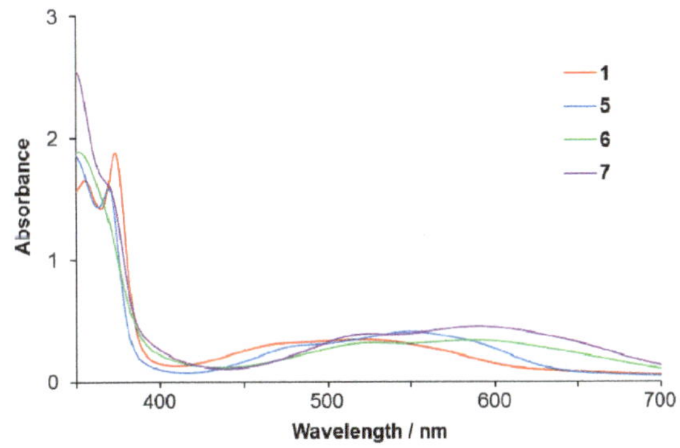

2.2.2. Molecular Structures

Single-crystal X-ray analyses of the nitrile complexes (**2**, **3**, and **8**) have been conducted to understand the geometry of series complexes in detail. The molecular structure of **1** has been reported elsewhere [3]. Selected parameters are summarized in Table 2, and the ORTEP diagrams for these cations are shown in Figures 3–5. All of the Ru–N distances of dnp are in the range 2.118(2)–2.167(6) Å with the exception of the central Ru–N bond, which is shorter (1.9660(14)–2.007(7) Å), as expected (Table 2). The phosphine ligands are in axial positions with respect to the tridentate dnp. The Ru–P bond distances of bis-phosphine complexes (2.4306(7) Å for **2**, 2.430(2) and 2.436(2) Å for **3**) are longer than that of the 2.3272(7) Å distance in **8**, which has a *trans* chlorido ligand. However, these distances are in the range previously observed in Ru(II)-phosphine complexes [15–18]. The nitrile units (N≡CR) are essentially linear and the Ru–N, N–C, and C–C bond distances are similar to typical values in other Ru(II)-N≡CR complexes [15]. The phenyl ring of benzonitrile in **3** is nearly coplanar with dnp; the dihedral angle between the central pyridyl ring and the phenyl ring is 13.0(4)°. This geometry is attributed to steric hindrance between the phenyl rings of PPh_3 and PhCN. In addition, the bond distance of Ru–Cl (2.4412(7) Å) in **8** is within the typical range for Ru(II)-chlorido complexes bearing tridentate pyridyl ligands [15].

Table 2. Selected bond distances (Å) and angles (°) for **2**, **3**, and **8**.

Parameter	2		3		$8{\cdot}2H_2O$	
Bond distances	Ru1-P1	2.4306(7)	Ru1-P1	2.436(2)	Ru1-P1	2.3272(7)
	Ru1-N2	2.118(2)	Ru1-P2	2.430(2)	Ru1-Cl1	2.4412(7)
	Ru1-N3	1.968(2)	Ru1-N2	2.167(6)	Ru1-N2	2.1206(15)
	Ru1-N4	2.051(2)	Ru1-N3	2.007(7)	Ru1-N3	1.9660(14)
	N4-C30	1.123(4)	Ru1-N4	2.163(6)	Ru1-N4	2.1442(15)
	C30-C31	1.464(5)	Ru1-N6	2.042(8)	Ru1-N6	2.0517(15)
	-	-	N6-C22	1.141(12)	N6-C40	1.130(3)
	-	-	C22-C23	1.438(14)	C40-C41	1.465(3)
Bond angles	Ru1-N4-C30	180.0	Ru1-N6-C22	179.3(6)	Ru1-N6-C40	174.2(3)
	N4-C30-C31	180.0	N6-C22-C23	176.3(9)	N6-C40-C41	178.2(3)
Dihedral angle[a]	-	-	-	13.0(4)	-	-

[a] The central pyridyl ring of dnp *vs.* the phenyl ring of benzonitrile.

Figure 3. Molecular structure of the cation in complex **2**. Counteranions and hydrogen atoms are omitted for clarity.

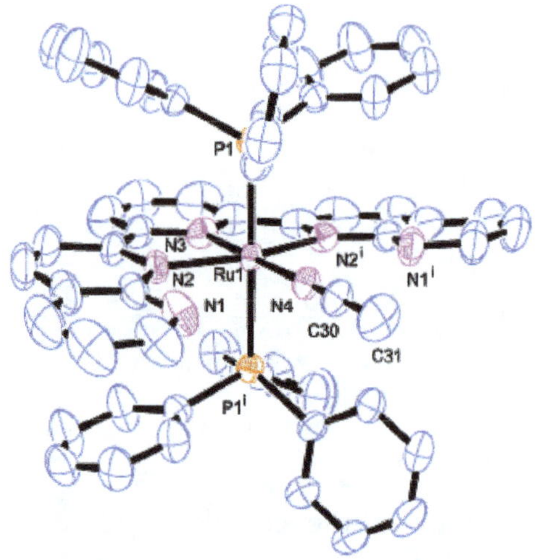

Figure 4. Molecular structure of the cation in complex **3**. Counteranions and hydrogen atoms are omitted for clarity.

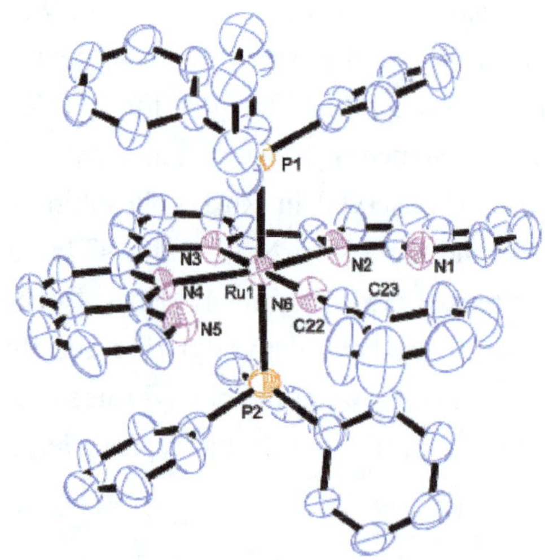

Figure 5. Molecular structure of the cation in complex **8**. The counteranion, solvent molecules, and hydrogen atoms are omitted for clarity.

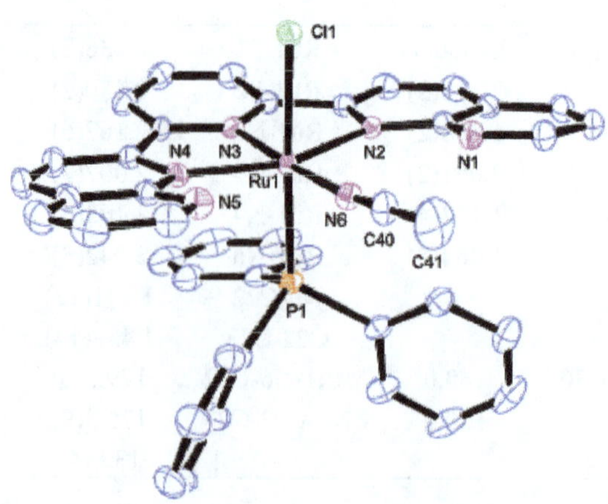

A remarkable structural feature is that the interatomic distances between noncoordinating nitrogen atoms of dnp are short. As described above, the complex system contains dnp that has two naphthyridine moieties on both sides of pyridine. The additional naphthyridine moieties could cause significant steric hindrance for Cl^- binding in **1** (or **2**). Thus, the presence of dnp seems to influence the selective coordination to the equatorial position. To examine this steric effect, density functional theory (DFT) calculations have been performed on **8** (and its geometrical isomer **8′**). The DFT-optimized structure for **8** is similar to its experimental structure. For example, for **8**, the calculated Ru–N and N–C bond distances and the angle of N–C–C in the acetonitrile ligand are 2.09 Å, 1.15 Å, and 179.5°, respectively, which are comparable to the experimental values (Table 2). Figure 6 depicts the optimized structures of **8** and **8′** and demonstrates greater steric hindrance at the equatorial ligand plane of **8′** than **8**. There is a large calculated energy difference between **8** and **8′**. **8** prefers *trans*-P, Cl geometry, which is 14.1 kcal/mol lower than that of **8′**. As shown in Figure 6, this geometrical preference is due to the narrower coordination space generated by the naphthyridine moieties in **8**, which prefers a linear CH_3CN coordination. This calculated result is strongly in accord with the fact that **8′** does not experimentally form.

Figure 6. Space-filling models of the lowest energy structures of geometrical isomers (**8** and **8′**) with the electronic energy difference.

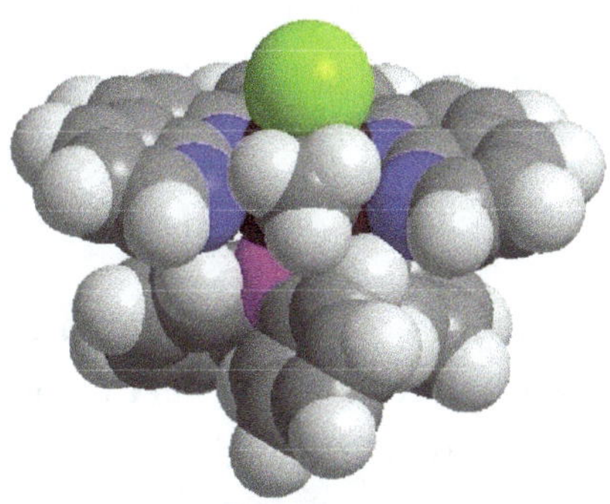

8

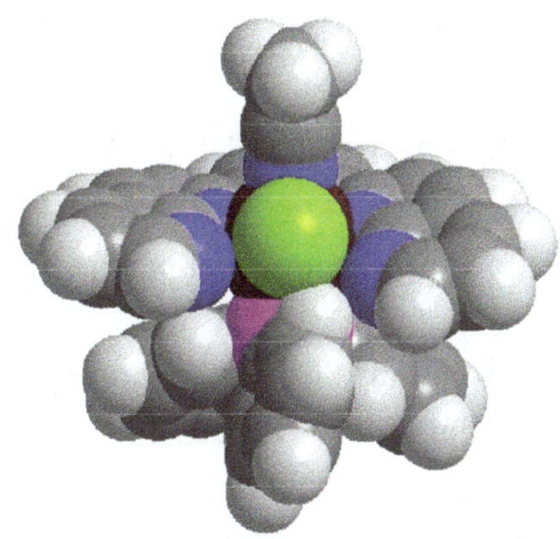

8'

(ΔE = +14.1 kcal/mol)

3. Experimental Section

3.1. Material and Methods

All solvents purchased for organic synthesis were anhydrous and used without further purification. Dnp [4,5] and $[RuCl_2(CH_3CN)_2(PPh_3)_2]$ [6] were prepared according to previously reported procedures.

Elemental analysis data were obtained on a Perkin Elmer 2400II series CHN analyzer. ^1H and $^{31}P\{^1H\}$ NMR spectra were recorded on a JEOL JMN-AL300 spectrometer (25 °C) operating at ^1H and ^{31}P frequencies of 300 and 121 MHz, respectively. ESI-MS data were obtained on a Bruker Daltonics microTOF. Electronic absorption spectra were obtained in 1-cm quartz cuvettes on a JASCO V-570 UV/VIS/NIR spectrophotometer. IR spectra were obtained using the KBr pellet method with a JASCO FT-IR 4100 spectrometer. DFT calculations were performed on **8** and **8′**. The ground-state, gas-phase structures of these complexes were optimized at the DFT level (B3LYP) [19,20]. The DGDZVP basis set was used for Ru atoms [21,22]. The 6-31G(d) basis set was employed for H, C, N, P, and Cl atoms [23,24]. A vibrational analysis was carried out to confirm the optimized geometry is a true minimum with no imaginary frequency. All of the calculations were performed using the *Gaussian 09W* program package [25].

3.2. Synthesis of the Complexes

3.2.1. Synthesis of $[Ru(dnp)(PPh_3)_2(OH_2)](PF_6)_2$ (**1**)

An ethanolic solution (60 mL) containing $[RuCl_2(CH_3CN)_2(PPh_3)_2]$ (200 mg, 0.257 mmol), dnp (94 mg, 0.268 mmol), and PPh_3 (68 mg, 0.257 mmol) was refluxed for 2 h. The solution was filtered and evaporated to 10 mL under reduced pressure. An aqueous solution of KPF_6 was added and the mixture was allowed to cool overnight. The precipitate was collected by filtration, washed with diethyl ether, and dried *in vacuo*. The crude product was purified by column chromatography on Al_2O_3 (eluent: acetone). Yield: 148 mg (45%). Anal. Calcd. for $[Ru(dnp)(PPh_3)_2(OH_2)](PF_6)_2{\cdot}CH_3CN$: $C_{59}H_{48}N_6OF_{12}P_4Ru$: C, 54.09; H, 3.69; N, 6.42. Found: C, 54.31; H, 3.79; N, 6.41%. ESI-MS (CH_3CN): *m/z* 349.5 ($[M–PPh_3–H_2O]^{2+}$), 358.5 ($[M–PPh_3]^{2+}$), 480.6 ($[M–H_2O]^{2+}$). ^1H NMR (acetone-d_6): δ 9.42 (dd, J = 4.5, 1.8 Hz, 2H, dnp), 8.69–8.59 (m, 6H, dnp), 8.43 (d, J = 8.7 Hz, 2H, dnp), 8.23–8.14 (m, 2H, dnp and H_2O), 8.02 (dd, J = 8.4, 4.2 Hz, 2H, dnp), 7.19 (t, J = 7.5 Hz, 6H, PPh_3), 6.91 (t, J = 9.0 Hz, 12H, PPh_3), and 6.79–6.73 (m, 12H, PPh_3) ppm. $^{31}P\{^1H\}$ NMR (acetone-d_6): δ 27.29 ppm.

3.2.2. Synthesis of $[Ru(dnp)(PPh_3)_2(RCN)](PF_6)_2$ (R = CH_3 (**2**); R = C_6H_5 (**3**))

An acetonitrile solution (30 mL) containing **1** (56 mg, 0.044 mmol) was stirred for 24 h at room temperature. The volume was reduced to 5 mL using a rotary evaporator, and orange crystals were precipitated by adding diethyl ether (20 mL). The precipitate was collected by filtration, washed with diethyl ether, and dried *in vacuo*. The crude product was finally recrystallized from acetonitrile and diethyl ether, yielding **2** as an orange powder. Yield: 49 mg (86%). Anal. Calcd. for $[Ru(dnp)(PPh_3)_2(CH_3CN)](PF_6)_2{\cdot}H_2O$: $C_{59}H_{48}N_6OF_{12}P_4Ru$: C, 54.09; H, 3.69; N, 6.42. Found: C, 53.77; H, 3.68; N, 6.57%. ESI-MS (CH_3CN): *m/z* 349.6 ($[M–PPh_3–CH_3CN]^{2+}$), 480.6 ($[M–CH_3CN]^{2+}$). ^1H NMR (acetonitrile-d_3): δ 9.35 (dd, J = 4.5, 1.8 Hz, 2H, dnp), 8.31–8.26 (m, 4H, dnp), 8.20–8.17 (m, 2H, dnp), 8.07 (t, J = 8.1 Hz, 1H, dnp), 7.89–7.81 (m, 4H, dnp), 7.13 (t, J = 7.6 Hz, 6H, PPh_3), and 6.85–6.70 (m, 24H, PPh_3) ppm [26]. $^{31}P\{^1H\}$ NMR (acetonitrile-d_3): δ 31.16 ppm.

A similar reaction between **1** (50 mg, 0.039 mmol) and benzonitrile gave **3** with an 82% (52 mg) yield. Anal. Calcd. for $[Ru(dnp)(PPh_3)_2(C_6H_5CN)](PF_6)_2$: $C_{64}H_{48}N_6F_{12}P_4Ru$: C, 56.77; H, 3.57; N, 6.21. Found: C, 56.88; H, 3.41; N, 6.26%. ESI-MS (CH_3CN): *m/z* 349.6 ($[M–PPh_3–C_6H_5CN]^{2+}$), 480.6 ($[M–C_6H_5CN]^{2+}$). ^1H NMR (acetone-d_6): δ 9.47 (dd, J = 3.9, 1.8 Hz, 2H, dnp), 8.69–8.53 (m, 6H, dnp), 8.33 (d, J = 8.7 Hz, 2H, dnp), 7.94 (dd, J = 8.1, 3.9 Hz, 2H, dnp), 7.81 (t, J = 7.5 Hz, 1H, dnp), 7.21–7.16 (m, 7H, PPh_3 and PhCN), 6.96–6.89 (m, 26H, PPh_3 and PhCN), and 6.77–6.75 (m, 2H, PhCN) ppm. $^{31}P\{^1H\}$ NMR (acetone-d_6): δ 30.02 ppm.

3.2.3. Synthesis of $[Ru(dnp)(PPh_3)_2(SCN)](PF_6)$ (**4**) and $[Ru(dnp)(PPh_3)(OH_2)X](PF_6)$ (X = NO_2 (**5**), Cl (**6**), Br (**7**))

In a typical preparation, a methanolic solution (40 mL) containing **1** (50 mg, 0.039 mmol) and 1.1 equiv. of NaSCN (10 mg) was refluxed for 30 min. The volume was reduced to 5 mL and a saturated solution of KPF_6 was added. The resulting solid was filtered and then washed with water followed by diethyl ether. The crude product was recrystallized from acetone/diethyl ether. The yield was 32 mg (69%) for **4**. ESI-MS (CH_3CN): *m/z* 757.0 ($[M–PPh_3]^+$), 1019.1 ($[M]^+$). ^1H NMR (acetone-d_6): δ 9.54 (dd, J = 3.9, 1.8 Hz, 2H, dnp), 8.42–8.33 (m, 6H, dnp), 8.15 (d, J = 8.7 Hz, 2H, dnp), 8.03 (t, J = 8.1 Hz, 1H, dnp), 7.86 (dd, J = 8.1, 3.9 Hz, 2H, dnp), 7.14–6.99 (m, 18H, PPh_3), and 6.83 (t, J = 6.9 Hz, 12H, PPh_3) ppm. $^{31}P\{^1H\}$ NMR (acetone-d_6): δ 27.38 ppm. IR (KBr): 2134 cm^{-1} (νN≡C).

A similar reaction between **1** and $NaNO_2$ under the same conditions described above gave **5** in 74% yield (26 mg). ESI-MS (CH_3CN): *m/z* 745.5 ($[M–H_2O]^+$). ^1H NMR (acetone-d_6): δ 9.50 (dd, J = 4.5, 1.8 Hz, 2H, dnp), 8.79–8.54 (m, 9H, dnp and H_2O), 8.13 (t, J = 8.1 Hz, 1H, dnp), 8.03 (dd, J = 4.8, 2.1 Hz, 2H, dnp), 7.20–7.16 (m, 3H, PPh_3), and 6.95–6.92 (m, 12H, PPh_3) ppm. $^{31}P\{^1H\}$ NMR (acetone-d_6): δ 38.33 ppm. IR (KBr): 1438, 1304 cm^{-1} (νNO_2).

A similar reaction between **1** and Et_4NCl under the same conditions described above gave **6** in 67% yield (23 mg). ESI-MS (CH_3CN): *m/z* 734.1 ($[M–H_2O]^+$). ^1H NMR (acetone-d_6): δ 9.46 (dd, J = 4.5, 1.8 Hz, 2H, dnp), 8.74–8.70 (m, 4H, dnp), 8.57 (dd, J = 8.4, 1.8 Hz, 4H, dnp), 8.48 (s, 1H, H_2O), 8.05–7.91 (m, 3H, dnp), 7.20–7.13 (m, 3H, PPh_3), and 6.95–6.91 (m, 12H, PPh_3) ppm. $^{31}P\{^1H\}$ NMR (acetone-d_6): δ 43.11 ppm.

A similar reaction between **1** and Bu_4NBr under the same conditions described above gave **7** in 94% yield (35 mg). ESI-MS (CH_3CN): *m/z* 780.1 ($[M–H_2O]^+$), 808.1 ($[M–H_2O+N_2]^+$). ^1H NMR (acetone-d_6): δ 9.46 (dd, J = 4.5, 2.1 Hz, 2H, dnp), 8.74–8.70 (m, 4H, dnp), 8.59 (dd, J = 8.7, 2.1 Hz, 4H, dnp), 8.38 (s, 1H, H_2O), 8.06–7.98 (m, 3H, dnp), 7.20–7.16 (m, 3H, PPh_3), and 6.95–6.91 (m, 12H, PPh_3) ppm. $^{31}P\{^1H\}$ NMR (acetone-d_6): δ 43.32 ppm.

3.2.4. Synthesis of $[Ru(dnp)(PPh_3)(CH_3CN)Cl]Cl$ (**8**)

An acetonitrile solution (30 mL) containing **1** (30 mg, 0.023 mmol) and Et_4NCl (8 mg, 0.048 mmol) was refluxed for 6 h. The volume was reduced to 5 mL using a rotary evaporator, and purple crystals were precipitated by adding diethyl ether (15 mL). The precipitate was collected by filtration, washed with diethyl ether, and dried *in vacuo*. The crude product was recrystallized from methanol/diethyl ether, yielding **8** as a violet powder. Yield: 15 mg (83%). Anal. Calcd. for $[Ru(dnp)(PPh_3)(CH_3CN)Cl]Cl\cdot 3H_2O$: $C_{42}H_{37}N_6O_3PRu$: C, 56.95; H, 4.31; N, 9.72. Found: C, 56.58; H, 4.23; N, 9.57%. ESI-MS (CH_3CN): *m/z* 734.1 ($[M–CH_3CN]^+$). ^1H NMR (acetonitrile-d_3): δ 9.34 (dd, J = 4.5, 1.8 Hz, 2H, dnp), 8.41–8.34 (m, 6H, dnp), 8.17 (d, J = 8.7 Hz, 2H, dnp), 8.03 (t, J = 8.1 Hz, 1H, dnp), 7.78 (dd, J = 8.1, 4.2 Hz, 2H, dnp), 7.15–7.08 (m, 3H, PPh_3), and 6.93–6.85 (m, 12H, PPh_3) ppm [26]. $^{31}P\{^1H\}$ NMR (acetonitrile-d_3): δ 46.95 ppm.

3.3. X-ray Crystallographic Analyses

Single crystals of the perchlorate salt of **2** and **8**·$2H_2O$ were obtained by diffusion of diethyl ether into an acetonitrile solution of each of the complexes. Single crystals of **3** were obtained by diffusion of diethyl ether into a benzonitrile solution of the complex. An orange crystal of **2** with dimensions 0.30 × 0.30 × 0.15 mm, an orange crystal of **3** with dimensions 0.20 × 0.10 × 0.05 mm, and a deep violet crystal of **8**·$2H_2O$ with dimensions 0.30 × 0.25 × 0.25 mm were mounted on a glass fiber. Measurements were made on a Rigaku RAXIS-RAPID diffractometer with graphite monochromated Mo Kα radiation (λ = 0.71075 Å). Data were collected to a maximum 2θ value of 55°. All calculations were conducted using the *CrystalStructure* crystallographic software package [27] except for refinement, which was performed using *SHELXL97* [28]. The structures were solved by direct methods [29,30] and were expanded using Fourier techniques. Multi-scan absorption corrections were applied [31]. The non-hydrogen atoms were refined anisotropically, and hydrogen atoms were included as raiding atoms. Crystallographic parameters of **2**, **3**, and **8**·$2H_2O$ are summarized in Table 3. Crystallographic data have been deposited with the Cambridge Crystallographic Data Centre as Supplementary Publication Nos. CCDC-962347 for **2**, 962348 for **3**, and 962349 for **8**·$2H_2O$.

Table 3. Crystallographic data for **2**, **3**, and **8**·$2H_2O$.

Parameter	**2**	**3**	**8·2H₂O**
Chemical formula	$C_{59}H_{46}N_6O_8Cl_2P_2Ru$	$C_{64}H_{48}N_6F_{12}P_4Ru$	$C_{41}H_{35}N_6O_2Cl_2PRu$
Formula weight	1200.97	1354.07	846.72
Temperature (K)	296(1)	296(1)	173(1)
Crystal system	monoclinic	monoclinic	Monoclinic
Space group	*C*2/*c*	*Cc*	*C*2/*c*
Unit cell parameters			
a (Å)	17.7754(3)	17.1743(4)	26.4297(5)
b (Å)	13.2405(2)	16.1763(4)	16.2228(3)
c (Å)	22.6972(4)	21.0790(4)	19.4965(4)
B (°)	96.4561(7)	97.8336(7)	114.9980(7)
V (Å^3)	5308.03(15)	5801.5(2)	7576.3(3)
Z	4	4	8
Calculated density (g cm^{-3})	1.503	1.550	1.485
μ (Mo Kα) (mm^{-1})	0.520	0.468	0.642
No. of measured reflections	25463	46099	34745
No. of observed reflections	6075	12645	8632
Refinement method		*Full-matrix least-squares on* F^2	
Parameters	356	785	474
$R1$ ($I > 2\sigma(I)$) [a]	0.0444	0.0580	0.0311
w$R2$ (all data) [b]	0.1292	0.2164	0.0803
S	1.050	1.126	1.038

[a] $R1 = \sum||F_o| - |F_c||/\sum|F_o|$; [b] $wR2 = \{\sum_w(F_o^2 - F_c^2)^2/\sum_w(F_o^2)^2\}^{1/2}$.

4. Conclusions

We have described the synthesis and X-ray crystallographic characterization of a new ruthenium complex system containing the highly functional dnp ligand. Computational analysis suggests that the incorporation of dnp not only stabilizes the complex but also enables this system to be synthesized with improved stereospecificity. The results provide useful information to elucidate this facile and selective synthetic approach to new transition metal complexes. For example, regulation of linkage isomerization in ruthenium complexes involving SCN^- could contribute to materials science in the development of photosensitizers for dye-sensitized solar cells (DSCs).

Acknowledgments

We thank the Foundation for Japanese Chemical Research (424-R) for financial support. Moreover, we are grateful to R. Naoi of Fukushima University for assistance with the spectroscopic measurements.

References

1. Chowdhury, A.D.; Das, A.; Irshad, K.; Mobin, S.M.; Lahiri, G.K. Isomeric complexes of $[Ru^{II}(trpy)(L)Cl]$ (trpy = 2,2':6',2''-terpyridine and HL = quinaldic acid): Preference of isomeric structural form in catalytic chemoselective epoxidation process. *Inorg. Chem.* **2011**, *50*, 1775–1785.
2. Dakkach, M.; López, M.I.; Romero, I.; Rodríguez, M.; Atlamsani, A.; Parella, T.; Fontrodona, X.; Llobet, A. New Ru(II) complexes with anionic and neutral N-donor ligands as epoxidation catalysts: An evaluation of geometrical and electronic effects. *Inorg. Chem.* **2010**, *49*, 7072–7079.
3. Oyama, D.; Yamanaka, T.; Fukuda, A.; Takase, T. Modulation of intramolecular hydrogen bonding strength by axial ligands in ruthenium(II) complexes. *Chem. Lett.* **2013**, *42*, 1554–1555.
4. Campos-Fernández, C.S.; Thomson, L.M.; Galán-Mascarós, J.R.; Ouyang, X.; Dunbar, K.R. Homologous series of redox-active, dinuclear cations $[M_2(O_2CCH_3)_2(pynp)_2]^{2+}$ (M = Mo, Ru, Rh) with the bridging ligand 2-(2-pyridyl)-1,8-naphthyridine (pynp). *Inorg. Chem.* **2002**, *41*, 1523–1533.
5. Tseng, H.-W.; Zong, R.; Muckerman, J.T.; Thummel, R. Mononuclear ruthenium(II) complexes that catalyze water oxidation. *Inorg. Chem.* **2008**, *47*, 11763–11773.
6. Al-Far, A.M.; Slaughter, L.M. *cis-cis-trans*-Bis(acetonitrile-κ*N*)dichloridobis(triphenylphosphine-κ*P*)ruthenium(II) acetonitrile disolvate. *Acta Crystallogr.* **2008**, *E64*, m184.
7. Kaveevivitchai, N.; Zong, R.; Tseng, H.-W.; Chitta, R.; Thummel, R.P. Further observations on water oxidation catalyzed by mononuclear Ru(II) complexes. *Inorg. Chem.* **2012**, *51*, 2930–2939.
8. Suen, H.F.; Wilson, S.W.; Pomerantz, M.; Walsh, J.L. Photosubstitution reactions of terpyridine complexes of ruthenium(II). *Inorg. Chem.* **1989**, *28*, 786–791.
9. Zong, R.; Naud, F.; Segal, C.; Burke, J.; Wu, F.; Thummel, R. Design and study of bi[1,8]naphthyridine ligands as potential photooxidation mediators in Ru(II) polypyridyl aquo complexes. *Inorg. Chem.* **2004**, *43*, 6195–6202.
10. Oyama, D.; Yamanaka, Y.; Watanabe, Y.; Takase, T. Fukushima University, Fukushima, Japan. Unpublished work, 2013.
11. Nakamoto, K. *Infrared and Raman Spectra of Inorganic and Coordination Compounds*, 4th ed.; John Wiley & Sons: New York, NY, USA, 1986; pp. 221–227.

12. Ooyama, D.; Nagao, N.; Nagao, H.; Miura, Y.; Hasegawa, A.; Ando, K.-I.; Howell, F.S.; Mukaida, M.; Tanaka, K. Redox- and thermally-induced nitro-nitrito linkage isomerizations of ruthenium(II) complexes having nitrosyl as a spectator ligand. *Inorg. Chem.* **1995**, *34*, 6024–6033.
13. Juris, A.; Balzani, V.; Barigelletti, F.; Campagna, S.; Belser, P.; von Zelewsky, A. Ru(II) polypyridine complexes: Photophysics, photochemistry, electrochemistry, and chemiluminescence. *Coord. Chem. Rev.* **1988**, *84*, 85–277.
14. Thummel, R.P.; Decloitre, Y. Polyaza cavity-shaped molecules. VIII. Ruthenium(II) complexes with 3,3'-annelated derivatives of 2-(2'-pyridyl)quinoline and 2-(2'-pyridyl)-1,8-naphthyridine. *Inorg. Chim. Acta* **1987**, *128*, 245–249.
15. Ooyama, D.; Saito, M. Synthesis, characterization and reactivity of polypyridyl ruthenium(II) carbonyl complexes with phosphine derivatives: ruthenium-carbon bond labilization based on steric and electronic effects. *Inorg. Chim. Acta* **2006**, *359*, 800–806.
16. Chen, J.; Szalda, D.J.; Fujita, E.; Creutz, C. Iron(II) and ruthenium(II) complexes containing P, N, and H ligands: structure, spectroscopy, electrochemistry, and reactivity. *Inorg. Chem.* **2010**, *49*, 9380–9391.
17. Ooyama, D.; Sato, M. A synthetic precursor for hetero-binuclear metal complexes, $[Ru(bpy)(dppy)_2(CO)_2](PF_6)_2$ (bpy = 2,2'-bipyridine, dppy = 2-(diphenylphosphino)pyridine). *Appl. Organomet. Chem.* **2004**, *18*, 380–381.
18. Ooyama, D.; Sato, M. Crystal structure of (2,2'-bipyridine-*N,N'*)(diphenyl-2-phosphinopyridine-*P*)chloro(dicarbonyl)ruthenium(II) hexafluorophosphate. *Anal. Sci.* **2003**, *19*, x39–x40.
19. Becke, A.D. Density-functional thermochemistry. III. The role of exact exchange. *J. Chem. Phys.* **1993**, *98*, 5648–5652.
20. Lee, C.; Yang, W.; Parr, R.G. Development of the Colle-Salvetti correlation-energy formula into a functional of the electron density. *Phys. Rev. B* **1988**, *37*, 785–789.
21. Sosa, C.; Andzelm, J.; Elkin, B.C.; Wimmer, E.; Dobbs, K.D.; Dixon, D.A. A local density functional study of the structure and vibrational frequencies of molecular transition-metal compounds. *J. Phys. Chem.* **1992**, *96*, 6630–6636.
22. Godbout, N.; Salahub, D.R.; Andzelm, J.; Wimmer, E. Optimization of Gaussian-type basis sets for local spin density functional calculations. Part I. Boron through neon, optimization technique and validation. *Can. J. Chem.* **1992**, *70*, 560–571.
23. Hehre, W.J.; Ditchfield, R.; Pople, J.A. Self-consistent molecular orbital methods. XII. Further extensions of Gaussian-type basis sets for use in molecular orbital studies of organic molecules. *J. Chem. Phys.* **1972**, *56*, 2257–2261.
24. Francl, M.M.; Pietro, W.J.; Hehre, W.J.; Binkley, J.S.; Gordon, M.S.; DeFrees, D.J.; Pople, J.A. Self-consistent molecular orbital methods. XXIII. A polarization-type basis set for second-row elements. *J. Chem. Phys.* **1982**, *77*, 3654–3665.
25. Frisch, M.J.; Trucks, G.W.; Schlegel, H.B.; Scuseria, G.E.; Robb, M.A.; Cheeseman, J.R.; Scalmani, G.; Barone, V.; Mennucci, B.; Petersson, G.A.; *et al. Gaussian 09W*, revision D.01; Gaussian, Inc.: Wallingford, CT, USA, 2009.
26. Methyl protons of both complexes could not be assigned due to their overlapping with solvent signals.
27. Rigaku. *CrystalStructure*, Version 3.8; Rigaku and Rigaku Americas: Woodlands, TX, USA, 2007.

28. Sheldrick, G.M. *SHELX*, SHELXL-97; University of Göttingen: Göttingen, Germany, 1997.
29. Altomare, A.; Cascarano, G.; Giacovazzo, C.; Guagliardi, A.; Burla, M.; Polidori, G.; Camalli, M. *SIR92*—A program for automatic solution of crystal structures by direct methods. *J. Appl. Crystallogr.* **1994**, *27*, 435.
30. Altomare, A.; Burla, M.; Camalli, M.; Cascarano, G.; Giacovazzo, C.; Guagliardi, A.; Molitern, A.; Polidori, G.; Spagna, R. *SIR97*: A new tool for crystal structure determination and refinement. *J. Appl. Crystallogr.* **1999**, *32*, 115–119.
31. Higashi, T. *ABSCOR*; Rigaku Corporation: Tokyo, Japan, 1995.

7

Bottom-Up, Wet Chemical Technique for the Continuous Synthesis of Inorganic Nanoparticles

Annika Betke and Guido Kickelbick *

Inorganic Chemistry, Saarland University, Am Markt Zeile 3, 66125 Saarbrücken, Germany; E-Mail: annika.betke@uni-saarland.de

* Author to whom correspondence should be addressed; E-Mail: guido.kickelbick@uni-saarland.de

---

Abstract: Continuous wet chemical approaches for the production of inorganic nanoparticles are important for large scale production of nanoparticles. Here we describe a bottom-up, wet chemical method applying a microjet reactor. This technique allows the separation between nucleation and growth in a continuous reactor environment. Zinc oxide (ZnO), magnetite (Fe_3O_4), as well as brushite ($CaHPO_4{\cdot}2H_2O$), particles with a small particle size distribution can be obtained continuously by using the rapid mixing of two precursor solutions and the fast removal of the nuclei from the reaction environment. The final particles were characterized by FT-IR, TGA, DLS, XRD and SEM techniques. Systematic studies on the influence of the different process parameters, such as flow rate and process temperature, show that the particle size can be influenced. Zinc oxide was obtained with particle sizes between 44 nm and 102 nm. The obtained magnetite particles have particle sizes in the range of 46 nm to 132 nm. Brushite behaves differently; the obtained particles were shaped like small plates with edge lengths between 100 nm and 500 nm.

Keywords: continuous synthesis; precipitation; inorganic nanoparticles; microjet reactor

---

1. Introduction

Nanoparticles have been identified as major compounds for many applications in the last decades. They are used in many different fields such as medicine, optics, electronics or catalysis [1]. For this reason, there is a lot of interest in facile synthetic routes for nanoparticles with controllable size and narrow size distribution. There are many different possibilities for the production of particles but due to its simplicity, low cost, and easy application in industrial production, stirred tank batch reactor processes are still the most employed wet chemical systems [2,3]. However, this method has often limitations for the preparation of nanoparticles on an industrial scale, because of long mixing times and thus uncontrolled nucleation and growth. To obtain nanoparticles with particle sizes smaller than 100 nm very short mixing times are required. Although there are methods which can be employed on a laboratory scale to obtain rapid mixing, their scale up is quite difficult [4,5]. In addition, it would be of great benefit if the synthesis could be performed continuously, because this would eliminate differences in nanoparticle properties from batch to batch [6–8]. There are continuous techniques, such as mixing starting material solutions on a macroscopic scale (e.g., tee shaped connectors) or microreactors [9]. These techniques have been used already for the synthesis of anatase-TiO_2, gold, and cobalt nanoparticles as well as tricalcium phosphate ceramics [10–13]. In general, all the above-mentioned techniques require an improvement with respect to the mixing time. There are several studies, which indicate that the effect of mixing has an important influence on the particle size distribution. Experimental, as well as theoretical, precipitation experiments using a tee-mixer and barium sulfate as exemplary material show that the particles becomes smaller with increasing mixing intensity [14,15]. In further studies, it was found that confined impinging jet reactors are more efficient than tee-mixers. Using such a reactor type, it is possible to generate polymer nanoparticles which have potential application in biomedicine; for example for controlled drug delivery [16,17]. Biocompatible nanomaterial for biomedicine and pharmaceutical applications can also be produced by using a novel continuous industrial reactor, called NETmix. This reactor is a network of interconnected chambers and channels creating zones of complete mixing and segregation. This technique is used already for the synthesis of hydroxyapatite nanoparticles with extremely high purity and crystallinity [18]. Fast mixing is also fulfilled when a microjet reactor (Figure 1) is used. The excellent improvement is that the reagent solutions are forced with high pressure through a narrow cone in a reaction chamber that is constantly flushed with a gas flow. As a result, the obtained product is removed directly from the reaction chamber so that clogging is avoided.

Figure 1. Schematic set-up of the microjet reactor.

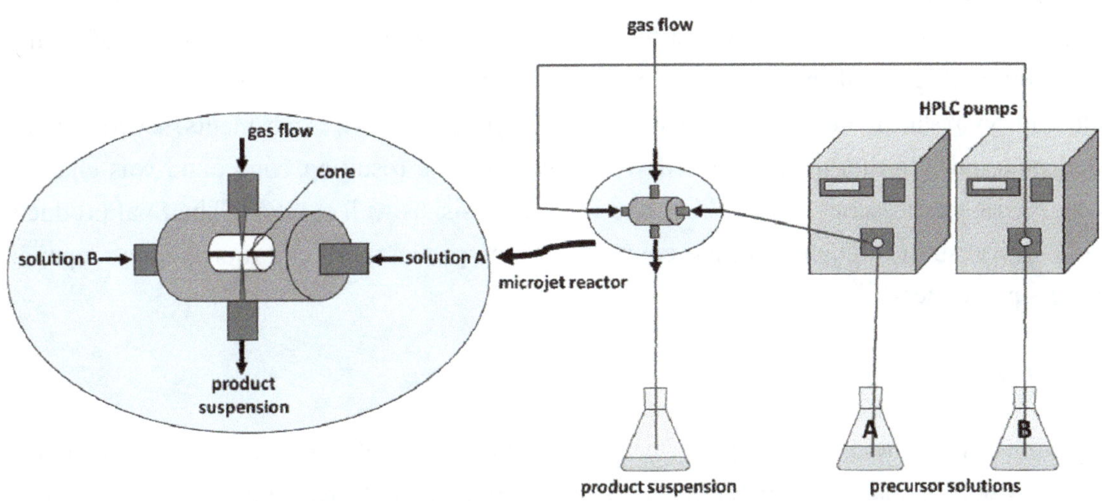

This technique can be used to perform different precipitation reactions continuously. Two pumps are used to transport the solutions containing the starting materials as dissolved substances to the microjet reactor. The collision takes place in a confined reaction environment, which is given by the format of the microjet reactor. Due to the fluid dynamics in the microjet reactor, there is a very short mixing time. Consequently, there is a maximum number of nuclei in an absolute homogeneous environment. In addition the nucleation rate is increased compared to the nucleus growth [19,20]. Directly after the nuclei are formed, they are transported from the reaction room to a tubing where the particle growth occurs. The length of this tubing up to the collection vessel influences the particle growth. The method has the advantage of being able to prepare nanoparticles even from a system with very low solubility products, e.g. barium sulfate [21]. However, it is also able to produce metal oxides, such as titania [22]. One great advantage is that the particles are obtained in a suspension, which could be used for a direct application, e.g. in spray coating or inkjet printing.

In this study, we investigated the microjet reactor technique for the synthesis of three types of inorganic nanoparticles, *i.e.*, zinc oxide (ZnO), magnetite (Fe_3O_4), and brushite ($CaHPO_4{\cdot}2H_2O$). These particles are not only interesting for academic reasons, but also have applications in optoelectronics and clinical, as well as biomedical devices [23–25]. The goal was to define the possibilities and limitations of this technique with regard to the different process parameters, such as flow rate and temperature, on the result of the precipitation reaction. Here, we do not focus on the importance of the time between nuclei formation and collection of the products, which also influences the size of the particles as this will be investigated in a further study.

2. Results and Discussion

The synthesis of inorganic nanoparticles was carried out using a microjet reactor system. In order to reveal the wide application field of the microjet reactor we performed the synthesis of three different species: zinc oxide (ZnO), magnetite (Fe_3O_4), as well as brushite ($CaHPO_4{\cdot}2H_2O$). All of these three systems show a low solubility under the applied reaction conditions. Two isocratic preparative HPLC pumps were used to transport the solutions containing the starting materials as dissolved substances. In the microjet-reactor, these two solutions clash and the precipitation reaction takes place. The precipitate was directly removed from the reactor by the use of a nitrogen gas flow. This process was performed with various flow rates of the precursor solutions (150 mL/min, 200 mL/min and 250 mL/min) and at different temperatures (25 °C, 50 °C and 75 °C). The reproducibility of the method was investigated by performing all experiments several times. The obtained material was thoroughly washed with water and the resulting compound was characterized applying FT-IR-spectroscopy, XRD, TGA and DLS analysis, as well as SEM. The final products were analyzed with respect to changes in the composition, the particle size and morphology depending on the process parameters.

2.1. Zinc Oxide

The synthesis of zinc oxide was performed by using aqueous solutions of zinc chloride and sodium hydroxide as starting materials (Equation 1) for the solution A and B, respectively.

$$ZnCl_2 + 2NaOH \rightarrow ZnO + 2NaCl + H_2O. \quad (1)$$

Under conventional precipitation conditions in an open vessel, the same starting materials deliver disc-like ZnO submicrometer structures with diameters between 300 nm and 500 nm [26].

Smaller particles could be obtained applying controlled double-jet-precipitation, which produces primary particles with diameters between 20–30 nm that tend to agglomerate in ellipsoidal particles with sizes of a few hundred nanometers [5].

The particles that were obtained by using the microjet reactor were characterized using various techniques. FT-IR-analysis of the prepared materials showed no specific signals with exception of OH and CO bands of adsorbed CO_2 and H_2O as well as a strong ZnO band around 400 cm^{-1}.

Thermogravimetric analysis (Figure 2) was performed to determine the amount of adsorbed CO_2 and H_2O. The results show that the samples contain between 5% and 6% adsorbed species. The development of the mass loss looks similar for all samples.

Figure 2. Thermogravimetric analysis of the final products. The samples show a mass loss of between 5 and 6%.

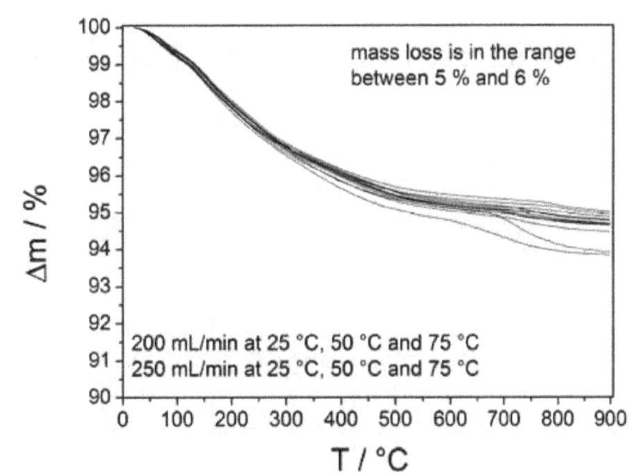

X-ray powder diffraction was applied to study the phase composition and crystallite size of the obtained material in dependence on the different synthesis parameters. The XRD pattern show that in each case pure zinc oxide is generated (Figure 3). The crystallite sizes decrease with increasing temperature and increasing flow rate (Table 1).

A further important parameter is the particle size including the size distribution, which is given by the standard deviation of the particle size. These parameters were determined using dynamic light scattering (Figure 4). The samples were dispersed in ethanol by the exposition to ultrasound for 5 min. The results show that the particles are easily dispersible and their size depends on the process parameters. The particle size decreases with increasing temperature and increasing flow rate, in addition the particle size distribution becomes smaller (Table 2).

Figure 3. X-ray powder pattern of the obtained particles. The results show that pure zinc oxide was obtained.

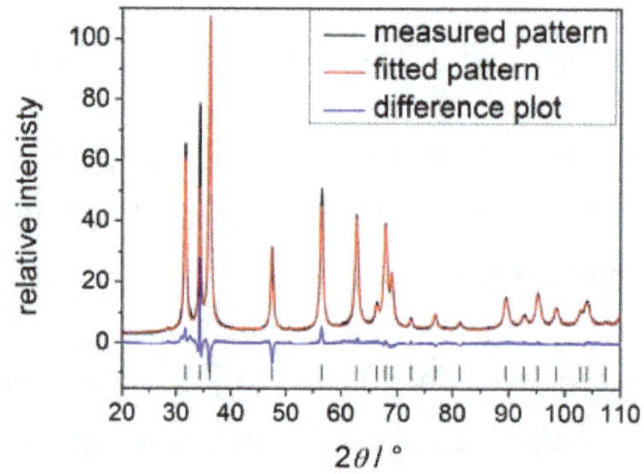

Table 1. Crystallite size in dependence on the process parameters.

Process parameters	Crystallite size
200 mL/min/25 °C	22.2 ± 0.5
200 mL/min/50 °C	20.4 ± 0.7
200 mL/min/75 °C	16.3 ± 0.3
250 mL/min/25 °C	22.7 ± 0.5
250 mL/min/50 °C	21.3 ± 0.4
250 mL/min/75 °C	16.9 ± 0.3

Figure 4. Particle sizes of the obtained zinc oxide particles determined by dynamic light scattering.

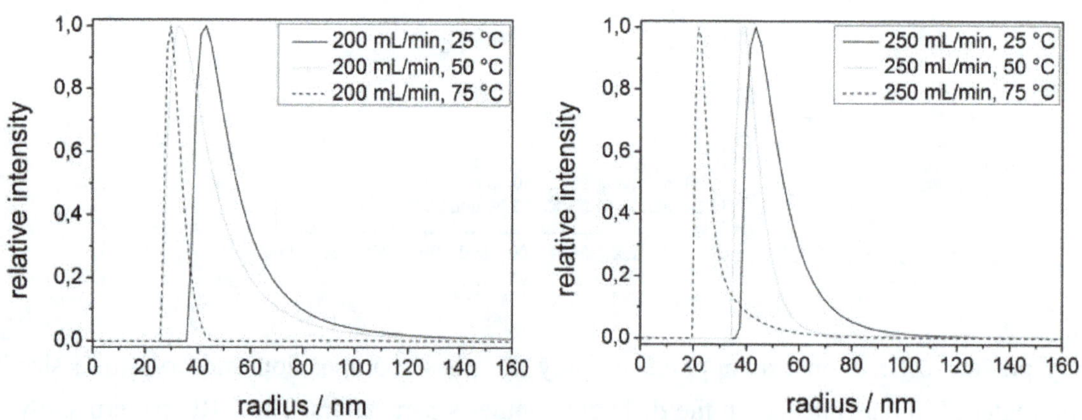

Table 2. Particle size determined using dynamic light scattering.

Process parameters	Radius/nm
200 mL/min/25 °C	51 ± 7
200 mL/min/50 °C	41 ± 8
200 mL/min/75 °C	27 ± 3
250 mL/min/25 °C	43 ± 6
250 mL/min/50 °C	36 ± 4
250 mL/min/75 °C	22 ± 2

The particle morphology was investigated by means of scanning electron microscopy (Figures 5 and 6). The images confirm that the particles are in the nanometer size range. However, SEM analysis does not show the pronounced differences in the particles sizes that were observed in the DLS measurements for the parameter variations. Generally, it can be observed in SEM that the particles are all sphere-shaped and the size distribution becomes more uniform with increasing temperature.

Figure 5. SEM images of zinc oxide synthesized with a flow rate of 200 mL/min at 25 °C (**left**) and at 75 °C (**right**).

Figure 6. SEM images of zinc oxide synthesized at 50 °C with a flow rate of 200 mL/min (**left**) and with a flow rate of 250 mL/min (**right**).

2.2. Magnetite

Systematic studies with respect to the synthesis of magnetite nanoparticles using a microjet reactor were performed with an aqueous solution of ferric and ferrous chloride as solution A and an ammonia solution B (Equation 2).

$$FeCl_2 + 2FeCl_3 + 8NH_4OH \rightarrow Fe_3O_4 + 8NH_4Cl + 4H_2O. \quad (2)$$

Ferrous chloride must be applied in excess because the synthesis was performed under air. Consequently, the Fe^{2+} ions tend to oxidize and form Fe^{3+} ions [27]. Magnetite nanoparticles were already obtained applying conventional synthetic routes using the same starting materials [28].

FT-IR analysis of the particles that were obtained in the microjet reactor does not show specific bands for the products with exception of the Fe-O band around 500 cm^{-1}.

Thermo gravimetric analysis (Figure 7) was performed to determine if the samples contain any adsorbed species like water. The particles show a mass loss between 4 and 7%, which can be explained by adsorbed water or the production of water by condensation of free hydroxyl groups.

Figure 7. Thermo gravimetric analysis of the final products. The samples show a mass loss of between 4 and 7%.

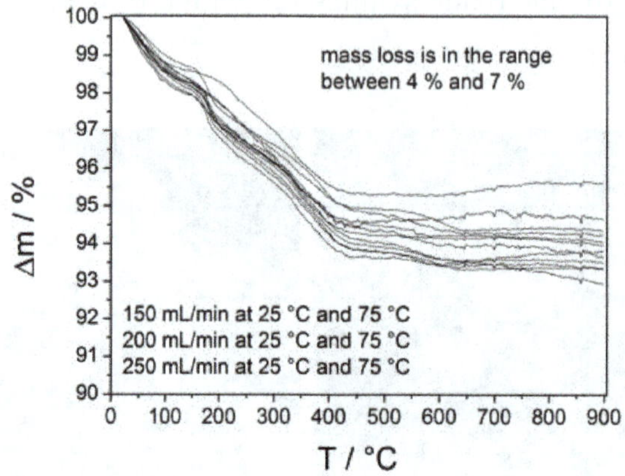

The phase composition and crystallite size of the obtained particles were determined by X-ray powder diffraction (Figure 8). The results show that magnetite with crystallite sizes around 10 nm were obtained in the microjet reactor. There are a few very weak reflections which cannot be correlated to magnetite. Consequently, there is a small fraction of an impurity phase. The reflection at approximately $2\theta = 32$ ° suggest this phase to be maghemite.

Figure 8. XRD pattern of the obtained particles. The results show that magnetite was obtained.

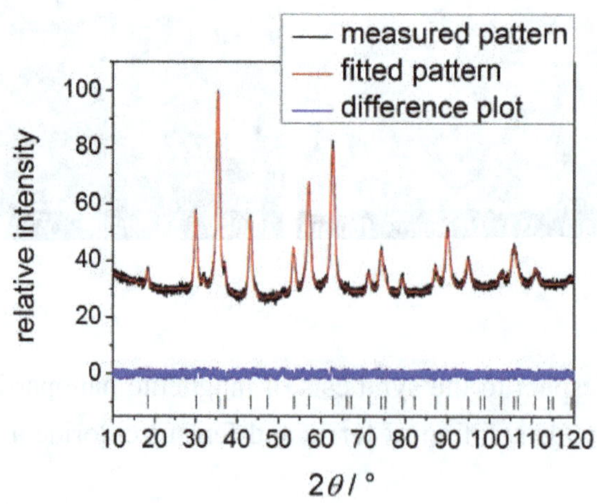

The particle size of the attained material was determined by using dynamic light scattering (Figure 9). The results reveal that particles between 46 nm and 112 nm were obtained. Just as in case of the ZnO particles the particle sizes including the size distribution decrease with increasing process temperature and flow rate (Table 3).

Previous studies applying rapid mixing techniques showed much smaller nanoparticle sizes around 7 nm [4]. The difference between the technique presented here and previous studies is the altered time for growth of the particles, which depends in our setup on the length of the tubing between microjet-reactor and collecting vessel. We are currently investigating these parameters in an additional study.

Figure 9. Particle size of the synthesized magnetite determined by dynamic light scattering.

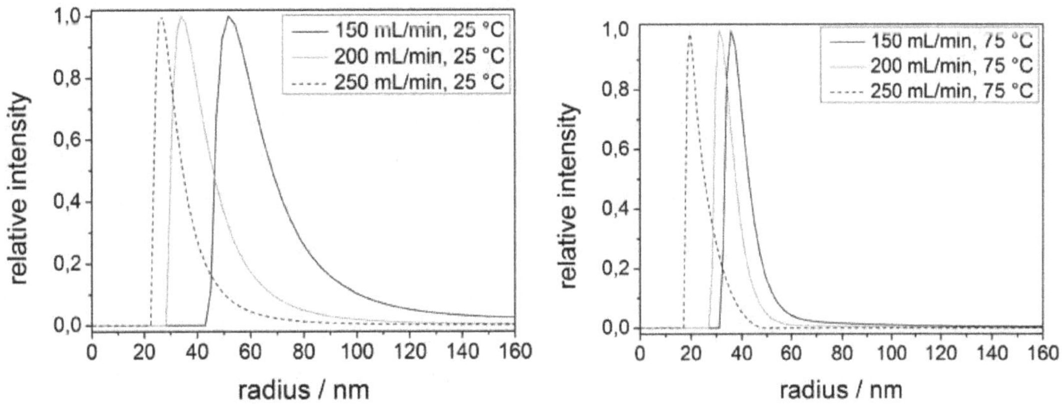

Table 3. Particle size determined using dynamic light scattering.

Process parameters	Radius/nm
150 mL/min/25 °C	66 ± 9
200 mL/min/25 °C	42 ± 7
250 mL/min/25 °C	32 ± 5
150 mL/min/75 °C	41 ± 4
200 mL/min/75 °C	35 ± 4
250 mL/min/75 °C	23 ± 3

SEM was applied to determine the morphology of the obtained magnetite particles (Figure 10). The results show that the particles are sphere-shaped and become smaller with increasing flow rate and temperature.

Figure 10. SEM images of the obtained magnetite particles synthesized at 25 °C with a flow rate of 150 mL/min (**left**) and at 75 °C with a flow rate of 250 mL/min (**right**).

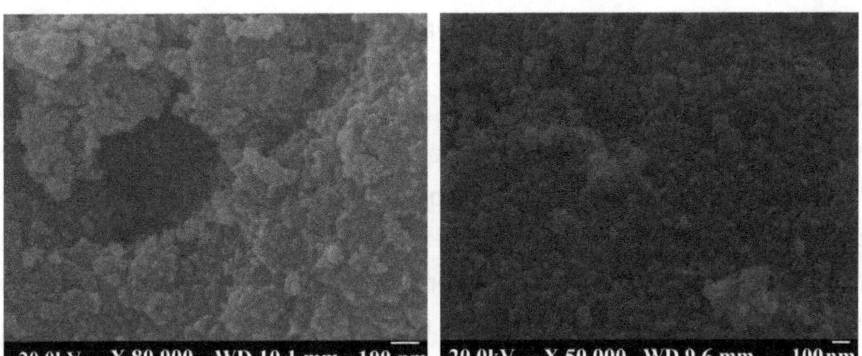

2.3. Calcium Hydrogen Phosphate/Brushite

As a third system the synthesis of calcium hydrogen phosphate particles using the mircojet-reactor was investigated. An aqueous solution of calcium nitrate (solution A) and di-ammonium hydrogen phosphate (solution B) were mixed in the microjet reactor (Equation 3). The pH of both solutions was adjusted to 5 in order to obtain the calcium hydrogen phosphate in the brushite phase.

$$Ca(NO_3)_2 \cdot 4H_2O + HPO_4(NH_4)_2 \rightarrow CaHPO_4 \cdot 2H_2O + 2NH_4NO_3 + 2H_2O. \quad (3)$$

Using the same starting materials, but applying a conventional batch reactor synthesis route, plate-like brushite particles in the micrometer range were obtained [29]. Employing the microjet reactor for this reaction it turned out that the precipitation reaction does not proceed at 25 °C. The material obtained using 75 °C as process temperature was analyzed with respect to composition and morphology. FT-IR analyses (Figure 11) of the obtained materials show that the synthesis of calcium hydrogen phosphate was successful at 75 °C. The spectra show absorption bands between 1200 cm^{-1} and 400 cm^{-1} which is the typical P-O region and indicates P-O vibrations as well as vibrations of a HPO_4 group. In addition, absorption bands relating to water are visible. At 1650 cm^{-1} there is the typical absorption band for the H_2O bending. Furthermore, between 3550 cm^{-1} and 3000 cm^{-1}, the absorption bands relating to the OH stretching of water are visible. This agrees well with FT-IR data of calcium hydrogen phosphate reported in the literature [30,31].

Figure 11. FT-IR spectra of the obtained materials. Absorption bands in the P-O region are observable as well as absorption bands relating to water. For each condition two spectra were shown in order to illustrate the reproducibility.

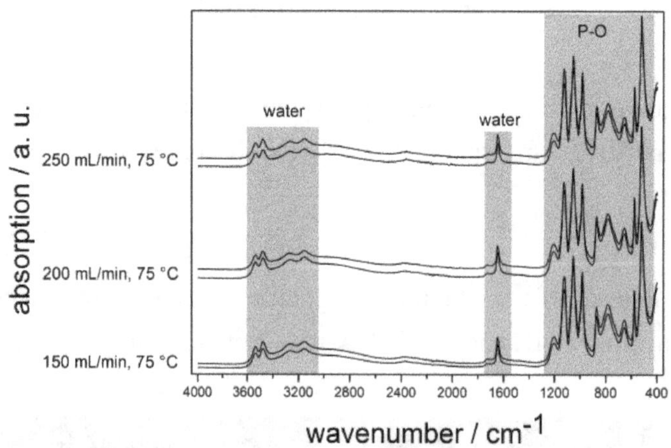

Thermo gravimetric analysis was performed to analyze the thermal decomposition of the obtained material (Figure 12). The results show that the decomposition takes place in two steps, which confirms that calcium hydrogen phosphate was obtained in the brushite phase. The first decomposition step proceeds between 100 °C and 350 °C degree and is relating to the loss of two molecules of crystal water. Consequently $CaHPO_4$ is formed. The second decomposition step takes place between 350 °C and 600 °C and is relating to the loss of 0.5 water molecules per Ca^{2+} ion and the formation of $Ca_2P_2O_7$. This is the typical pattern for decomposition of brushite.

Figure 12. Thermal decomposition of the obtained material. The decomposition takes place in two steps, which is typical for the thermal decomposition of calcium hydrogen phosphate in the brushite phase.

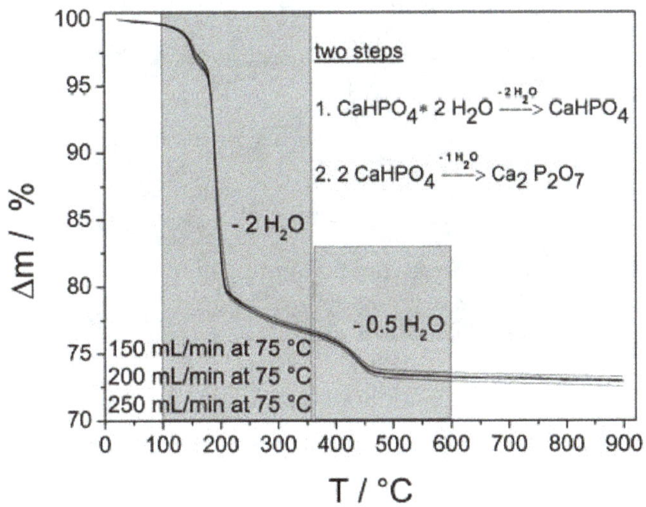

The results of the X-ray powder diffraction measurements (Figure 13) confirm that pure brushite was obtained. The particles show a high anisotropy in their crystallite sizes. An average crystallite size could be determined which is in the range of 200 to 300 nm depending on the reaction conditions.

Figure 13. X-ray powder pattern. Pure calcium hydrogen phosphate in the brushite phase was obtained.

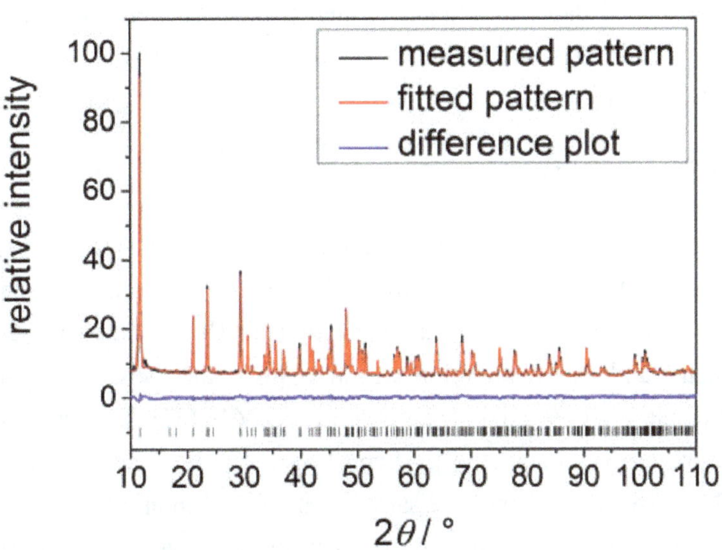

SEM was applied to analyze the morphology of the obtained material (Figure 14). The results show that the particles are shaped like small plates based on the layered structure of brushite. The layers grow faster in two directions so that plates are formed. The edge lengths of these plates were determined to be between 100 nm and 500 nm.

Figure 14. SEM image of the obtained material. Brushite synthesized at 75 °C with a flow rate of 250 mL/min.

All three examples showed that the continuous production of nanoparticles, in case of ZnO and Fe_3O_4, or submicron sized plates in case of brushite is possible with the wet chemical microjet reaction technique. All samples showed that flow rates as well as temperatures have an important influence on the size and size-distribution of the particles. In the microjet environment the nuclei are formed and the particle growth most likely occurs in the tube between the reactor and the collecting vessel. We did not investigate the effect of this distance on the particle growth yet. This will be the focus of a further study.

3. Experimental Section

3.1. Materials

Zinc chloride, sodium hydroxide and ferric chloride were purchased from Grüssing GmbH (Filsum, Germany). Calcium nitrate and di-ammonium hydrogen phosphate were purchased from Fluka (Buchs, Switzerland). Ferrous chloride was purchased from Alfa Aesar (Karlsruhe, Germany). Ammonia was purchased from VWR (Fontenay-sous-Bois, France). All chemicals were used without further purification.

3.2. Instruments and Characterization

FT-IR measurements were performed under ambient air (40 scans at a resolution of 4 cm^{-1}) in attenuated total reflectance (ATR) mode on a Bruker Vertex 70 spectrometer. X-ray powder diffraction was carried out on two different diffractometers a Panalytical X'Pert and a Bruker D8 Advance-system, Bragg-Brentano geometry and CuKα radiation was used in both cases. The quantitative analysis was carried out by the Rietveld-method using the program TOPAS [32] and crystallographic data for the different species (zinc oxide [33], magnetite [34] and brushite [35]). Thermogravimetric analysis (TGA) was carried out on a Netzsch Iris TG 209. The sample was placed in an alumina crucible which was then heated from room temperature to 900 °C under nitrogen atmosphere with a rate of 20 K min^{-1}.

Scanning electron microscopy (SEM) images were recorded on a JEOL SEM-7000 microscope. The SEM samples were prepared by placing some grains on a specimen stub with attached carbon adhesive foil followed by deposition of a gold layer. Dynamic light scattering (DLS) measurements were carried out on an ALV/CGS-3 compact goniometer system with an ALV/LSE-5003 multiple τ correlator at a wavelength of 632.8 nm (He-Ne Laser) and at a 90° goniometer angle. The particle radius was then determined by the analysis of the correlation-function via the g2(t) method followed by a logarithmic number-weighting (n.w.) of the distribution function.

3.3. Synthesis

The following experiments were performed in a microjet reactor construction. Two LaPrep P110 preparative HPLC pumps (VWR) are responsible for the transport of the solutions containing the starting materials as dissolved substances. The reaction takes place in a microjet reactor (Synthesechemie, Heusweiler, Germany and MJR PharmJet, Homburg, Germany) with a cone diameter of 300 micrometers. The precipitate was directly removed from the reactor by the use of a nitrogen gas flow. The length of the tube between the microjet reactor and the collecting vessel was 150 cm. Zinc oxide was synthesized using a 1 M aqueous zinc chloride and a 2 M aqueous sodium hydroxide solution as starting materials. Magnetite was obtained using a 0.4 M aqueous ammonia solution and an aqueous solution containing 0.0625 M ferric and ferrous chloride respectively. Brushite was synthesized by using 0.2 M aqueous solution of calcium nitrate and a 0.1 M aqueous solution of di-ammonium hydrogen phosphate, the pH of both solutions was adjusted to 5 using acetic acid or ammonia. The synthesis of each species was performed using different process parameters. The temperature was varied between 25 °C and 75 °C, the flow rate was varied between 150 mL/min and 250 mL/min.

4. Conclusions

The microjet-reactor technique is a promising method for the continuous synthesis of inorganic nanoparticles. Systematic studies on the effect of the process parameters, such as flow rate and process temperature, show that the particle size is controllable by means of these parameters. Our aim was the continuous synthesis of inorganic nanoparticles with controllable particle size. In our work we successfully demonstrated that it is possible to obtain pure zinc oxide nanoparticles using the microjet-reactor system. The prepared particles are uniformly shaped with a particle size of between 44 nm and 102 nm depending on the reaction conditions. The same could be shown for the synthesis of magnetite nanoparticles that were obtained with particle sizes between 46 nm and 132 nm. In both cases, the size distributions were quite small and decreased with increasing reaction temperature. Furthermore, we demonstrated the synthesis of calcium hydrogen phosphate (brushite) by using the microjet-reactor. In this case brushite as a layered structure formed small plates.

References

1. Kango, S.; Kalia, S.; Celli, A.; Njuguna, J.; Habibi, Y.; Kumar, R. Surface modification of inorganic nanoparticles for development of organic-inorganic nanocomposites - A review. *Prog. Polym. Sci.* **2013**, *38*, 1232–1261.
2. Castro, F.; Kuhn, S.; Jensen, K.; Ferreira, A.; Rocha, F.; Vicente, A.; Teixeira, J.A. Process intensification and optimization for hydroxyapatite nanoparticles production. *Chem. Eng. Sci.* **2013**, *100*, 352–359.

3. Ying, Y.; Chen, G.; Zhao, Y.; Li, S.; Yuan, Q. A high throughput methodology for continuous preparation of monodispersed nanocrystals in microfluidic reactors. *Chem. Eng. J.* **2008**, *135*, 209–215.
4. Ström, V.; Olsson, R.T.; Rao, K.V. Real-time monitoring of the evolution of the magnetism during precipitation of superparamagnetic nanoparticles for bioscience application. *J. Mater. Chem.* **2010**, *20*, 4168–4175.
5. Van den Rul, H.; Mondelaers, D.; van Bael, M.K.; Mullens, J. Water-based wet chemical synthesis of (doped) ZnO nanostructures. *J. Sol-Gel Sci. Tech.* **2006**, *39*, 41–47.
6. Jones, A.; Rigopoulos, S.; Zauner, R. Crystallisation and precipitation engineering. *Comput. Chem. Eng.* **2005**, *29*, 1159–1166.
7. Chen, J.; Zheng, C.; Chen, G. Interaction of macro- and micromixing on particle size distribution in reactive precipitation. *Chem. Eng. Sci.* **1996**, *51*, 1957–1966.
8. Mersmann, A. Crystallization and precipitation. *Chem. Eng. Process.* **1999**, *38*, 345–353.
9. Luo, G.; Du, L.; Wang, Y.; Lu, Y.; Xu, J. Controllable preparation of particles with microfluidics. *Particuology* **2011**, *9*, 545–558.
10. Chen, G.; Luo, G.; Yang, X.; Sun, Y.; Wang, J. Anatase-TiO_2 nano-particle preparation with a micro-mixing technique and its photocatalytic performance. *Mater. Sci. Eng. A* **2004**, *380*, 320–325.
11. Wagner, J.; Kirner, T.; Mayer, G.; Albert, J.; Köhler, J.M. Generation of metal nanoparticles in a microchannel reactor. *Chem. Eng. J.* **2004**, *101*, 251–260.
12. Zinoveva, S.; De Silva, R.; Louis, R.D.; Datta, P.; Kumar, C.S.S.R.; Goettert, J.; Hormes, J. The wet chemical synthesis of Co nanoparticles in a microreactor system: A time-resolved investigation by X-ray absorption spectroscopy. *Nucl. Instrum. Methods Phys. Res. Sect. A* **2007**, *582*, 239–241.
13. Du, L.; Wang, Y.J.; Lu, Y.C.; Luo, G.S. Preparation of highly purified β-tricalcium phosphate ceramics with a microdispersion process. *Chem. Eng. J.* **2013**, *221*, 55–61.
14. Schwarzer, H.C.; Schwertfirm, F.; Manhart, M.; Schmid, H.J.; Peukert, W. Predictive simulation of nanoparticle precipitation based on the population balance equation. *Chem. Eng. Sci.* **2006**, *61*, 167–181.
15. Schwarzer, H.C.; Peukert, W. Experimental Investigation into the Influence of Mixing on Nanoparticles Precipitation. *Chem. Eng. Technol.* **2002**, *25*, 6, 657–661.
16. Lince, F.; Marchisio, D.L.; Barresi, A.A. A comparative study for nanoparticle production with passive mixers via solvent-displacement: Use of CFD models for optimization and design. *Chem. Eng. Process.* **2011**, *50*, 356–368.
17. Lince, F.; Marchisio, D.L.; Barresi, A.A. Smart mixers and reactors for the production of pharmaceutical nanoparticles: Proof of concept. *Chem. Eng. Res. Des.* **2009**, *87*, 543–549.
18. Silva, V.M.T.M.; Quadros, P.A.; Laranjeira, P.E.M.S.C.; Dias, M.M.; Lopes, J.C.B. A novel Continuous Industrial Process for Producing Hydroxyapatite Nanoparticles. *J. Dispersion Sci. Technol.* **2008**, *29*, 542–547.
19. Penth, B. (K)ein Fall für die Fällung. *Chem. Tech.* **2004**, *3*, 18–20.
20. Penth, B. Kontinuierliche Produktion in Mikroreaktoren. German Patent DE 102006004350 A1, August 2006.

21. Rüfer, A.; Räuchle, K.; Krahl, F.; Reschetilowski, W. Kontinuierliche Darstellung von Bariumsulfat-Nanopartikeln im MicroJet-Reaktor. *Chem. Ing. Tech.* **2009**, *81*, 1949–1954.
22. Dittert, B.; Gavrilovic, A.; Schwarz, S.; Angerer, P.; Steiner, H.; Schöftner, R. Phase content controlled TiO_2 nanoparticles using the MicroJetReactor technology. *J. Eur. Ceram. Soc.* **2011**, *31*, 2475–2480.
23. Djutisic, A.B.; Ng, A.M.C.; Chen, X.Y. ZnO nanostructures for optoelectronics: Material properties and device applications. *Prog. Quantum Electron.* **2010**, *34*, 191–259.
24. Sugawara, A.; Asaoka, K.; Ding, S.J. Calcium phosphate-based cements: Clinical needs and recent progress. *J. Mater. Chem. B* **2013**, *1*, 1081–1089.
25. Vékás, L.; Tombácz, E.; Turcu, R.; Morjan, I.; Avdeev, M.V.; Krasia-Christoforou, T.; Socoliuc, V. Synthesis of Magnetite Nanoparticles and Magnetic Fluids for Biomedical Applications. Nanomedicine-Basic and Clinical Applications in Diagnostics and Therapy. *Else Kröner-Fresenius Symp. Basel Karger* **2011**, *2*, 35–52.
26. Samanta, P.K.; Mishra, S. Wet chemical growth and optical property of ZnO nanodiscs. *Optik* **2013**, *124*, 2871–2873.
27. RuizMoreno, R.G.; Martínez, A.I.; Falcony, C.; Castro-Rodriguez, R.; Bartolo-Pérez, P.; Castro-Román, M. One pot synthesis of water compatible and monodisperse magnetite nanoparticles. *Mater. Lett.* **2013**, *92*, 181–183.
28. Fang, M.; Ström, V.; Olsson, R.T.; Belova, L.; Rao, K.V. Particle size and magnetic properties dependence on growth temperature for rapid mixed co-precipitated magnetite nanoparticles. *Nanotechnology* **2012**, *23*, 145601.
29. Arifuzzaman, S.M.; Rohani, S. Experimental study of brushite precipitation. *J. Cryst. Growth* **2004**, *267*, 624–634.
30. Petrov, I.; Soptrajanov, B.; Fuson, N.; Lawson, J.R. Infra-red investigation of dicalcium phosphates. *Spectrochim. Acta* **1967**, *23A*, 2637–2646.
31. Trpkovska, M.; Soptrajanov, B.; Malkov, P. FTIR reinvestigation of the spectra of synthetic brushite and its partially deuterated analogues. *J. Mol. Struct.* **1999**, *480–481*, 661–666.
32. *Topas*, V4.2; General profile and structure analysis software for powder diffraction data, User Manual; Bruker AXS: Karlsruhe, Germany, 2008.
33. Khan, A.A. X-ray determination of thermal expansion of zinc oxide. *Acta Cryst. A* **1968**, *24*, 403.
34. Okudera, H.; Kihara, K.; Matsumoto, T. Temperature Dependence of Structure Parameters in Natural Magnetite: Single Crystal X-ray Studies from 126 to 773 K. *Acta Cryst. B* **1996**, *52*, 450–457.
35. Beevers, C.A. The Crystal Structure of Dicalcium Phosphate Dihydrate, $CaHPO_4{\cdot}2H_2O$. *Acta Cryst.* **1958**, *11*, 273–277.

Synthesis and Characterisation of Lanthanide *N*-Trimethylsilyl and -Mesityl Functionalised Bis(iminophosphorano)methanides and -Methanediides

George Marshall, Ashley J. Wooles, David P. Mills, William Lewis, Alexander J. Blake and Stephen T. Liddle *

School of Chemistry, University of Nottingham, University Park, Nottingham, NG7 2RD, UK; E-Mails: pcxgm4@nottingham.ac.uk (G.M.); ashley.wooles@nottingham.ac.uk (A.J.W.); david.mills@manchester.ac.uk (D.P.M.); pczwl@exmail.nottingham.ac.uk (W.L.); a.j.blake@nottingham.ac.uk (A.J.B.)

* Author to whom correspondence should be addressed; E-Mail: stephen.liddle@nottingham.ac.uk

---

Abstract: We report the extension of the series of $\{BIPM^{TMS}H\}^-$ ($BIPM^R = C\{PPh_2NR\}_2$, TMS = trimethylsilyl) derived rare earth methanides by the preparation of $[Ln(BIPM^{TMS}H)(I)_2(THF)]$ (Ln = Nd, Gd, Tb), **1a–c**, in 34–50% crystalline yields via the reaction of $[Ln(I)_3(THF)_{3.5}]$ with $[Cs(BIPM^{TMS}H)]$. Similarly, we have extended the range of $\{BIPM^{Mes}H\}^-$ (Mes = 2,4,6-trimethylphenyl) derived rare earth methanides with the preparation of $[Gd(BIPM^{Mes}H)(I)_2(THF)_2]$, **3**, (49%) and $[Yb(BIPM^{Mes}H)(I)_2(THF)]$, **4**, (26%), via the reaction of $[Ln(I)_3(THF)_{3.5}]$ with $[\{K(BIPM^{Mes}H)\}_2]$. Attempts to prepare dysprosium and erbium analogues of **3** or **4** were not successful, with the ion pair species $[Ln(BIPM^{Mes}H)_2][BIPM^{Mes}H]$ (Ln = Dy, Er), **5a–b**, isolated in 31–39% yield. The TMEDA (*N',N',N'',N''*-tetramethylethylenediamine) adducts $[Ln(BIPM^{Mes}H)(I)_2(TMEDA)]$ (Ln = La, Gd), **6a–b**, were prepared in quantitative yield via the dissolution of $[La(BIPM^{Mes}H)(I)_2(THF)]$ or **3** in a TMEDA/THF solution. The reactions of $[Ln(BIPM^{Mes}H)(I)_2(THF)]$ [Ln = La, Ce, Pr, and Gd (**3**)] or **6a–b** with a selection of bases did not afford $[La(BIPM^{Mes})(I)(S)_n]$ (S = solvent) as predicted, but instead led to the isolation of the heteroleptic complexes $[Ln(BIPM^{Mes})(BIPM^{Mes}H)]$ (Ln = La, Ce, Pr and Gd), **7a–d**, in low yields due to ligand scrambling.

Keywords: lanthanide; methanide; methanediide

1. Introduction

Since Cavell *et al.* reported the first rare earth bis(iminophosphorano)methanediide complex, namely [Sm(BIPM^{TMS})(NCy$_2$)(THF)] (BIPM^{TMS} = {C(PPh$_2$NSiMe$_3$)$_2$}$^{2-}$) [1], a wide range of BIPM^{R}- derived rare earth complexes have been reported with the BIPM^{R} ligand being either mono-deprotonated ({BIPM^{R}H}$^{-}$) and coordinating as a methanide, or doubly deprotonated ({BIPM^{R}}$^{2-}$) and coordinating as either a methanediide or carbene [2–4]. These complexes have shown extensive utility in a range of areas, including hydroamination/cyclisation, hydrosilylation and polymerisation reactions [5,6], and our work towards: (a) the preparation of rare earth heterobimetallics [7]; (b) unusual reactivity towards small molecules [8–10]; and (c) the stabilisation of the Ce(IV) oxidation state [11]. In our studies we have employed primarily the BIPM^{TMS} and BIPM^{Mes} ({C(PPh$_2$NMes)$_2$}$^{2-}$; Mes = 2,4,6-trimethylphenyl) ligand frameworks [7,12–16], however, during our studies we have encountered several synthetic limitations in our attempts to prepare BIPM^{R}-derived rare earth complexes for the complete rare earth series. We have employed both alkane elimination and salt metathesis strategies towards the preparation of methanediide complexes of the general formula [Ln(BIPM^{R})(X)(THF)$_n$] (R = TMS or Mes; X = I, CH$_2$SiMe$_3$ or CH$_2$Ph), with each approach having advantages and limitations.

The reaction of a range of rare earth tri-benzyl complexes [Ln(CH$_2$Ph)$_3$(THF)$_3$] (Ln = La, Ce, Pr, Nd, Sm, Gd, Dy, Er, Y) or [Y(CH$_2$SiMe$_3$)$_3$(THF)$_3$] with the protonated pro-ligand BIPM^{TMS}H$_2$ afforded different products depending on the size of the rare earth centre [16–18]. In the case of the smaller rare earths, the methanediide alkyl complexes [Ln(BIPM^{TMS})(R)(THF)] (R = CH$_2$Ph: Ln = Dy, Er, Y; R = CH$_2$SiMe$_3$: Ln = Y) were isolated [16–18], however for the larger rare earths the only isolable products were the bis-BIPM complexes [Ln(BIPM^{TMS})(BIPM^{TMS}H)] (Ln = La, Ce, Pr, Nd, Sm, Gd) [16]. This variation in reactivity was ascribed to the lanthanide contraction [19], with ligand scrambling occurring in the attempted preparations of the larger rare earth analogues despite the high steric demands of the BIPM^{TMS} ligand [16]. By a similar methodology we prepared [Ln(BIPM^{TMS})(I)(THF)$_2$] (Ln = Y, Er) via the reaction of [Ln(CH$_2$Ph)$_2$(I)(THF)$_3$] with BIPM^{TMS}H$_2$ [15], but unfortunately this methodology could not be applied across the rare earth series due to the inaccessibility of [Ln(CH$_2$Ph)$_2$(I)(THF)$_3$] for rare earths larger than erbium [12].

Salt metathesis strategies were first employed in the preparation of [Y(BIPM^{TMS}H)(I)$_2$(THF)], which was isolated in 64% yield via the reaction of [{K(BIPM^{TMS}H)}$_2$] with [Y(I)$_3$(THF)$_{3.5}$] in refluxing THF over 4 hours [7]. Fine-tuning of the reaction conditions was required with excessive reaction times leading to decomposition. Perhaps surprisingly, applying a similar strategy to La with the reaction of [La(I)$_3$(THF)$_4$] and [{K(BIPM^{TMS}H)}$_2$] does not afford [La(BIPM^{TMS}H)(I)$_2$(THF)] as expected; no reaction is observed at ambient temperature and decomposition is observed at elevated temperatures [12]. Gratifyingly, however, the reaction of [Ln(I)$_3$(THF)$_4$] (Ln = La, Ce) with

[Rb(BIPM^{TMS}H)(THF)$_n$] affords [Ln(BIPM^{TMS}H)(I)$_2$(THF)] (Ln = La, Ce) in high yield with no forcing conditions required [11,13]. This variation exemplifies the importance of selecting the correct synthetic reagent for the reaction in hand. The reaction of [Ln(BIPM^{TMS}H)(I)$_2$(THF)] (Ln = Ce, Y) with [K(Bn)] affords the rare earth carbene complexes [Ln(BIPM^{TMS})(I)(S)$_n$] (Ln = Y, S = THF, n = 2; Ln = Ce, S = 1,2-dimethoxyethane, n = 1) in good yield [7,11].

To fully investigate the importance of the steric demands of the *N*-SiMe$_3$ group in BIPM^{TMS}, we employed a different BIPM ligand, namely the *N*-aryl substituted BIPM^{Mes} system. In contrast to the difficulties experienced when employing salt metathesis strategies with the BIPM^{TMS} framework, we previously prepared [Ln(BIPM^{Mes}H)(I)$_2$(THF)$_2$] (Ln = La, Ce, Pr, Nd, Sm) via the straight-forward reaction of [Ln(I)$_3$(THF)$_n$] with [{K(BIPM^{Mes}H)}$_2$] [12,14]. Similarly to the BIPM^{TMS} system [La(BIPM^{Mes}H)(I)$_2$(THF)$_2$] was converted to the rare earth carbene complex [La(BIPM^{Mes})(I)(THF)$_3$] in 53% yield via reaction with [K(Bn)] [12].

Whilst a range of complexes of the formula [{Ln(BIPM^{TMS}H)(Cl)(μ-Cl)}$_2$] have been reported previously by the reaction of [Ln(Cl)$_3$] and [{K(BIPM^{TMS}H)}$_2$] [20], we have solely utilised iodide precursors. This is to reduce the likelihood of salt occlusion occurring as KI, RbI and CsI have a much lower propensity for salt occlusion in comparison to LiCl or KCl, and as salt occlusion can often negatively affect reactivity and promote unwanted side reactions this is a major advantage of utilising [Ln(I)$_3$(THF)$_n$] precursors. As ligand scrambling and Schlenk-type equilibra proved problematic in alkane elimination strategies, we focused on salt metathesis strategies in our attempts to complete the range of rare earth BIPM^{R}H (R = TMS, Mes) complexes of the general formula [Ln(BIPM^{R}H)(I)$_2$(THF)$_n$]. These complexes would provide an insight into the structure and bonding of rare earth complexes as well as having utility in the preparation of methanediide complexes of the general formula [Ln(BIPM^{R})(I)(THF)$_n$]. With these complexes in hand we would be able to investigate what, if any, effect the varying metal size would have on their structure and reactivity.

2. Results and Discussion

As rare earth ions are predominantly paramagnetic, any NMR spectroscopic studies performed are often difficult to interpret, which makes it problematic to both follow reaction progress and to identify the products of reactions. For this reason single crystal X-ray diffraction studies are essential to unambiguously identify the outcome of reactions involving paramagnetic rare earth metals. Unfortunately, we often observe that some complexes do not crystallise readily, despite repeated attempts, one reason for which may be that they do not possess an optimal metal/ligand size ratio for crystal growth [21]. For this reason we commonly employ a range of rare earths of varying sizes, Table 1, in our studies to ensure we have the best chance at fully identifying the products of reactions. Due to the difficulties in identifying reaction products without X-ray diffraction studies, we have only included structurally authenticated complexes in this report.

Table 1. Six-coordinate ionic radii and covalent radii (Å) of Ln(III) complexes [22,23].

Metal	Ionic radius	Covalent radius	Metal	Ionic radius	Covalent radius
Sc	0.745	1.70	Gd	0.938	1.96
Y	0.900	1.90	Tb	0.923	1.94
La	1.032	2.07	Dy	0.912	1.92
Ce	1.01	2.04	Ho	0.901	1.92
Pr	0.99	2.03	Er	0.890	1.89
Nd	0.983	2.01	Tm	N/A	1.90
Pm	N/A	1.99	Yb	0.868	1.87
Sm	0.958	1.98	Lu	0.861	1.87
Eu	0.947	1.98			

2.1. Preparation of BIPM^{TMS}H Rare Earth Methanides

As previously discussed, we have reported the preparation of [Ln(BIPM^{TMS}H)(I)$_2$(THF)] (Ln= La, Ce) in high yields (77% and 89%, respectively) via the reaction of [Ln(I)$_3$(THF)$_4$] with the heavy group 1 ligand transfer reagent [Rb(BIPM^{TMS}H)(THF)$_n$] under ambient conditions [11,13]. As this strategy was successful for the two largest rare earth metals (see Table 1), we extended this methodology across the rare earth series using the cesium ligand transfer reagent to maximise our chances of success. The reactions of [Ln(I)$_3$(THF)$_{3.5}$)] (Ln= Nd, Gd, Tb) with [Cs(BIPM^{TMS}H)] afforded, following filtration and workup, [Ln(BIPM^{TMS}H)(I)$_2$(THF)] (Ln= Nd, **1a**, 37%; Gd, **1b**, 50%; Tb, **1c**, 34%) with concomitant elimination of CsI (Scheme 1). As Nd(III), Gd(III) and Tb(III) are highly paramagnetic, NMR spectroscopic studies were inconclusive, and due to this each complex was crystallised from either THF or toluene to ensure purity of the sample. Unfortunately this also led to lower isolated yields (36-54%) compared to the diamagnetic La (77%) or weakly paramagnetic Ce (89%) analogues which could both be identified by NMR spectroscopy.

Scheme 1. Preparation of **1a–c**.

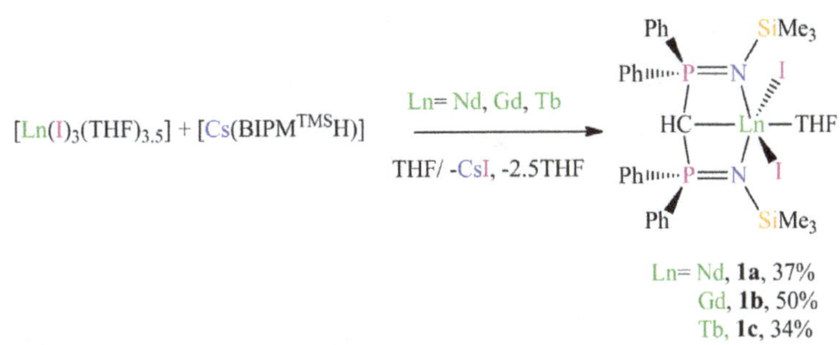

The identities of **1a–c** were confirmed by single crystal X-ray diffraction studies, elemental analysis and solution magnetic studies (*vide infra*, Table 2). The solid-state structures of **1a** C_7H_8 and **1c** OC_4H_8 are shown in Figures 1 and 2, respectively, with selected bond lengths and angles shown in Table S1. Each complex exhibits a distorted octahedral geometry with the metal centre coordinated by tridentate {BIPM^{TMS}H}$^-$ through the methanide carbon atom and the two imino nitrogens [N-Ln-N angles: **1a**, 116.28(14)°; **1b**, 93.6(2)°; **1c**, 93.40(10)°], two iodide ligands and a THF molecule. Despite exhibiting the same overall structure there is a variation in conformation of the iodides and THF molecules in

1a–c. Complex **1a** is isostructural to [Ln(BIPM^{TMS}H)(I)$_2$(THF)] (Ln = La, Ce) [11,13], with the coordinated THF molecule being *trans* to the methanide carbon [O1-Nd1-C1 angle: 151.54(13)°] with the two iodides being mutually *trans* [I1-Nd1-I2 angle: 156.167(13)°]. In contrast, **1b–c** are isostructural to [Ln(BIPM^{TMS}H)(I)$_2$(THF)] (Ln = Dy, Y) [7,12], and exhibit the THF molecule *cis* to the methanide carbon rather than *trans* [O1-Gd1-C1: 86.5(2)°; O1-Tb1-C1: 87.30(10)°], with the two iodide ligands being mutually *cis* [I1-Gd1-I2: 97.20(2)°: I1-Tb1-I2: 96.364(9)°]. The variation of conformation may be due to the varying size of the metal centre present as the larger rare earths La, Ce and Nd favour the *trans* conformation with the smaller rare earths Gd, Tb, Dy and Y favouring the *cis* conformation (for radii see Table 1). However, this could also be due to solvent effects on crystallisation favouring different conformations during crystal packing as **1a** and [Ln(BIPM^{TMS}H)(I)$_2$(THF)] (Ln = La, Ce) [11,13] were crystallised from toluene and adopt the *trans* geometry, whereas **1b-c** and [Ln(BIPM^{TMS}H)(I)$_2$(THF)] (Ln = Dy, Y) [7,12] were crystallised from THF and adopt the *cis* conformation. There is a general shortening of the Ln-C, Ln-N, Ln-I and Ln-O bond distances between **1a–1c** which is in agreement with the lanthanide contraction (see Table 1) [19], and is consistent with the bond distances reported for [Ln(BIPM^{TMS}H)(I)$_2$(THF)] (Ln = La, Ce, Dy, Y) [7,11–13]. The bond distances about the metal centres in **1a–c** are all within the range of previously reported bond distances [24], and are unremarkable. The mean endocyclic P-C and P-N bond distances in **1a–c** and the previously reported [Ln(BIPM^{TMS}H)(I)$_2$(THF)] (Ln = La, Ce, Dy, Y) [7,11–13], are statistically indistinguishable suggesting the {BIPM^{TMS}H}$^-$ coordinates to the metal centre in a similar way despite the variation in metal centre.

Figure 1. Molecular structure of **1a** C_7H_8 with selected atom labelling. Displacement ellipsoids are drawn at 50% probability, with lattice solvent and non-methanide hydrogen atoms omitted for clarity.

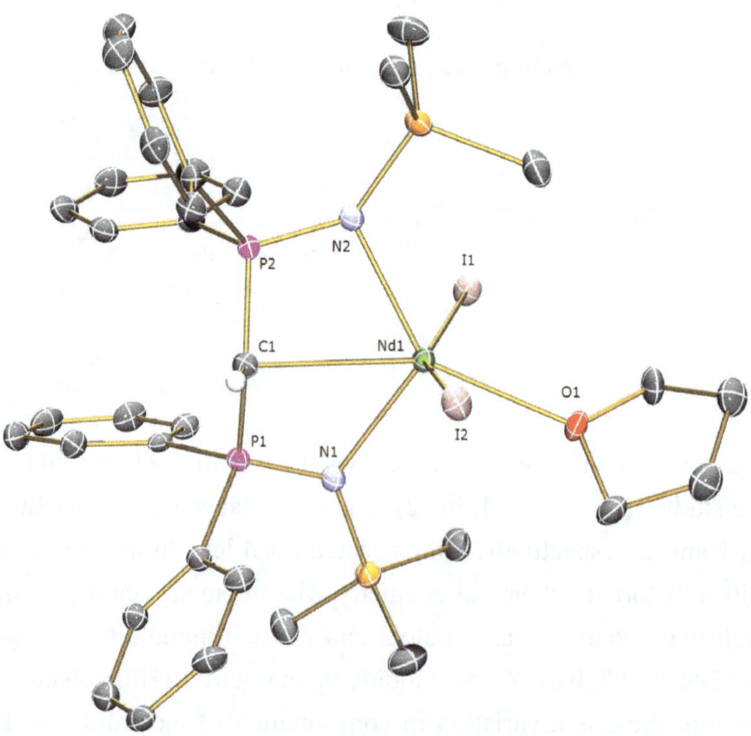

Figure 2. Molecular structure of **1c** OC_4H_8 with selected atom labelling. Displacement ellipsoids are drawn at 50% probability, with disordered components, lattice solvent and non-methanide hydrogen atoms omitted for clarity. The structure of **1b** is very similar.

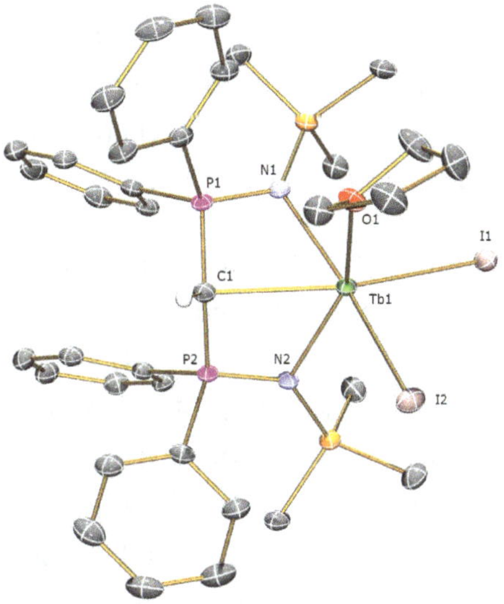

Scheme 2. Preparation of **2**.

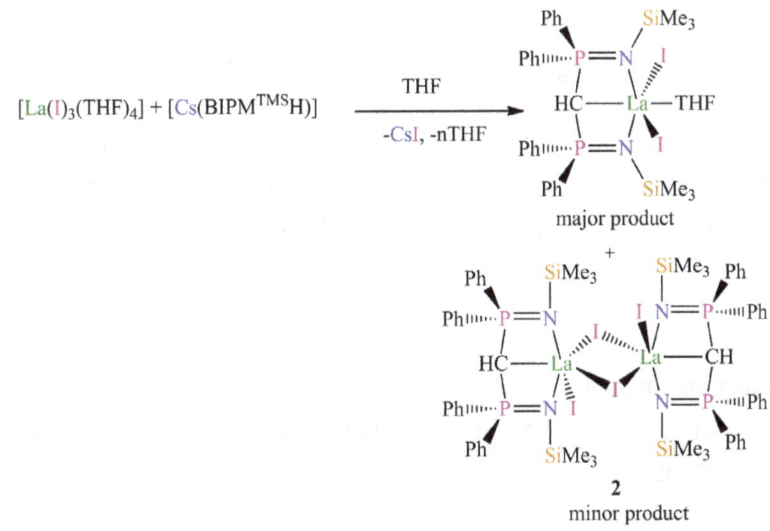

During our studies we have prepared many samples of [Ln(BIPM^{TMS}H)(I)$_2$(THF)] and on one occasion during preparation of [La(BIPM^{TMS}H)(I)$_2$(THF)] we isolated, from toluene solution, a small crop of colourless crystals which appeared to be of a different morphology to our previously reported [La(BIPM^{TMS}H)(I)$_2$(THF)] [13]. To ensure purity of the sample a single crystal X-ray diffraction study was performed which confirmed its identity as [{La(BIPM^{TMS}H)(I)(μ-I)}$_2$], **2** $2C_7H_8$ (Scheme 2). Complex **2** is dimeric in the solid-state with each lanthanum centre adopting a heavily distorted octahedral geometry and being coordinated by a {BIPM^{TMS}H}$^-$ in a tridentate fashion [N1-La1-N2 angle: 118.3(2)°], by one terminal iodide and by two iodides that are bridging the two lanthanide centres. The dimeric motif of **2** is likely caused by a scarcity of coordinating solvent present during crystallisation with the dimeric form allowing greater saturation of the coordination sphere of the lanthanide centres compared to a solvent free monomeric form. Once isolated, **2** has a very low solubility in arene solvents (e.g., d_6-benzene and d_8-toluene), which precluded solution NMR spectroscopic studies, while dissolution in more polar solvents to increase solubility (e.g., d_8-THF) led to solvated monomeric complexes such as [La(BIPM^{TMS}H)(I)$_2$(d_8-THF)]. Although NMR spectroscopy was not conclusive, the identity of **2** was confirmed by elemental analysis. The dimeric form of **2** is analogous to the chloride congeners reported by Roesky *et al.*, namely [{Ln(BIPM^{TMS}H)(Cl)(μ-Cl)}$_2$] (Ln = Sm ,Dy, Er, Yb, Lu, Y), which despite being prepared in THF solution adopt solvent free dimeric conformations when crystallised from toluene [20].

The solid-state structure of **2** $2C_7H_8$ is depicted in Figure 3 with selected bond lengths and angles shown in Table S1. There is a variation in La1-I bond distances in **2**, with the two bridging iodides being bound asymmetrically, with bond lengths of 3.2253(8) Å (La1-I2) and 3.3114(6) Å (La1-I2a), but both are still longer than the La1-I1 distance of 3.1144(7) Å due to the bridging *vs* terminal nature of the iodide ligands. Although the mean La1-N distance in **2** [2.443(6) Å] is statistically indistinguishable to the La1-N distances in [La(BIPM^{TMS}H)(I)$_2$(THF)] (mean: 2.438(6) Å), there is a large variation in the La1-C1 bond distance, with the La1-C1 distance in **2** [2.778(8) Å] being shorter than the La-C distance in [La(BIPM^{TMS}H)(I)$_2$(THF)] [2.859(8) Å] [13]. This shortening of the La1-C1 bond distance is likely due to the absence of O-donor solvents in **2**, compared to [La(BIPM^{TMS}H)(I)$_2$(THF)] [13], as, according to the HSAB principle, the hard lanthanum centre is predicted to exhibit a stronger interaction with the oxygen of THF rather than the relatively soft methanide carbon, leading to the longer La-C distance in [La(BIPM^{TMS}H)(I)$_2$(THF)] [13].

Despite the change in coordination mode between **2** and [La(BIPM^{TMS}H)(I)$_2$(THF)] the {BIPM^{TMS}H}$^-$ ligand appears to be bound to lanthanum in a similar manner in each complex, as evidenced by the mean endocyclic P-C and P-N bond distances and P1-C1-P2 angle in **2** [1.720(8), 1.607(7) Å and 138.3(5)°, respectively] being statistically indistinguishable to the corresponding values in [La(BIPM^{TMS}H)(I)$_2$(THF)] [1.734(8), 1.610(7) Å and 135.9(5)°, respectively] [13].

Figure 3. Molecular structure of **2** $2C_7H_8$ with selected atom labelling. Displacement ellipsoids are drawn at 50% probability, with lattice solvent and non-methanide hydrogen atoms omitted for clarity.

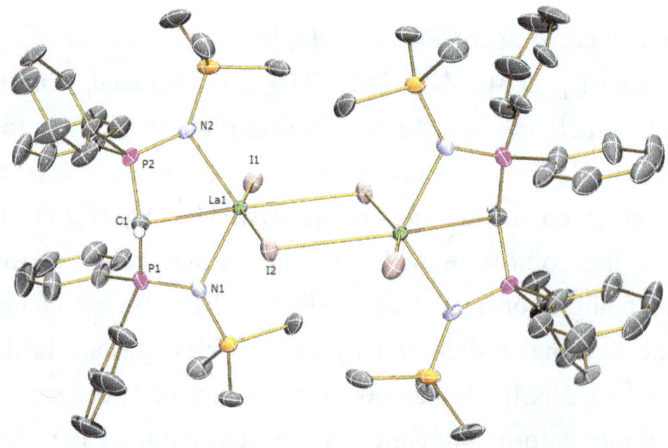

We have previously reported the conversion of [Ln(BIPM^{TMS}H)(I)$_2$(THF)] (Ln = Ce, Y) to [Ce(BIPM^{TMS})(I)(DME)] and [Y(BIPM^{TMS})(I)(THF)$_2$], respectively, via reaction with [K(Bn)]. This approach provides access to rare earth {BIPM^{TMS}}$^{2-}$ derived methanediide complexes for both the larger and smaller rare earths (Ce, and Y, respectively, Table 1). To fully investigate the nature of the {BIPM^{R}}$^{2-}$ ligand on the stability of these systems we have investigated the *N*-Mesityl variant of {BIPM^{TMS}}$^{2-}$, namely {BIPM^{Mes}}$^{2-}$.

2.2. Preparation of {BIPM^{Mes}H} Rare Earth Methanides

We have previously reported the preparation of [Ln(BIPM^{Mes}H)(I)$_2$(THF)$_2$] (Ln = La, Ce, Pr, Nd, Sm) in variable yield (40–78%) via the reaction of [Ln(I)$_3$(THF)$_n$] with [{K(BIPM^{Mes}H)}$_2$] [12,14]. In contrast to the preparation of the {BIPM^{TMS}H}$^{-}$ analogues (*vide supra*), no forcing conditions or use of heavier group 1 ligand transfer reagents (e.g., [Cs(BIPM^{Mes}H)]) was required. This is likely due to the decreased steric bulk of the {BIPM^{Mes}H}$^{-}$ ligand compared to {BIPM^{TMS}H}$^{-}$ leading to more reactive ligand transfer reagents as the alkali metal is less sterically shielded.

As we had previously prepared [Ln(BIPM^{Mes}H)(I)$_2$(THF)$_2$] for the larger rare earths (Ln = La, Ce, Pr, Nd, Sm; See Table 1), we attempted to extend this series to investigate any changes in structure or reactivity imparted on the complex by varying the metal centre. Analogous to our previous reports, the reaction of half an equivalent of [{K(BIPM^{Mes}H)}$_2$] with [Ln(I)$_3$(THF)$_{3.5}$] (Ln = Gd, Yb) afforded [Gd(BIPM^{Mes}H)(I)$_2$(THF)$_2$], **3**, and [Yb(BIPM^{Mes}H)(I)$_2$(THF)], **4**, respectively (Scheme 3). Each complex was identified by elemental analyses, solution magnetic studies (*vide infra*) and by single crystal X-ray diffraction studies. However, while the X-ray diffraction data for **3** 3C_7H_8 is of good quality (R_{int} = 0.034, R = 0.0419, R_w = 0.109) and confirmed its identity as [Gd(BIPM^{Mes}H)(I)$_2$(THF)$_2$] 3C_7H_8, the data-set obtained for **4** 3C_7H_8 is poor (R_{int} = 0.0744, R = 0.1097, R_w = 0.2865), and despite exhaustive attempts more satisfactory data could not be obtained. While the data-set collected on crystals of **4** 3C_7H_8 is of poor quality and precludes any assessment of the metrical parameters of the complex, the connectivity is clear-cut, and together with elemental analysis and comparison to lighter rare earth analogues, we are confident in our assignment of **4**. The difficulty in obtaining crystals of **4** of suitable quality for single crystal X-ray diffraction studies is perhaps surprising as [Ln(BIPM^{Mes}H)(I)$_2$(THF)$_2$] (Ln = La, Ce, Pr, Nd, Sm, Gd) all crystallise readily from toluene solutions to afford large crystals suitable for single crystal X-ray diffraction studies. The change in crystallinity is possibly due to **4** possessing only one THF molecule coordinated to ytterbium, compared to the two THF molecules coordinated in [Ln(BIPM^{Mes}H)(I)$_2$(THF)$_2$] (Ln = La, Ce, Pr, Nd, Sm, Gd) [12,14]. This variation in coordination number and ligand environment may lead to less efficient crystal packing, however, the lack in crystallinity may also be due to **4** not containing the optimal metal/ligand size ratio for crystal growth [21].

The solid-state structure of **3** 3C_7H_8 is shown in Figure 4, with selected bond lengths and angles shown in Table S1. As **3** 3C_7H_8 crystallises with the same cell settings as [Ln(BIPM^{Mes}H)(I)$_2$(THF)$_2$] (Ln = La, Ce, Pr, Nd, Sm) [12,14], they adopt the same distorted pentagonal bipyramidal structure, with the gadolinium centre being coordinated by tridentate {BIPM^{Mes}H}$^{-}$ [N1-Gd1-N2 angle:

113.23(10)°], two iodides and two THF molecules. The two iodides can be considered as occupying the axial sites, with the I1-Gd1-I2 angle of 155.765(9)° revealing the degree of distortion away from idealised pentagonal bipyramidal geometry due to the coordination of the bulky $\{BIPM^{Mes}H\}^-$ ligand. As expected, due to the lanthanide contraction, the bond distances about the gadolinium centre in **3** are generally shorter than those observed in $[Ln(BIPM^{Mes}H)(I)_2(THF)_2]$ (Ln = La, Ce, Pr, Nd, Sm), due to the smaller radius of Gd(III) (Table 1), and are each within the range of previously reported distances [24], but are otherwise unremarkable. The $\{BIPM^{Mes}H\}^-$ ligand in **3** appears to be bound to the metal centre in an analogous manner to $[Ln(BIPM^{Mes}H)(I)_2(THF)_2]$ (Ln = La, Ce, Pr, Nd, Sm) [12,14], as evidenced by the mean endocyclic P-C, P-N distances and P1-C1-P2 angle in **3** [1.723(4) Å, 1.626(3) Å and 134.3(2)°, respectively] being statistically indistinguishable to corresponding values reported for $[Ln(BIPM^{Mes}H)(I)_2(THF)_2]$ (Ln = La, Ce, Pr, Nd, Sm) [ranges P-C: 1.719(4)–1.731(4) Å; P-N: 1.612(3)–1.634(3) Å and P-C-P: 133.7(3)–134.1(2)°] [12,14].

Whilst we were successful in the preparation of $[Ln(BIPM^{Mes}H)(I)_2(THF)_n]$ for the larger rare earths (Ln = La, Ce, Pr, Nd, Sm, Gd) and smaller rare earth ytterbium, we had difficulty in the preparation of analogues for intermediate sized rare earths, namely dysprosium and erbium (Table 1). It is difficult to follow reaction progress between $[Ln(I)_3(THF)_{3.5}]$ (Ln = Dy, Er) and half an equivalent of $[\{K(BIPM^{Mes}H)\}_2]$, as the highly paramagnetic nature of Dy(III) ($4f^7$) and Er(III) ($4f^9$) lead to inconclusive NMR spectra, and also because $[Ln(I)_3(THF)_{3.5}]$ has low solubility in THF so the elimination of KI cannot be monitored easily as a suspension is observed throughout the reaction. As we followed the same preparative method that yielded $[Ln(BIPM^{Mes}H)(I)_2(THF)_n]$ (*vide supra*), we worked up the reaction mixture in a similar manner, namely filtration and recrystallisation from hot toluene solutions. In the cases of dysprosium and erbium this did not afford $[Ln(BIPM^{Mes}H)(I)_2(THF)_n]$ (Ln = Dy, Er) as expected, but instead led to the isolation of the separated ion pair species $[Ln(BIPM^{Mes}H)_2][BIPM^{Mes}H]$ (Ln = Dy, **5a**; Er, **5b**) in moderate yields (**5a**: 39%; **5b**: 31%, based on BIPM) (Scheme 3). Our hypothesis for the isolation of **5a–b** is that the reaction of $[Ln(I)_3(THF)_{3.5}]$ (Ln = Dy, Er) with half an equivalent of $[\{K(BIPM^{Mes}H)\}_2]$ proceeds sluggishly under ambient conditions, primarily due to the inherent insolubility of $[Ln(I)_3(THF)_{3.5}]$ (Ln = Dy, Er) in THF [25]. During workup, filtration separates insoluble $[Ln(I)_3(THF)_{3.5}]$ (Ln = Dy, Er) and yields a solution of $[\{K(BIPM^{Mes}H)\}_2]$ in excess to $[Ln(I)_3(THF)_{3.5}]$ (Ln = Dy, Er), which, when heated in toluene during attempted recrystallisation, would lead to a rapid reaction, where $[Ln(I)_3(THF)_{3.5}]$ (Ln = Dy, Er) would effectively react with three equivalents of $[\{K(BIPM^{Mes}H)\}_2]$ to afford **5a–5b**, with concomitant elimination of KI.

Repeated attempts to prepare $[Ln(BIPM^{Mes}H)(I)_2(THF)_n]$ (Ln = Dy, Er) by varying reaction times and solvents (THF, 1,2-dimethoxyethane, Et_2O) were not successful, and when forcing conditions were employed the reaction proceeded rapidly rather than in a slow and controlled manner, leading to isolation of **5a–b**. Complexes **5a–b** were identified by solution magnetic moment measurements (*vide infra*), elemental analyses and by single crystal X-ray diffraction studies.

Scheme 3. Preparation of **3–5a–b**.

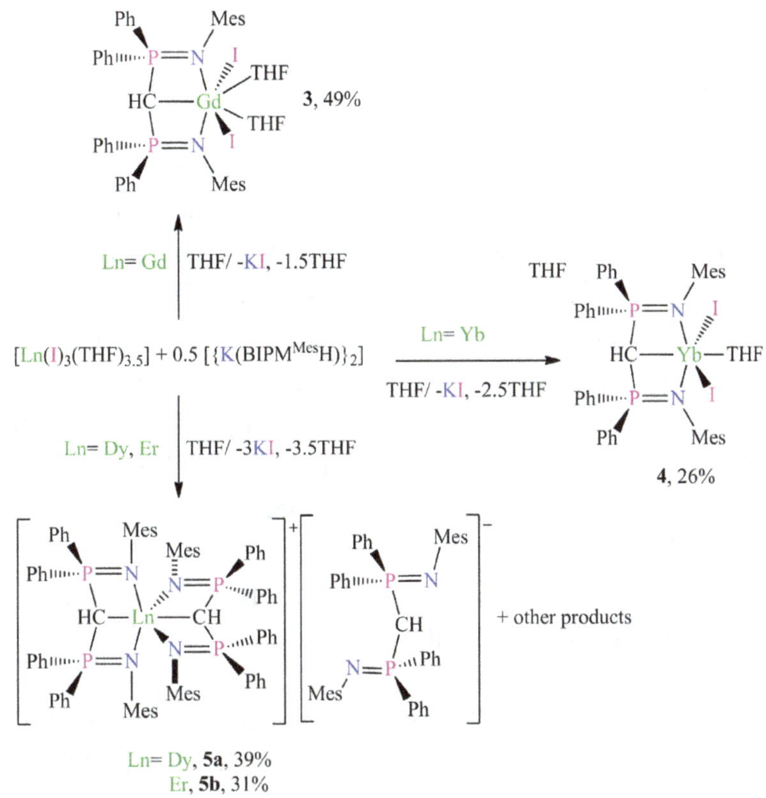

Figure 4. Molecular structure of **3** $3C_7H_8$ with selected atom labelling. Displacement ellipsoids are drawn at 50% probability, with lattice solvent and non-methanide hydrogen atoms omitted for clarity.

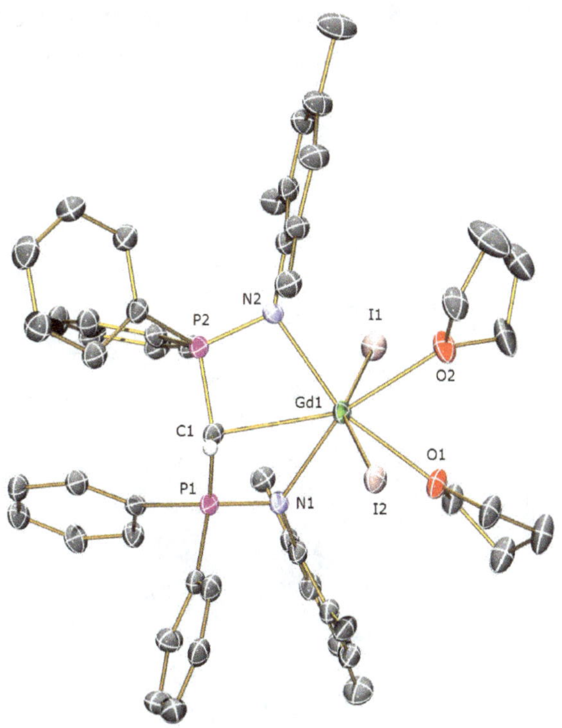

As complexes **5a** $4C_7H_8$ and **5b** $5C_7H_8$ have very similar solid-state structures and metrical parameters, only **5a** $4C_7H_8$ is discussed in detail. Selected bond lengths and angles of **5a** $4C_7H_8$ and **5b** $5C_7H_8$ are shown in Table S1. The cationic $[Dy(BIPM^{Mes}H)_2]^+$ component of **5a** is shown in Figure 5, while, for clarity, the anionic $[BIPM^{Mes}H]^-$ component is shown separately in Figure 6.

The dysprosium centre in **5a** is coordinated by two mutually orthogonal $\{BIPM^{Mes}H\}^-$ ligands, each exhibiting a tridentate binding mode, leading to a distorted octahedral geometry. Each $\{BIPM^{Mes}H\}^-$ is coordinated to dysprosium in a similar mode, as shown by the statistically indistinguishable Dy1-C1 [2.676(2) Å] and Dy1-C44 [2.679(2) Å] distances, but there is a variation in Dy-N distances, with each $\{BIPM^{Mes}H\}^-$ exhibiting one shorter Dy-N distance [Dy1-N2: 2.3445(18) Å; Dy1-N3: 2.3584(17) Å] and one longer Dy-N distance [Dy1-N1: 2.3849(19) Å; Dy1-N4: 2.3866(19) Å]. The mean endocyclic P-C and P-N distances, and P-C-P angle in the cationic component of **5a** [1.742(2), 1.632(2) Å and 128.65(13)°, respectively] are similar to the corresponding distances in $[Ln(BIPM^{Mes}H)(I)_2(THF)_2]$ (Ln = La, Ce, Pr, Nd, Sm) [12,14], but there is a slight elongation of the P-C bonds [**5a**: 1.742(2) Å compared to range 1.719(4)–1.731(4) Å] and lowering of the P-C-P angle [**5a**: 128.65(13)° compared to range 133.7(3)–134.1(2)°], which is likely a reflection of the coordination of two bulky $\{BIPM^{Mes}H\}^-$ ligands to one metal centre leading to increased steric hindrance about the endocyclic ring.

The anionic component of **5a** $[BIPM^{Mes}H]^-$ exhibits shortened endocyclic P-C and P-N distances [mean values of 1.709(2) and 1.592 Å, respectively] and a larger P-C-P angle [P5-C87-P6: 133.89(15)°] in comparison to the $\{BIPM^{Mes}H\}^-$ ligands in the cationic component, due to the anion not being coordinated to a metal centre. The P-C and P-N bonds in the N-P-C-P-N framework of the anionic $[BIPM^{Mes}H]^-$ fragment are shorter, and longer, respectively, than the distances reported in $BIPM^{Mes}H_2$ [26], suggesting a degree of delocalisation of charge about the N-P-C-P-N framework.

Figure 5. Cationic component of **5a** $4C_7H_8$. Displacement ellipsoids are drawn at 50% probability, with lattice solvent and non-methanide hydrogen atoms omitted for clarity.

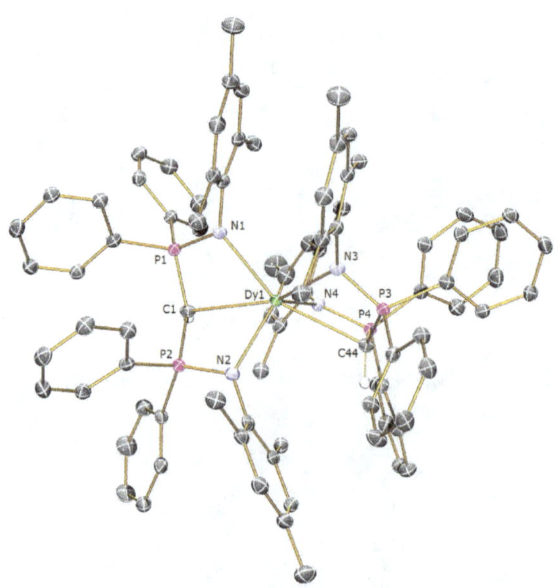

Figure 6. Anionic component of **5a** $4C_7H_8$. Displacement ellipsoids are drawn at 50% probability, with lattice solvent and non-methanide hydrogen atoms omitted for clarity.

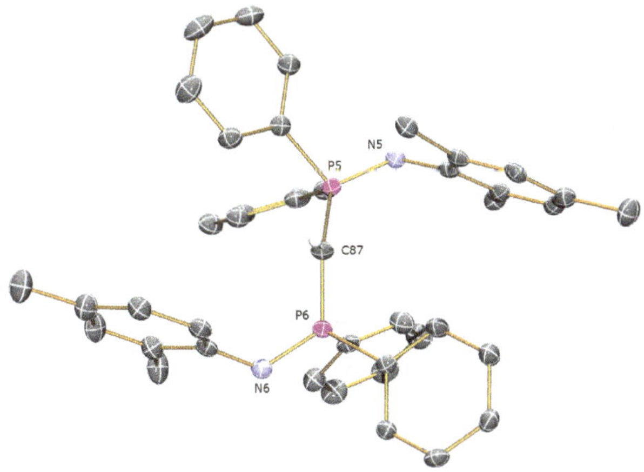

With $[Ln(BIPM^{Mes}H)(I)_2(THF)_2]$ (Ln = Ce, Pr, Nd, Sm) and **3** in hand, we attempted to prepare a range of $\{BIPM^{Mes}\}^{2-}$ derived rare earth methanediides in an analogous manner to our previously reported $[La(BIPM^{Mes})(I)(THF)_3]$ [12], which was prepared via the reaction of $[La(BIPM^{Mes}H)(I)_2(THF)_2]$ with [K(Bn)]. Unfortunately this preparative route proved to be capricious (*vide infra*), with a potential cause of the reaction not proceeding smoothly being the ineffectiveness of the labile coordinated THF molecules to fully saturate and stabilise the rare earth centre during the reaction. Replacement of these THF molecules with a bulkier bidentate donor solvent, namely TMEDA (*N',N',N'',N''*-tetramethylethylenediamine) was predicted to provide increased saturation of the rare earth metal centre and lead to an improved route to complexes of the formula $[Ln(BIPM^{Mes})(I)(S)_n]$ (S = donor solvent).

We initially investigated the lanthanum congener, as its La(III) $4f^0$ state is diamagnetic allowing us to follow the reaction progress easily by NMR spectroscopy. The straightforward reaction of $[La(BIPM^{Mes}H)(I)_2(THF)_2]$ [12] with TMEDA in THF solution over 18 h afforded $[La(BIPM^{Mes}H)(I)_2(TMEDA)]$, **6a**, in essentially quantitative yield (Scheme 4).

NMR spectroscopy performed on **6a** revealed it to have similar resonances to $[La(BIPM^{Mes}H)(I)_2(THF)_2]$, with the ^1H and $^{13}C\{^1H\}$ NMR spectra of **6a** exhibiting resonances for the methanide hydrogen and carbon at 3.91 ppm ($^2J_{PH}$ = 8.8 Hz), and 8.44 ppm (J_{PC} = 138 Hz), respectively, which are both slightly downfield compared to the equivalent resonances reported for $[La(BIPM^{Mes}H)(I)_2(THF)_2]$ (3.36 ppm, $^2J_{PH}$ = 14.5 Hz and 5.73 ppm, J_{PC} = 136 Hz).[12] The $^{31}P\{^1H\}$ NMR spectrum of **6a** exhibits a resonance at 15.70 ppm which is slightly downfield to the corresponding resonance reported for $[La(BIPM^{Mes}H)(I)_2(THF)_2]$ (11.46 ppm) [12].

While elemental analysis and NMR spectroscopy supported the formulation of **6a**, a single crystal X-ray diffraction study was performed to confirm its structure. The solid-state structure of **6a** $3C_7H_8$ is shown in Figure 7, with selected bond lengths and angles complied in Table S1. The lanthanum centre in **6a** adopts a heavily distorted pentagonal bipyramidal geometry with the two iodides occupying the axial sites [I1-La1-I2 angle: 158.395(12)°]. The tridentate $\{BIPM^{Mes}H\}^-$ [N1-La1-N2 angle:110.72(13)°] and the two amine nitrogens of TMEDA occupy the equatorial sites. The La1-N_{TMEDA} distances [La1-N3: 2.852(5) Å; La1-N4: 2.844(4) Å] are far longer than the La1-N_{BIPM} distances [La1-N1: 2.564(4) Å; La1-N2: 2.526(4) Å] consistent with the N^{-}-P^{+}-C^{-}-P^{+}-N^{-} charge distribution about the $\{BIPM^{Mes}H\}^{-}$ framework leading to an increased electrostatic interaction between La1 and the imino nitrogens. Despite the variation of donor solvent coordinated to lanthanum, the mean La1-N_{BIPM}, La1-I and La1-C1 bond distances [2.545(4), 3.2229(5) and 2.822(5) Å, respectively] are very similar to, if not statistically indistinguishable, from the corresponding distances in [La($BIPM^{Mes}H$)$(I)_2$$(THF)_2$] [2.537(3), 3.2042(4) and 2.833(4) Å, respectively] [12]. Similarly, the $\{BIPM^{Mes}H\}^{-}$ ligand exhibits statistically indistinguishable endocyclic P-C and P-N distances [mean distances: 1.732(5) and 1.621(5) Å, respectively], and P1-C1-P2 angle [134.1(3)°] to the THF congener [1.726(4) Å, 1.617(3) Å and 133.8(2)°, respectively] [12].

Following the successful isolation of **6a**, we prepared a heavier rare earth analogue, namely [Gd($BIPM^{Mes}H$)$(I)_2$(TMEDA)], **6b**, via the same methodology. As Gd(III) is smaller than La(III) (Table 1) this would provide us with a range of metal sizes to fully investigate if the size of the metal centre has any effect on the successful preparation of our target $\{BIPM^{Mes}\}^{2-}$ methanediide complexes (*vide infra*). Complex **6b** was prepared and utilised *in situ* in attempted deprotonation/salt elimination reactions, but a small sample was recrystallised from toluene to afford single crystals suitable for X-ray diffraction studies. Although **6b** crystallises in a different cell setting to **6a** (*C*mca *vs. P*-1), leading to the TMEDA molecule possessing positional disorder about the mirror plane, the overall structures are very similar with the gadolinium centre in **6b** adopting a heavily distorted pentagonal bipyramidal geometry. As expected due to the lanthanide contraction the bond distances about the gadolinium centre in **6b** are each shortened by *ca.* 0.07–0.14 Å compared to the corresponding distances about lanthanum in **6a**, and are all well within the range of previously reported bond distances [24].

With [Ln($BIPM^{Mes}H$)$(I)_2$$(THF)_2$] (Ln = La, Ce, Pr, Gd) and [Ln($BIPM^{Mes}H$)$(I)_2$(TMEDA)] (Ln = La, Gd) secured, we turned our focus to the preparation of $\{BIPM^{Mes}\}^{2-}$ derived rare earth methandediide complexes.

Scheme 4. Preparation of **6a–b**.

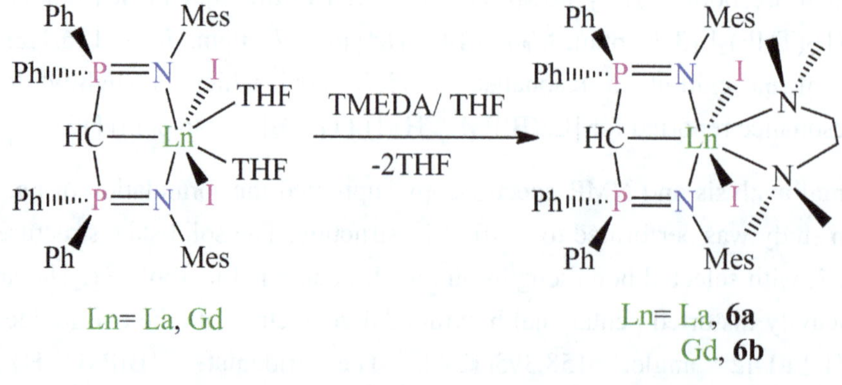

Figure 7. Molecular structure of **6a** $3C_7H_8$ with selected atom labelling. Displacement ellipsoids are drawn at 50% probability, with lattice solvent and non-methanide hydrogen atoms omitted for clarity. The molecular structure of **6b** $2C_7H_8$ is very similar.

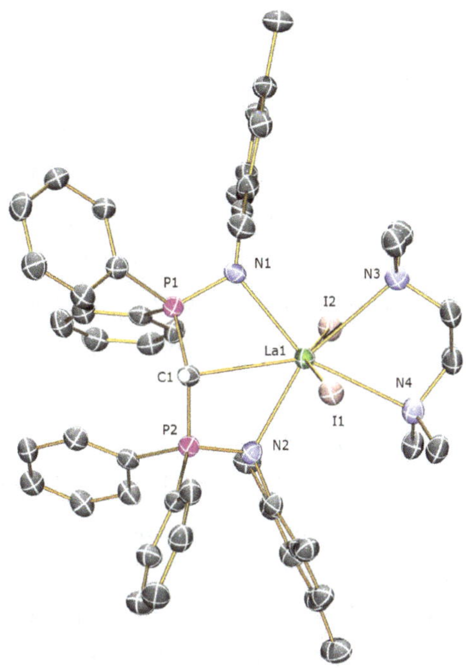

2.3. Attempted Preparation of BIPM^{Mes} Rare Earth Carbene Complexes

We have previously reported that the reaction of [La(BIPM^{Mes}H)(I)$_2$(THF)] with [K(Bn)] afforded the methanediide complex [La(BIPM^{Mes})(I)(THF)$_3$] in 53% yield [12]. Unfortunately, this reaction appears to be capricious in nature, and despite following the reported methodology the reaction often yields a mixture of products from which [La(BIPM^{Mes})(I)(THF)$_3$] cannot be isolated cleanly. As NMR spectroscopy revealed a mixture of products, of which none could be unambiguously identified, the reaction mixture was recrystallised from toluene which afforded [La(BIPM^{Mes})(BIPM^{Mes}H)], **7a**, in 8% yield. The low yield of **7a** is a reflection of the reaction not proceeding smoothly and affording a mixture of products, of which only **7a** was crystalline and able to be extracted cleanly. Complex **7a** was identified by single crystal X-ray diffraction studies (*vide infra*), NMR studies and elemental analysis. The $^{31}P\{^1H\}$ NMR spectrum of **7a** exhibits two singlet resonances, at 2.88 and 10.51 ppm, which are assigned to the methanediide and methanide centres, respectively, by comparison to the resonances reported for [La(BIPM^{Mes})(I)(THF)$_3$] (4.84 ppm) and [La(BIPM^{Mes}H)(I)$_2$(THF)$_2$] (11.46 ppm) [12]. The ^1H and $^{13}C\{^1H\}$ NMR spectra of **7a** reveal methanide resonances of 2.26 (no coupling observed) and 4.59 ppm (J_{PC} = 127 Hz), which are both upfield of the values reported for [La(BIPM^{Mes}H)(I)$_2$(THF)$_2$] (3.36 ppm, $^2J_{PH}$ = 14.5 Hz and 5.73 ppm, J_{PC} = 136 Hz, respectively), while the methanediide is observed in the $^{13}C\{^1H\}$ NMR spectrum of **7a** at 45.06 ppm, which is also upfield to the value reported for [La(BIPM^{Mes})(I)(THF)$_3$] (58.76 ppm, J_{PC} = 148.4 Hz) [12].

The isolation of **7a** is reminiscent of our attempts to prepare [La(BIPM^{TMS})(Bn)(THF)$_n$] via the reaction of [La(Bn)$_3$(THF)$_3$] and BIPM^{TMS}H$_2$ which instead afforded [La(BIPM^{TMS})(BIPM^{TMS}H)], the

BIPM^{TMS} analogue of **7a** [16]. In the cases of both **7a** and [La(BIPM^{TMS})(BIPM^{TMS}H)], the observed products are likely due to the instability of the intermediate complex [La(BIPM^{Mes}H)(Bn)(I)(THF)$_n$] or [La(BIPM^{TMS}H)(Bn)$_2$(THF)$_n$], which would each be susceptible to ligand scrambling due to the highly labile nature of lanthanide alkyl bonds.

As the preparation of [La(BIPM^{Mes})(I)(THF)$_3$] via the reaction of [La(BIPM^{Mes}H)(I)$_2$(THF)$_2$] with [K(Bn)] proved to be unreliable we attempted the reaction with other rare earth metals to investigate of the size of the metal present affected the result of the reaction. Also, in an attempt to stabilise the postulated "[Ln(BIPM^{Mes}H)(Bn)(I)(THF)$_n$]" intermediate, which we hypothesise is undergoing ligand scrambling during reaction, we varied both the alkyl and coordinated solvent present. The alkyl could be substituted by utilising a base that is bulkier than [K(Bn)], namely [K(CHPh$_2$)] or [K(CPh$_3$)], while the coordinated solvent was varied by utilising **6a–b** in place of [Ln(BIPM^{Mes}H)(I)$_2$(THF)$_2$].

Unfortunately the reactions of [Ln(BIPM^{Mes}H)(I)$_2$(THF)] (Ln = La, Ce, Pr, Gd) or **6a–b** with [K(R)] (R = Bn, CHPh$_2$, CPh$_3$) did not afford [Ln(BIPM^{Mes})(I)(S)$_n$] as anticipated, but instead led to mixtures of products in each case. The only isolable and identifiable products from these reactions were **7a**, [Ce(BIPM^{Mes})(BIPM^{Mes}H)], **7b**, [Pr(BIPM^{Mes})(BIPM^{Mes}H)], **7c**, or [Gd(BIPM^{Mes})(BIPM^{Mes}H)], **7d**, each in low yields (Scheme 5). Complexes **7b–d** were each identified by single crystal X-ray diffraction studies (*vide infra*), and while a clean sample of **7b** was isolated which allowed full characterisation (elemental analysis and solution magnetic studies), isolated samples of **7c** and **7d** were contaminated with impurities which precluded full analysis. Despite the lack of full characterisation for **7c** and **7d** we are confident our formulation is correct as they are analogous to **7a–b**, and it appears that varying the size of the metal present has little effect on the outcome of the reaction.

Scheme 5. Preparation of **7a–d**.

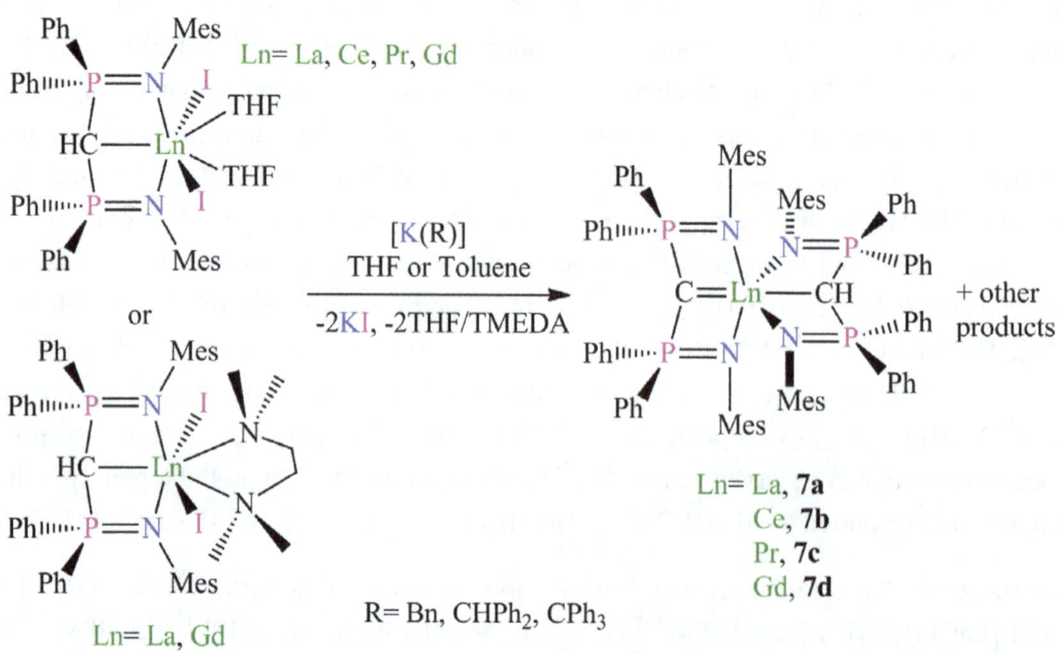

Figure 8. Molecular structure of **7a** $3C_7H_8$ with selected atom labelling. Displacement ellipsoids are drawn at 50% probability, with lattice solvent and non-methanide hydrogen atoms omitted for clarity. The molecular structures of **7b–d** are very similar.

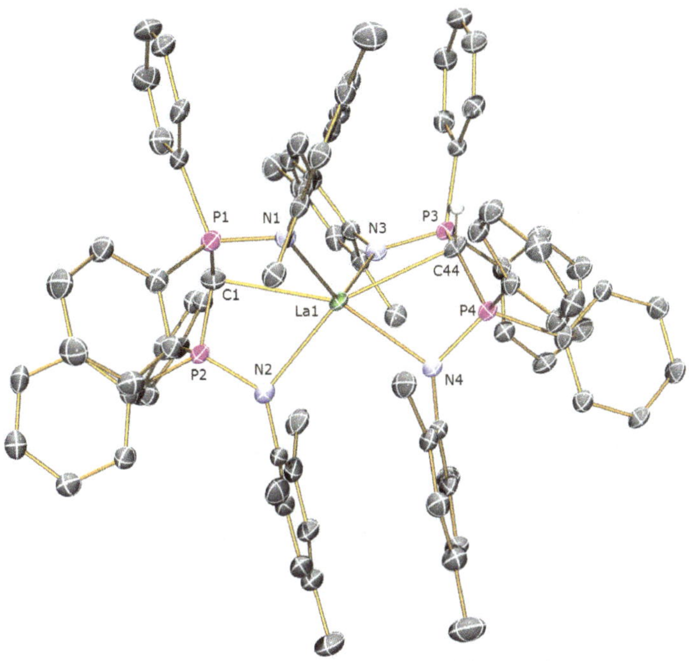

The complexes **7a–d** exhibit very similar solid-state structures, with the rare earth metal coordinated by mutually orthogonal $\{BIPM^{Mes}\}^{2-}$ and $\{BIPM^{Mes}H\}^{-}$ ligands, each bound in a tridentate fashion. A representative structure of **7a** $3C_7H_8$ is shown in Figure 8 with selected bond lengths and angles for each complex shown in Table S1. Despite exhibiting the same overall structure there is a variation in metrical parameters between **7a**, **c–d** and **7b**, which we propose is due to a variation in solvent of crystallisation. As each of **7a–d** exhibit one $\{BIPM^{Mes}H\}^{-}$ ligand bound to the metal centre as a methanide and one $\{BIPM^{Mes}\}^{2-}$ ligand bound as a methanediide, varying Ln-C distances would be expected as the metal centre would have an increased electrostatic interaction to the dianionic methanediide centre over the monoanionic methanide centre. This is the case for **7b**, which exhibits a longer Ce1-C1 distance of 2.819(11) Å, and a shorter Ce1-C44 distance of 2.681(11) Å, which leads to the assignment of C1 as being the methanide centre, and C44 being the methanediide centre. This is also analogous to the previously reported $BIPM^{TMS}$ congeners $[Ln(BIPM^{TMS})(BIPM^{TMS}H)]$ (Ln = La, Ce, Pr, Sm, Gd), which each exhibit one long and one short Ln-C interaction [16].

However, in the cases of **7a** and **7c–d**, each $\{BIPM^{Mes}\}$ ligand appears to be bound to the metal in an identical manner, which is exemplified by statistically indistinguishable Ln-C bond distances in each case [**7a**: La1-C1: 2.725(5) Å, La1-C44: 2.731(4) Å; **7c**, Pr1-C1: 2.667(6) Å, Pr1-C44: 2.662(6) Å; **7d**, Gd1-C1: 2.605(11) Å, Gd1-C44: 2.573(10) Å]. We propose that this variation in coordination is simply an artefact of crystallisation, with the crystal packing in **7a,c–d** being random, with the resulting data-set revealing an averaged geometry about the metal centre leading to equivalent Ln-C bond distances. This is supported by the mean La1-C distances in **7a** of 2.728(5) Å being intermediate to the Ce1-C distances in **7b** of 2.681(11) and 2.819(11) Å, which would be expected if the two bond distances were averaged out. The other possibility is that each {$BIPM^{Mes}$} ligand in **7a** is bound as a methanide leading to a formal oxidation state assignment of La(II), analogous to the previously reported [Sm($BIPM^{Mes}$H)$_2$] [27], but this is easily ruled out by **7a** being diamagnetic and its $^{31}P\{^{1}H\}$ NMR spectrum exhibiting two resonances consistent with the presence of two different ligand environments.

As expected due to the lanthanide contraction (Table 1) [19], there is a shortening of the mean Ln-C and Ln-N bond distances across the series from **7a** to **7d**, but these distances are otherwise unremarkable. In each complex the {$BIPM^{Mes}$} ligands are bound in a similar manner [endocyclic P-C distances ranging 1.671(12)–1.740(11) Å and P-N distances ranging 1.618(9)–1.684(9) Å], with each adopting a *pseudo*-boat conformation [P-C-P angles: range: 130.2(7)–134.3(7)°]. This conformation is in contrast to the {$BIPM^{TMS}$}$^{2-}$ ligands in [Ln($BIPM^{TMS}$)($BIPM^{TMS}$H)], which exhibit near planar geometries, with P-C-P angles of *ca.* 166° [16].

2.4. Solution State Magnetic Properties of $BIPM^{TMS}$ and $BIPM^{Mes}$ Rare Earth Complexes

The room temperature solution magnetic moments of **1a–c**, **3**, **4**, **5a–b** and **7b** were determined utilising the Evans method [28]. These are compiled in Table 2 along with their ground state terms and theoretical magnetic moments. The theoretical magnetic moments of rare earth complexes can be approximated by $\mu_J = gJ\sqrt{J(J+1)}$ [where $gJ = 3/2 + [S(S+1) - L(L+1)]/2J(J+1)$] according to the Van Vleck equation for magnetic susceptibility [29], assuming the $^{2S+1}L_J$ ground state is well separated from excited states and crystal field splitting is negligible. This is generally true for rare earths, with the exception of Sm(III) and Eu(III), which exhibit low lying excited states of $^{6}H_{7/2}$ and $^{7}F_1$, respectively, which each contribute to the room temperature magnetic moments [19]. As shown in Table 2 there is reasonable agreement between the theoretical and observed magnetic moments for each complex reported in this work.

Table 2. Theoretical and observed solution magnetic moments.

Complex	Rare earth ion	Ground state term	μ_J (μ_B) [a]	μ_{eff} (μ_B)
1a	Nd(III)	$^{4}I_{9/2}$	3.68	4.14
1b	Gd(III)	$^{8}S_{7/2}$	7.94	8.47
1c	Tb(III)	$^{7}F_6$	9.72	9.46
3	Gd(III)	$^{8}S_{7/2}$	7.94	8.00
4	Yb(III)	$^{2}F_{7/2}$	4.53	4.29
5a	Dy(III)	$^{6}H_{15/2}$	10.63	9.81
5b	Er(III)	$^{4}I_{15/2}$	9.59	10.25
7b	Ce(III)	$^{2}F_{5/2}$	2.54	1.88

[a] $\mu_J = gJ\sqrt{J(J+1)}$ where $gJ = 3/2 + [S(S+1) - L(L+1)]/2J(J+1)$.

3. Experimental Section

All manipulations were carried out using standard Schlenk techniques, or an MBraun UniLab glovebox, under an atmosphere of dry nitrogen. Solvents were dried by passage through activated alumina towers and degassed before use. All solvents were stored over potassium mirrors (with the exception of THF which was stored over activated 4 Å molecular sieves). Deuterated solvents were distilled from potassium, degassed by three freeze-pump-thaw cycles and stored under nitrogen.

[Cs(BIPM^{TMS}H)] [13], [Ln(I)$_3$(THF)$_{3.5}$] (Ln = Nd, Gd, Dy, Er, Yb) [25], [{K(BIPM^{Mes}H)}$_2$] [6], [Ln(BIPM^{Mes}H)(I)$_2$(THF)$_2$] (Ln = La, Ce, Pr) [12,14], [La(BIPM^{TMS}H)(I)$_2$(THF)] [13], and [K(Bn)] [30] were prepared according to published procedures. [Tb(I)$_3$(THF)$_{3.5}$] was prepared according to a modified literature procedure [25]. [K(CHPh$_2$)] and [K(CPh$_3$)] were prepared by the reaction of KOBu^t/Bu^{n}Li with either diphenylmethane or triphenylmethane in hexanes, analogously to the preparation of [K(Bn)] [30]. All other materials were purchased and dried before use.

^{1}H, ^{13}C and ^{31}P NMR spectra were recorded on a Bruker 400 spectrometer operating at 400.2, 100.6 and 162.0 MHz, respectively; chemical shifts are quoted in ppm and are relative to tetramethylsilane (^{1}H, ^{13}C) or external 85% H_3PO_4 (^{31}P). FTIR spectra were recorded on a Bruker Tensor 27 spectrometer. Elemental microanalyses were carried out by Mr. Stephen Boyer at the Microanalysis Service, London Metropolitan University, UK.

Preparation of [Nd(BIPM^{TMS}H)(I)$_2$(THF)] (1a): THF (40 mL) was added to a precooled (−78 °C) mixture of [Cs(BIPM^{TMS}H)] (0.69 g, 1.00 mmol) and [Nd(I)$_3$(THF)$_{3.5}$] (0.78 g, 1.00 mmol) and the resulting mixture was allowed to warm to room temperature with stirring over 18 h. The mixture was filtered to remove CsI and all volatiles were removed *in vacuo* to afford a pale blue solid. Recrystallisation from toluene (5 mL) afforded **1a** C_7H_8 as pale blue crystals. Yield 0.42 g, 37%. Anal. Calcd for $C_{42}H_{55}I_2N_2NdOP_2Si_2$: C, 45.05; H, 4.95; N, 2.50. Found: C, 44.89; H, 4.86; N, 2.58. FTIR ν/cm^{-1} (Nujol): 1377 (m), 1366 (w), 1214 (w), 1064 (m), 992 (w), 839 (m), 770 (w), 746 (w). μ_{eff} (Evans method, 298 K, C_6D_6): 4.14 μ_B.

Preparation of [Gd(BIPM^{TMS}H)(I)$_2$(THF)] (1b): THF (40 mL) was added to a precooled (−78 °C) mixture of of [Cs(BIPM^{TMS}H)] (0.69 g, 1.00 mmol) and [Gd(I)$_3$(THF)$_{3.5}$] (0.79 g, 1.00 mmol) and the resulting mixture was allowed to warm to room temperature with stirring over 18 h. The mixture was filtered to remove CsI and the solution reduced in volume to *ca.* 5 mL. Cooling of the solution to −30 °C afforded **1b** OC_4H_8 as yellow crystals. Yield 0.56 g, 50%. Anal. Calcd for $C_{39}H_{55}GdI_2N_2O_2P_2Si_2$: C, 42.09; H, 4.98; N, 2.52. Found: C, 42.07; H, 4.95; N, 2.61. FTIR ν/cm^{-1} (Nujol): 1437 (m), 1152 (m), 1114 (s), 1057 (s), 920 (w), 839 (m), 752 (w), 745 (w), 725 (w), 712 (w), 695 (w), 660 (w), 551 (m), 504 (w). μ_{eff} (Evans method, 298 K, C_6D_6): 8.47 μ_B.

Preparation of [Tb(BIPM^{TMS}H)(I)$_2$(THF)] (1c): THF (40 mL) was added to a precooled (−78 °C) mixture of [Tb(I)$_3$(THF)$_{3.5}$] (0.79 g, 1.00 mmol) and [Cs(BIPM^{TMS}H)] (0.69 g, 1.00 mmol) and the resulting mixture was allowed to warm to room temperature with stirring over 18 h. The mixture was filtered to remove CsI and the resulting solution reduced in volume to *ca.* 5 mL. Cooling of the solution to 4 °C afforded **1c** OC_4H_8 as yellow crystals. Yield 0.38 g, 34%. Anal. Calcd for $C_{39}H_{55}I_2N_2O_2P_2Si_2Tb$: C, 42.02; H, 4.97; N, 2.51. Found: C, 41.92; H, 4.88; N, 2.62. FTIR ν/cm^{-1} (Nujol): 1437 (m), 1154 (m), 1114 (m), 1056 (m), 938 (w), 920 (w), 840 (m), 771 (m), 753 (w), 725 (w), 712 (w), 695 (w), 551 (m). μ_{eff} (Evans method, 298 K, C_6D_6): 9.46 μ_B.

Preparation of [{La(BIPM^{TMS}H)(I)(μ-I)}$_2$] (2): During our studies we prepared many samples of [La(BIPM^{TMS}H)(I)$_2$(THF)], and on one occasion, following routine workup and recrystallisation, isolated a crop of crystals of **2** $2C_7H_8$ suitable for single crystal X-ray diffraction studies. Anal. Calcd for $C_{76}H_{94}I_4La_2N_4P_4Si_4$: C, 43.78; H, 4.54; N, 2.69. Found: C, 43.77; H, 4.49; N, 2.53. FTIR ν/cm^{-1} (Nujol): 3054 (w), 1438 (m), 1415 (w), 1261 (m), 1214 (w), 1181 (w), 1116 (m), 1070 (m), 991 (m),

839 (m), 770 (m), 745 (m), 695 (m) 407 (m). The low solubility of **2** in arene solvents precluded NMR spectroscopy being performed in d_6-benzene or d_8-toluene, while dissolution in donor solvents such as d_8-THF afforded the solvated complex [La(BIPM^{TMS}H)(I)$_2$(d_8-THF)] [13].

Preparation of [Gd(BIPM^{Mes}H)(I)$_2$(THF)$_2$] (3): THF (40 mL) was added to a precooled (−78 °C) mixture of [{K(BIPM^{Mes}H)}$_2$] (0.69 g, 0.50 mmol) and [Gd(I)$_3$(THF)$_{3.5}$] (0.79 g, 1.00 mmol) and the resulting mixture was raised to room temperature with stirring over 18 h. The mixture was filtered to remove KI and all volatiles were removed *in vacuo* to afford a yellow solid. Recrystallisation from hot toluene (3 ml) afforded **3a** 3C_7H_8 as yellow crystals on cooling to 4 °C. Yield 0.72 g, 49%. Anal. Calcd for $C_{72}H_{83}GdI_2N_2O_2P_2$: C, 58.37; H, 5.65; N, 1.89. Found: C, 58.17; H, 5.57; N, 1.74. FTIR ν/cm^{-1} (Nujol): 1435 (m), 1213 (m), 1155 (m), 940 (w), 854 (m), 741 (m), 695 (m), 663 (w), 590 (w), 560 (w), 535 (m). μ_{eff} (Evans method, 298 K, C_6D_6): 8.00 μ_B.

Preparation of [Yb(BIPM^{Mes}H)(I)$_2$(THF)] (4): THF (40 mL) was added to a precooled (−78 °C) mixture of [Yb(I)$_3$(THF)$_{3.5}$] (1.21 g, 1.50 mmol) and [{K(BIPM^{Mes}H)}$_2$] (1.05 g, 0.75 mmol), and the resulting orange mixture was raised to room temperature with stirring over 18 h. The mixture was filtered to remove KI and all volatiles removed *in vacuo* to afford an orange solid. Recrystallisation from hot toluene (15 ml) afforded **4** 3C_7H_8 as orange crystals on cooling to −30 °C. Yield 0.55 g, 26%. Anal. Calcd for $C_{47}H_{51}I_2N_2OP_2Yb$: C, 49.14; H, 4.48; N, 2.44. found: C, 49.34; H, 4.58; N, 2.10. FTIR ν/cm^{-1} (Nujol): 1588 (w), 1302 (w), 1261 (m), 1214 (m), 1156 (m), 1106 (s), 1014 (s), 859 (w), 792 (w), 771 (m), 733 (m), 588 (w), 572 (w), 536 (w). μ_{eff} (Evans method, 298 K, THF): 4.29 μ_B.

Preparation of [Dy(BIPM^{Mes}H)$_2$][BIPM^{Mes}H] (5a): THF (20 mL) was added to a precooled (−78 °C) mixture of [{K(BIPM^{Mes}H)}$_2$] (0.69 g, 0.50 mmol) and [Dy(I)$_3$(THF)$_{3.5}$] (0.80 g, 1.00 mmol) and the resulting mixture was allowed to raise to room temperature with stirring over 18 h. The mixture was filtered to remove KI and volatiles removed *in vacuo* to yield a pale yellow solid. Recrystallisation from a hot toluene solution (3 ml) afforded **5a** 4C_7H_8 as yellow crystals on cooling to ambient temperature. Yield 0.32 g, 39%. Anal. Calcd for $C_{157}H_{161}DyN_6P_6$: C, 76.03; H, 6.54; N, 3.39. Found: C, 69.03; H, 6.72; N, 3.58. Despite repeated attempts more satisfactory elemental analyses could not be obtained, with the low carbon value ascribed to carbide formation [31] causing incomplete combustion during analysis. FTIR ν/cm^{-1} (Nujol): 1604 (w), 1587 (w), 1573 (w), 1435 (m), 1332 (w), 1208 (m), 1154 (w), 972 (w), 940 (w), 853 (w) 739 (m), 694 (m). μ_{eff} (Evans method, 298 K, C_6D_6): 9.81 μ_B.

Preparation of [Er(BIPM^{Mes}H)$_2$][BIPM^{Mes}H] (5b): THF (20 mL) was added to a precooled (−78 °C) mixture of [Er(I)$_3$(THF)$_{3.5}$] (1.20 g, 1.50 mmol) and [{K(BIPM^{Mes}H)}$_2$] (1.05 g, 0.75 mmol), and the resultant beige mixture allowed to raise to room temperature with stirring over 72 h. The resulting suspension was filtered and volatiles were removed *in vacuo* to afford a pink solid. The solid was washed with hexanes and recrystallised from hot toluene (15 mL) to afford **5b** 5C_7H_8 as pale yellow crystals on cooling to −30 °C. Yield 0.40 g, 31%. Anal. Calcd for $C_{164}H_{169}ErN_6P_6$: C, 76.43; H, 6.61; N, 3.26. Found: C, 76.32: H, 6.48: N, 3.37. FTIR ν/cm^{-1} (Nujol): 1261 (s), 1205 (w), 1098 (s), 1020 (s), 971 (w), 853 (w), 800 (s), 694 (m), 525 (m). μ_{eff} (Evans method, 298 K, THF): 10.25 μ_B.

Preparation of [La(BIPM^{Mes}H)(I)$_2$(TMEDA)] (6a) and [Gd(BIPM^{Mes}H)(I)$_2$(TMEDA)] (6b): Complexes **6a** and **6b** were prepared by the dissolution of [Ln(BIPM^{Mes}H)(I)$_2$(THF)] (Ln = La, Gd) in a THF/TMEDA solution followed by stirring for 18 h. In each case all volatiles were removed *in vacuo* to afford **6a** (Ln = La) or **6b** (Ln = Gd) as off-white and yellow solids, respectively, in quantitative yield. Recrystallisation of a small portion of each sample from toluene afforded crops of crystals of **6a** $3C_7H_8$ and **6b** $2C_7H_8$, respectively, suitable for single crystal X-ray diffraction studies. Data for **6a** $3C_7H_8$: Anal. Calcd for $C_{70}H_{83}I_2LaN_4P_2$: C, 58.59; H, 5.83; N, 3.90. Found: C, 58.47; H, 5.93; N, 4.02. ^{1}H NMR (298 K, C_6D_6): 1.65–2.40 (34 H, m, br, Mes-CH_3, NCH_2, NCH_3), 3.91 (1 H, t, $^2J_{PH}$= 8.8 Hz, HCP$_2$), 6.55–7.15 (24 H, m, br, Ar-H). ^{13}C{^{1}H} NMR (298 K, C_6D_6): 8.44 (CH, t, J_{PC} =138 Hz, HCP$_2$), 20.51 (Mes-CH$_3$), 20.65 (Mes-CH$_3$), 21.18 (Mes-CH$_3$), 21.44 (Mes-CH$_3$), 49.84 (NCH$_3$), 57.87 (NCH$_2$), 125.44 (Ar-CH), 125.82 (Ar-C), 129.08 (Ar-CH), 129.81 (Ar-CH), 130.61 (Ar-CH), 131.11 (Ar-CH), 131.31 (Ar-CH), 132.17 (Ar-C), 132.58 (Ar-CH), 135.53 (Ar-C), 137.64 (Ar-C), 142.50 (Ar-C). ^{31}P{^{1}H} NMR (298 K, C_6D_6): 15.70 (s). FTIR ν/cm^{-1} (Nujol): 3053 (w), 1436 (m), 1215 (m), 1158 (m), 401 (m). As full spectroscopic data supported the formulation of **6a**, full spectroscopic data was not obtained for paramagnetic **6b** which was subsequently prepared and utilised *in situ*. During our investigations we attempted many preparations of [Ln(BIPM^{Mes})(I)(THF)$_n$] (Ln = La, Ce, Pr, Gd) utilising a range of methanide precursors including [Ln(BIPM^{Mes})(I)$_2$(THF)$_2$] (Ln = La, Ce, Pr, Gd) and [Ln(BIPM^{Mes})(I)$_2$(TMEDA)] (Ln = La, **6a**; Gd, **6b**) and bases including [K(Bn)], [K{CHPh$_2$}] and [K{C(Ph)$_3$}]. In each case [Ln(BIPM^{Mes})(I)(THF)$_n$] was not isolated cleanly with mixtures of products observed in each case. The sole isolable products from these reactions were [Ln(BIPM^{Mes}H)(BIPM^{Mes})] (Ln = La, **7a**; Ce, **7b**; Pr, **7c**; Gd, **7d**), with representative preparations of **7a** and **7b** given below.

Preparation of [La(BIPM^{Mes}H)(BIPM^{Mes})] (7a): THF (15 mL) was added to a pre-cooled (−78 °C) mixture of [La(BIPM^{Mes}H)(I)$_2$(THF)] 3toluene (0.71 g, 0.49 mmol) and [K(CHPh$_2$)] (0.10 g, 0.49 mmol). The resulting orange suspension was stirred at this temperature for 10 m and then raised to room temperature with stirring over 18 h. The resulting deep brown suspension was filtered to remove KI and all volatiles removed *in vacuo* to afford a brown solid. Recrystallisation from toluene (4 mL) afforded colourless crystals of **7a** $3C_7H_8$. Yield 70 mg, 8%. Anal. Calcd for $C_{107}H_{109}LaN_4P_4$: C, 74.99; H, 6.41; N, 3.27. Found: C, 74.77; H, 6.78; N, 3.17. ^{1}H NMR (d_6-benzene, 298 K): δ 2.10 (12 H, s, br, o-Mes-CH_3), 2.26 (1 H, s, HCP$_2$), 2.37 (12 H, s, br, o-Mes-CH_3), 2.46 (6 H, s, p-Mes-CH_3), 2.57 (6 H, s, p-Mes-CH_3), 6.6-7.1 (28 H, m, br, Ar-CH), 7.37 (8 H, s, br, Ar-CH), 7.6–7.9 (12 H, m, br, Ar-CH). ^{13}C{^{1}H} NMR (d_6-benzene, 298 K): δ 4.59 (t, J_{PC} = 127 Hz, HCP$_2$), 20.67 (Mes-CH$_3$), 20.83 (Mes-CH$_3$), 21.20 (Mes-CH$_3$), 22.80 (Mes-CH$_3$), 45.06 (s, CP$_2$), 125.45 (Ar-C), 126.09 (Ar-C), 126.75 (Ar-CH), 128.09 (m, Ar-CH), 129.03 (Ar-CH), 129.64 (Ar-CH), 130.34 (Ar-CH), 130.57 (Ar-CH), 131.12 (Ar-CH), 132.17 (Ar-CH), 134.54 (Ar-C), 135.18 (br, Ar-C), 137.65 (Ar-C), 143.07 (Ar-C). ^{31}P{^{1}H} NMR (d_6-benzene, 298 K): 2.88 (CP_2), 10.51 (HCP_2). FTIR ν/cm-1 (Nujol): 1589 (w), 1261 (s), 1098 (s), 1020 (s), 854 (w), 800 (s), 663 (m).

Preparation of [Ce(BIPM^{Mes}H)(BIPM^{Mes})] (7b): THF (25 mL) was added to a pre-cooled (−78 °C) mixture of [Ce(BIPM^{Mes}H)(I)$_2$(THF)] (3.00 g, 2.54 mmol) and [K(Bn)] (0.33 g, 2.54 mmol) and the resulting mixture allowed to warm to room temperature with stirring over 20 h. The resulting mixture was filtered and all volatiles removed *in vacuo* to afford a brown solid. The solid was washed with hexanes and the resulting brown powder was recrystallised from toluene (5 mL) to afford **7b** $3C_7H_8$ as yellow crystals, as identified by unit cell check [32]. Yield 0.10 g, 3%. On standing, the hexanes extract yielded **7b** $2C_6H_{14}$ as yellow crystals suitable for single crystal X-ray diffraction studies. Yield 0.14 g, 4%. Anal. Calcd for $C_{86}H_{85}CeN_4P_4$: C, 71.80; H, 5.96; N, 3.89. Found: C, 71.67; H, 5.88; N, 3.79. FTIR ν/cm-1 (Nujol): 1261 (m), 1212 (m), 1153 (m), 1099 (s), 1019 (s), 800 (s), 721 (w), 694 (w), 519 (w). μ_{eff} (Evans method, 298 K, THF): 1.88 μ_B.

X-ray Crystallography

Crystal data for compounds **1-7d** are given in Table S2. Bond lengths and angles are listed in Table S1. Crystals were examined variously on a Bruker APEX CCD area detector diffractometer using graphite-monochromated MoKα radiation (λ = 0.71073 Å), or on an Oxford Diffraction SuperNova Atlas CCD diffractometer using mirror-monochromated CuKα radiation (λ = 1.5418 Å). Intensities were integrated from data recorded on 0.3 (APEX) or 1° (SuperNova) frames by *ω* rotation. Cell parameters were refined from the observed positions of all strong reflections in each data set. Semi-empirical absorption correction based on symmetry-equivalent and repeat reflections (APEX) or Gaussian grid face-indexed absorption correction with a beam profile correction (Supernova), were applied. The structures were solved variously by direct and heavy atom methods and were refined by full-matrix least-squares on all unique F^2 values, with anisotropic displacement parameters for all non-hydrogen atoms, and with constrained riding hydrogen geometries; U_{iso}(H) was set at 1.2 (1.5 for methyl groups) times U_{eq} of the parent atom. The largest features in final difference syntheses were close to heavy atoms and were of no chemical significance. The data-set obtained for **4** is of low quality, and while the connectivity is clear, no assessment could be made of the geometric parameters, and despite exhaustive attempts, a better data-set could not be obtained. Highly disordered solvent molecules of crystallisation in **4** and **7a–d** could not be modelled and were treated with the Platon SQUEEZE procedure [33]. Programs were Bruker AXS SMART [34] and CrysAlisPro (control) [35], Bruker AXS SAINT [34] and CrysAlisPro (integration) [35], and SHELXTL [36] and OLEX2 [37] were employed for structure solution and refinement and for molecular graphics. Crystal data have been deposited with the Cambridge Structural Database CCDC numbers 970500-970513.

4. Conclusions

We have successfully extended the series of $\{BIPM^{TMS}H\}^-$ derived rare earth methanides by the preparation of $[Ln(BIPM^{TMS}H)(I)_2(THF)]$ (Ln = Nd, Gd, Tb), **1a–c**, in moderate to good yields. Similarly, we have successfully extended the range of $\{BIPM^{Mes}H\}^-$ derived rare earth methanides with the preparation of $[Gd(BIPM^{Mes}H)(I)_2(THF)_2]$, **3**, and $[Yb(BIPM^{Mes}H)(I)_2(THF)]$, **4**, but solubility issues of $[Ln(I)_3(THF)_{3.5}]$ prevented the preparation of dysprosium and erbium analogues, with $[Ln(BIPM^{Mes}H)_2][(BIPM^{Mes}H)]$, (Ln = Dy, **5a**; Er, **5b**) being the only isolable products of these reactions. Dissolution of $[Ln(BIPM^{Mes}H)(I)_2(THF)]$ (Ln = La, Gd) in a TMEDA/THF solution afforded the TMEDA adduct complexes $[Ln(BIPM^{Mes}H)(I)_2(TMEDA)]$ (Ln = La, **6a**; Gd, **6b**) in essentially quantitative yields. Attempts to prepare $[La(BIPM^{Mes})(I)(S)n]$ and other rare earth analogues were not successful, with the reactions of $[Ln(BIPM^{Mes}H)(I)_2(S)]$ (Ln = La, Ce, Pr, Gd; S = THF or TMEDA) with a range of bases yielding mixtures of products, of which, only $[Ln(BIPM^{Mes})(BIPM^{Mes}H)]$ (Ln = La, **7a**; Ce, **7b**; Pr, **7c**; Gd, **7d**) could be identified.

Acknowledgments

We thank the Royal Society for a University Research Fellowship (S.T.L.), and the EPSRC, European Research Council, and the University of Nottingham for generously supporting this work.

References and Notes

1. Aparna, K.; Ferguson, M.; Cavell, R.G. A Monomeric Samarium Bis(Iminophosphorano) Chelate Complex with a Sm=C Bond. *J. Am. Chem. Soc.* **2000**, *122*, 726–727.
2. Liddle, S.T.; Mills, D.P.; Wooles, A.J. Bis(phosphorus-stabilized)methanide and methandiide derivatives of Group 1–5 and f-element metals. *Organomet. Chem.* **2010**, *36*, 29–55.
3. Liddle, S.T.; Mills, D.P.; Wooles, A.J. Early metal bis(phosphorus-stabilized)carbene chemistry. *Chem. Soc. Rev.* **2011**, *40*, 2164–2176.
4. Roesky, P.W. Bis(phosphinimino)methanides as ligands in the chemistry of the rare earth elements. *Z. Anorg. Allg. Chem.* **2006**, *632*, 1918–1926.
5. Rastätter, M.; Zulys, A.; Roesky, P.W. A bis(phosphinimino)methanide lanthanum amide as catalyst for the hydroamination/cyclisation, hydrosilylation and sequential hydroamination/hydrosilylation catalysis. *Chem. Commun.* **2006**, 874–876.
6. Ahmed, S.A.; Hill, M.S.; Hitchcock, P.B. Synthesis and M−Cγ Hemilability of Group 2 Bis(phosphinimino)methanides. *Organometallics* **2006**, *25*, 394–402.
7. Liddle, S.T.; Mills, D.P.; Gardner, B.M.; McMaster, J.; Jones, C.; Woodul, W.D. A Heterobimetallic Gallyl Complex Containing an Unsupported Ga-Y Bond. *Inorg. Chem.* **2009**, *48*, 3520–3522.
8. Mills, D.P.; Soutar, L.; Lewis, W.; Blake, A.J.; Liddle, S.T. Regioselective C-H Activation and Sequential C-C and C-O Bond Formation Reactions of Aryl Ketones Promoted by an Yttrium Carbene. *J. Am. Chem. Soc.* **2010**, *132*, 14379–14381.
9. Mills, D.P.; Soutar, L.; Cooper, O.J.; Lewis, W.; Blake, A.J.; Liddle, S.T. Reactivity of the Yttrium Alkyl Carbene Complex $[Y(BIPM)(CH_2C_6H_5)(THF)]$ (BIPM = $\{C(PPh_2NSiMe_3)_2\})^{2-}$: From Insertions, Substitutions, and Additions to Nontypical Transformations. *Organometallics* **2013**, *32*, 1251–1264.
10. Mills, D.P.; Lewis, W.; Blake, A.J.; Liddle, S.T. Reactivity Studies of a T-Shaped Yttrium Carbene: C–F and C–O Bond Activation and C=C Bond Formation Promoted by $[Y(BIPM)(I)(THF)_2]$ (BIPM = $C(PPh_2NSiMe_3)_2$). *Organometallics* **2013**, *32*, 1239–1250.
11. Gregson, M.; Lu, E.; McMaster, J.; Lewis, W.; Blake, A.J.; Liddle, S.T. A Cerium(IV)–Carbon Multiple Bond. *Angew. Chem. Int. Ed.* **2013**, *52*, 13016–13019.
12. Wooles, A.J.; Cooper, O.J.; McMaster, J.; Lewis, W.; Blake, A.J.; Liddle, S.T. Synthesis and Characterization of Dysprosium and Lanthanum Bis(iminophosphorano)methanide and -methanediide Complexes. *Organometallics* **2010**, *29*, 2315–2321.
13. Wooles, A.J.; Gregson, M.; Cooper, O.J.; Middleton-Gear, A.; Mills, D.P.; Lewis, W.; Blake, A.J.; Liddle, S.T. Group 1 Bis(iminophosphorano)methanides, Part 1: N-Alkyl and Silyl Derivatives of the Sterically Demanding Methanes $H_2C(PPh_2NR)_2$ (R = Adamantyl and Trimethylsilyl). *Organometallics* **2011**, *30*, 5314–5325.

14. Wooles, A.J.; Gregson, M.; Robinson, S.; Cooper, O.J.; Mills, D.P.; Lewis, W.; Blake, A.J.; Liddle, S.T. Group 1 Bis(iminophosphorano)methanides, Part 2: *N*-Aryl Derivatives of the Sterically Demanding Methanes $H_2C(PPh_2NR)_2$ (R = 2,4,6-trimethylphenyl or 2,6-diisopropylphenyl). *Organometallics* **2011**, *30*, 5326–5337.
15. Mills, D.P.; Wooles, A.J.; McMaster, J.; Lewis, W.; Blake, A.J.; Liddle, S.T. Heteroleptic $[M(CH_2C_6H_5)_2(I)(THF)_3]$ Complexes (M = Y or Er): Remarkably Stable Precursors to Yttrium and Erbium T-Shaped Carbenes. *Organometallics* **2009**, *28*, 6771–6776.
16. Wooles, A.J.; Mills, D.P.; Lewis, W.; Blake, A.J.; Liddle, S.T. Lanthanide tri-benzyl complexes: structural variations and useful precursors to phosphorus-stabilised lanthanide carbenes. *Dalton Trans.* **2010**, *39*, 500–510.
17. Liddle, S.T.; McMaster, J.; Green, J.C.; Arnold, P.L. Synthesis and structural characterisation of an yttrium-alkyl-alkylidene. *Chem. Commun.* **2008**, 1747–1749.
18. Mills, D.P.; Cooper, O.J.; McMaster, J.; Lewis, W.; Liddle, S.T. Synthesis and reactivity of the yttrium-alkyl-carbene complex $[Y(BIPM)(CH_2C_6H_5)(THF)]$ (BIPM = $\{C(PPh_2NSiMe_3)_2\}$). *Dalton Trans.* **2009**, 4547–4555.
19. Cotton, S. *Lanthanide and Actinide Chemistry*; Wiley: Chichester, UK, 2006.
20. Gamer, M.T.; Dehnen, S.; Roesky, P.W. Synthesis and Structure of Yttrium and Lanthanide Bis(phosphinimino)methanides. *Organometallics* **2001**, *20*, 4230–4236.
21. Evans, W.J.; Lee, D.S.; Rego, D.B.; Perotti, J.M.; Kozimor, S.A.; Moore, E.K.; Ziller, J.W. Expanding Dinitrogen Reduction Chemistry to Trivalent Lanthanides via the LnZ_3/Alkali Metal Reduction System: Evaluation of the Generality of Forming $Ln_2(\mu-\eta^2:\eta^2-N_2)$ Complexes via LnZ_3/K. *J. Am. Chem. Soc.* **2004**, *126*, 14574–14582.
22. Cordero, B.; Gomez, V.; Platero-Prats, A.E.; Reves, M.; Echeverria, J.; Cremades, E.; Barragan, F.; Alvarez, S. Covalent radii revisited. *Dalton Trans.* **2008**, 2832–2838.
23. Shannon, R.D. Revised effective ionic radii and systematic studies of interatomic distances in halides and chalcogenides. *Acta Crystallogr. Sect. A* **1976**, *A32*, 751–767.
24. Allen, F.H. The Cambridge Structural Database: A quarter of a million crystal structures and rising. *Acta Crystallogr. Sect. B* **2002**, *58*, 380–388.
25. Izod, K.; Liddle, S.T.; Clegg, W. A Convenient Route to Lanthanide Triiodide THF Solvates. Crystal Structures of $LnI_3(THF)_4$ [Ln = Pr] and $LnI_3(THF)_{3.5}$ [Ln = Nd, Gd, Y]. *Inorg. Chem.* **2004**, *43*, 214–218.
26. Hill, M.S.; Hitchcock, P.B.; Karagouni, S.M. A. Group 1 and 13 complexes of aryl-substituted bis(phosphinimino)methyls. *J. Organomet. Chem.* **2004**, *689*, 722–730.
27. Hill, M.S.; Hitchcock, P.B. Synthesis of a homoleptic Sm(II) bis(phosphinimino)methanide. *Dalton Trans.* **2003**, 4570–4571.
28. Evans, D.F. The determination of the paramagnetic susceptibility of substances in solution by nuclear magnetic resonance. *J. Chem. Soc. (Resumed)* **1959**, 2003–2005.
29. Vleck, J.H.V. *Theory of Electric and Magnetic Susceptibilities*; Oxford University Press: Oxford, UK, 1932.
30. Schlosser, M.; Hartmann, J. Transmetalation and Double Metal Exchange: A Convenient Route to Organolithium Compounds of the Benzyl and Allyl Type. *Angew. Chem. Int. Ed. Engl.* **1973**, *12*, 508–509.

31. Hitchcock, P.B.; Lappert, M.F.; Maron, L.; Protchenko, A.V.; Lanthanum Does Form Stable Molecular Compounds in the +2 Oxidation State. *Angew. Chem. Int. Ed.* **2008**, *47*, 1488–1491.
32. Crystals of **7b** $3C_7H_8$ were identified by unit cell check comparisons to La, Pr and Gd analogues.
33. Spek, A.L. *Platon*, version 1.17; University of Utrecht, Utrecht, The Netherlands, 2000.
34. SMART and SAINT Bruker AXS Inc.: Madison, WI, USA, 2001.
35. *CrysAlis PRO*; version1.171.35.19; Agilent Technologies: Yarnton, UK, 2010.
36. Sheldrick, G.M. A short history of SHELX. *Acta Crystallogr. Sect. A* **2008**, *A64*, 112–122.
37. Dolomanov, O.V.; Bourhis, L.J.; Gildea, R.J.; Howard, J.A.K.; Puschmann, H. OLEX2: A complete structure solution, refinement and analysis program. *J. Appl. Crystallogr.* **2009**, *42*, 339–341.

9

Supercritical Fluid Synthesis of $LiCoPO_4$ Nanoparticles and their Application to Lithium Ion Battery

Murukanahally Kempaiah Devaraju *, Quang Duc Truong, Hiroshi Hyodo, Takaaki Tomai and Itaru Honma *

Institute of Multidisciplinary Research for Advanced Materials, Tohoku University, 2-1-1, Katahira, Aoba-ku, Sendai 980-8577, Japan; E-Mails: tqduc@mail.tagen.tohoku.ac.jp (Q.D.T.); hyodo@tagen.tohoku.ac.jp (H.H.); tomai@tagen.tohoku.ac.jp (T.T.)

* Authors to whom correspondence should be addressed;
E-Mails: devaraju113@gmail.com (M.K.D.); i.honma@tagen.tohoku.ac.jp (I.H.)

Abstract: In this work, $LiCoPO_4$ nanoparticles were synthesized by supercritical fluid method using cobalt nitrate hexahydrate ($Co(NO_3)_2$ $6H_2O$) and cobalt acetate tetrahydrate ($C_4H_6CoO_4$ $4H_2O$) as starting materials. The effect of starting materials on particle morphology, size, and the crystalline phase were investigated. The as-synthesized samples were systematically characterized by XRD, TEM, STEM, EDS, BET, and TG and charge-discharge measurements. In addition, Rietveld refinement analysis was performed. The electrochemical measurements of $LiCoPO_4$ nanoparticles have shown differences in capacities depending on the starting materials used in the synthesis and the results have been discussed in this paper.

Keywords: supercritical fluid; starting materials; high voltage; nanoparticles; $LiCoPO_4$ cathode

1. Introduction

In recent years, rechargeable lithium ion batteries have received great attention due to their importance in power sources for electric and hybrid electric vehicles and they offer many more opportunities in electric and electronic domain with higher energy densities than the commercially available rechargeable batteries [1,2]. The crucial factors for the success of new cathode materials depend on the synthesis method, starting materials, particle size, cation order, and other experimental parameters [3]. Since the demonstration of electrochemical properties of $LiFePO_4$ cathode material by Padhi *et al.* [4] olivine-type $LiMPO_4$ (M = Fe, Mn, Co and Ni) and Nasicon-type $Li_3V_2(PO_4)_3$ have received great interest owing to their low cost, high reversible capacity, and good stability [5–8]. The potential of the M^{3+}/M^{2+} redox couple *versus* Li/Li^+ of $LiMPO_4$ is as follows; 3.5 V for $LiFePO_4$, 4.1 V for $LiMnPO_4$, 5.2–5.4 V for $LiNiPO_4$, and 4.8 V for $LiCoPO_4$. Among these, $LiCoPO_4$ is appealing since it offers both a flat high potential (~4.8 V *vs.* Li/Li^+) and good theoretical capacity (167 mA h/g) [9]. The cyclic and rate performances of $LiCoPO_4$ material is affected by electrolyte degradation which is due to high working voltage, low electronic conductivity, and low intrinsic conductivity. Thus far, the first cycle discharge capacities of 30–140 mA h/g with poor or moderate cyclic performances have been observed for $LiCoPO_4$ synthesized via different synthetic routes. In recent days, effort has been made to improve the cyclic performances by cationic doping and decreasing the particle size. Particularly, $LiCoPO_4$ nanoparticles have shown improved electrochemical performances, which is due to a small Li ion diffusion length. In addition, nanoparticles facilitate fast charge and mass transport because of large surface to volume ratio, as well as enhancing the close contact with the electrolyte and which could speed up the reaction kinetics.

$LiCoPO_4$ cathode material can be synthesized by various methods, such as solid-state reaction [10–17], hydrothermal synthesis [18–20], sol–gel method [11,21,22], co-precipitation [23], optical floating zone method [24], radio frequency magnetron sputtering [25], electrostatic spray deposition technique [26], and microwave heating method [27]. In addition, some approaches have been applied in order to improve the electronic conductivity of the cathode, such as carbon coating [11], making composite with carbon [10,28]. However, most of the above methods required long duration time to synthesize $LiCoPO_4$ particles, and often it is not easy to control the shape and size of the synthesized particles. Among these methods, hydrothermal and solvothermal process have proved to be beneficial in terms of cost, time and energy savings [29,30]. Very recently, we reported supercritical fluid process for controlled synthesis of plate-like $LiCoPO_4$ nanoparticles within a short reaction time [31]. Supercritical fluid method possess unique properties, such as gas like diffusivity, low viscosity, and the density of the reactants closer to that of liquid, these properties leads to the preparation of high quality materials. In recent days, varieties of electrode materials have been successfully synthesized via supercritical fluid process [32–38]. Recently, our group has reported supercritical fluid synthesis of nanosize lithium metal phosphates [30,38,39], lithium metal silicates [40–44], and flurophosphates [45].

Herein, we report supercritical fluid synthesis of $LiCoPO_4$ nanoparticles using two kinds of starting materials, such as cobalt nitrate hexahydrate ($Co(NO_3)_2$ $6H_2O$) and cobalt acetate tetrahydrate ($C_4H_6CoO_4$ $4H_2O$). The reaction was carried out at 400 °C within a short reaction period (7 min). The effect of starting materials on the crystalline structure and particles size were investigated and their electrochemical properties have been characterized.

2. Results and Discussion

2.1. One Pot Synthesis

Supercritical fluid process is a one pot synthesis process for $LiCoPO_4$ nanoparticles as shown in Figure 1. Two kinds of precursor solution were used for the synthesis of $LiCoPO_4$ nanoparticles. For the synthesis, 5 ml of precursor solution was fed into the supercritical reactor and heated at 400 °C for 7 min to obtain well-crystallized $LiCoPO_4$ nanoparticles.

Figure 1. One pot synthesis of $LiCoPO_4$ nanoparticles using two kinds of starting materials via supercritical fluid process.

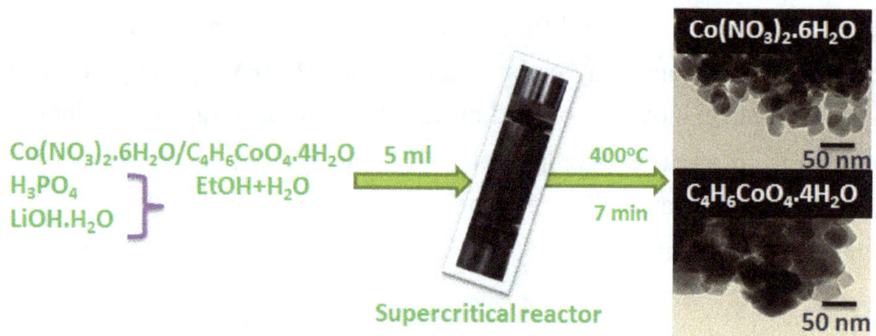

2.2. X-ray Powder Diffraction Analysis

The XRD patterns of as-synthesized $LiCoPO_4$ particles at 400 °C for 7 min using $Co(NO_3)_2\ 6H_2O$ and $C_4H_6CoO_4\ 4H_2O$ as starting materials via supercritical fluid process are shown in Figure 2.

Figure 2. XRD patterns of $LiCoPO_4$ nanoparticles synthesized using (**a**) $Co(NO_3)_2\ 6H_2O$ and (**b**) $C_4H_6CoO_4\ 4H_2O$ via supercritical fluid process; (**c**) crystal structure of $LiCoPO_4$.

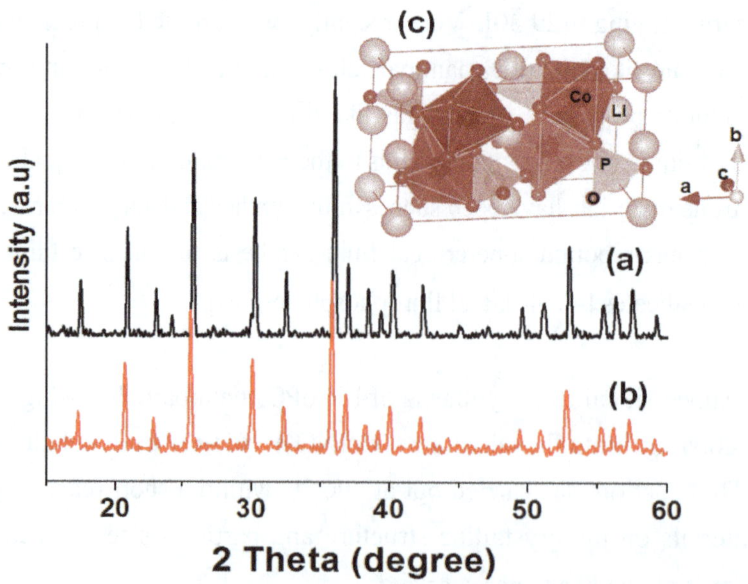

The observed diffraction peaks of both $LiCoPO_4$ particles are well matched with JCPDS file (# 00-085-0002), and belongs to orthorhombic crystal system with *Pnma* space group. The XRD patterns are in agreement with the reported pattern [23,31,46–49]. The calculated cell parameters and the Rietveld reliability factors for $LiCoPO_4$ particles synthesized using $Co(NO_3)_2$ $6H_2O$ and $C_4H_6CoO_4$ $4H_2O$ are a = 10.222Å, b = 5.9273, c = 4.7088, R_{WP} = 29.42%, R_I = 13.19% and a = 10.2046Å, b = 5.9183, c = 4.6977, R_{WP} = 27.63%, R_I = 16.22%, respectively. The cell parameters are almost consistent with the reported values [50–52]. The Rietveld refinement results obtained from XRD pattern are shown in Table 1, the results showed almost same structural parameters for $LiCoPO_4$ nanoparticles synthesized using $Co(NO_3)_2$ $6H_2O$ and $C_4H_6CoO_4$ $4H_2O$. The experimental, calculated and observed XRD pattern are shown in Figure S1(a,b) (Supplementary information), which show no significant changes and impurity phases. From the XRD results, it is confirmed that single phase of $LiCoPO_4$ was synthesized using $Co(NO_3)_2$ $6H_2O$ (Figure 1(a)) and $C_4H_6CoO_4$ $4H_2O$ (Figure 1(b)) as starting materials via supercritical fluid process. The XRD pattern did not shows much difference except variations in the peak intensity. Due to higher solubility of $Co(NO_3)_2$ $6H_2O$ in water-ethanol, the particles are well crystallized, so that the peak intensity of $LiCoPO_4$ particles synthesized using $Co(NO_3)_2$ $6H_2O$ is higher than that of $LiCoPO_4$ particles synthesized using $C_4H_6CoO_4$ $4H_2O$. The model of $LiCoPO_4$ crystal structure is shown in Figure 2(c), which shows the arrangement of cobalt octahedral and phosphor tetrahedral linked with oxygen either by corner sharing or edge sharing. It is also proved that, regardless of using different source materials in the present synthesis, supercritical fluid process is capable of producing high quality $LiCoPO_4$ nanoparticles.

Table 1. Rietveld refinement results of XRD pattern of $LiCoPO_4$ synthesized using $Co(NO_3)_2$ $6H_2O$ and $C_4H_6CoO_4$ $4H_2O$.

Formula Crystal system Space group		**$LiCoPO_4$ (Using $Co(NO_3)_2$ $6H_2O$) Orthorhombic *Pnma***					**$LiCoPO_4$ (Using $C_4H_6CoO_4$ $4H_2O$) Orthorhombic *Pnma***				
Atom	**Site**	**X**	**Y**	**Z**	**g**	**B/Å^2**	**X**	**Y**	**Z**	**g**	**B/Å^2**
Li	4a	0.5	0.5	0.5	1	0.8	0.5	0.5	0.5	1	0.8
Co	4c	0.2778	0.25	0.9819	1	1.0	0.2783	0.25	0.9801	1	1.0
P	4c	0.096	0.25	0.420	1	1.5	0.096	0.25	0.416	1	1.5
O(1)	4c	0.091	0.25	0.736	1	1.7	0.085	0.25	0.728	1	1.7
O(2)	4c	0.453	0.25	0.240	1	0.6	0.445	0.25	0.230	1	0.6
O(3)	8d	0.160	0.029	0.278	1	0.9	0.157	0.0276	0.275	1	0.9

2.3. TEM and HRTEM Analysis

The as-synthesized particles morphology, size and their crystalline nature are observed by TEM and HRTEM analysis and the results are shown in Figure 3(a–d) and Figure 3(e–h). The particles synthesized at 400 °C for 7 min using $Co(NO_3)_2$ $6H_2O$ (Figure 3(a–d)) and $C_4H_6CoO_4$ $4H_2O$ (Figure 3(e–h)) as starting materials via supercritical fluid process show mixed type morphologies, such as sphere, plate, and rod. The particles size are ranging from 20–50 nm in diameter for as-synthesized $LiCoPO_4$ particles using $Co(NO_3)_2$ $6H_2O$ (Figure 3(a,b)) and $C_4H_6CoO_4$ $4H_2O$ (Figure 3(e,f)) as starting materials. The as-synthesized particles are well distributed and softly agglomerated. The ED pattern and HRTEM images of as-synthesized particles using $Co(NO_3)_2$ $6H_2O$ and $C_4H_6CoO_4$ $4H_2O$ are shown in Figure 3(c,d) and Figure 3(g,h), respectively. The SAED images of both samples (Figure 3(c,g)) are exhibited dot like pattern, which clearly shows the single crystalline nature of $LiCoPO_4$ particles. However, as-synthesized particles using $Co(NO_3)_2$ $6H_2O$ (Figure 4(c)) show weaker diffraction than the SAED pattern of as-synthesized particle using $C_4H_6CoO_4$ $4H_2O$ (Figure 3(g)). The HRTEM images show well resolved lattice fringes exhibiting interplanar spacing along [001] and [100] axis as shown in Figure 3(d,h). The interplanar spacing along *a* axis show 1.02 nm which is consistent with the unit cell parameter of olivine structured $LiCoPO_4$ along *a*-axis (a = 10.02 A°). $LiCoPO_4$ nanoparticles synthesized using two kinds of starting materials within a short reaction time via supercritical fluid process show well crystalline nanosize particles. In order to investigate effect of temperature and time on the particles formation, synthesis was carried out using $Co(NO_3)_2$ $6H_2O$ and $C_4H_6CoO_4$ $4H_2O$ as starting materials and the results are shown in Table 2. When the reaction temperature was 300 °C, sphere like particles were observed at 7, 10, and 15 min of reaction time with particles size ranging from 10–30 nm, the XRD pattern of this particles show nearly amorphous phase and particles were covered with lot of carbonaceous materials. When the reaction temperature was increased to 350 °C for about 7, 10, and 15 min of reaction time, the obtained particles were sphere and rod like particles with 20–70 nm in diameter. The XRD of these particles show mixed phase of $LiCoPO_4$, and Li_3PO_4 phase. Single phase $LiCoPO_4$ particles were obtained only at 400 °C for about 7, 10, and 15 min of reaction time, which were the optimum experimental conditions for the formation of $LiCoPO_4$ under supercritical fluid conditions.

Figure 3. TEM, HRTEM, SAED images of $LiCoPO_4$ nanoparticles synthesized using (**a–d**) $Co(NO_3)_2$ $6H_2O$ and (**e–h**)$C_4H_6CoO_4$ $4H_2O$ as starting materials.

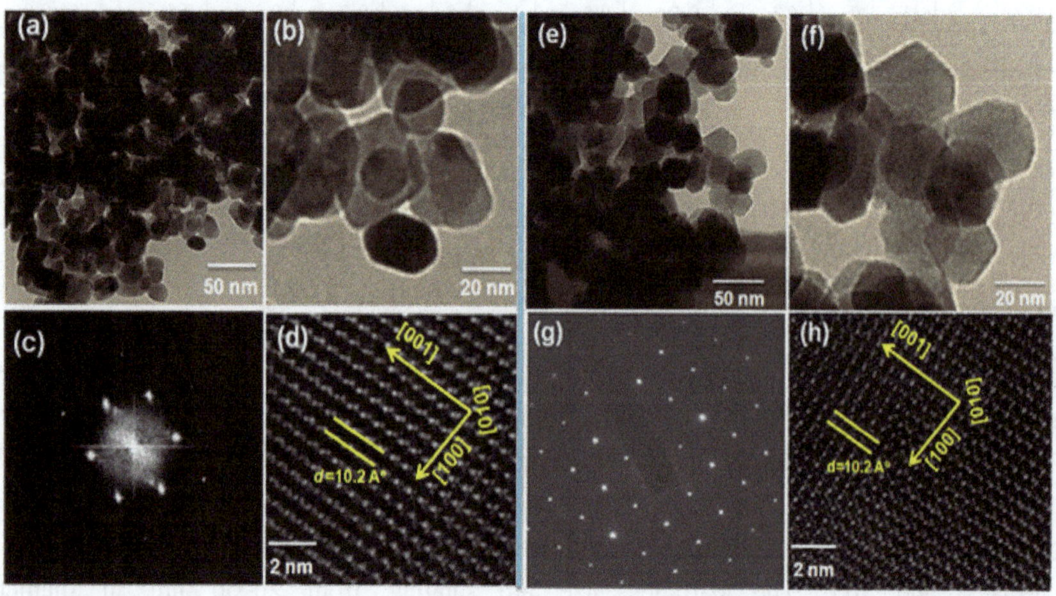

Table 2. Synthesis conditions for $LiCoPO_4$ particles using $Co(NO_3)_2\ 6H_2O$ and $C_4H_6CoO_4\ 4H_2O$ as starting materials.

Synthesis Temperature (°C)	Reaction Time (min)	Morphology	Particle size (nm)	XRD
300	7,10 &15	Sphere	10-30	nearly amorphous
350	7,10 &15	Sphere & rods	10-30	$LiCoPO_4$, Li_3PO_4
400	7,10 &15	Sphere, plates & rods	10-30	$LiCoPO_4$

2.4. Elemental Mapping

The as-synthesized particles using $Co(NO_3)_2\ 6H_2O$ and $C_4H_6CoO_4\ 4H_2O$ as starting materials were subjected to STEM analysis to verify the purity of $LiCoPO_4$ nanoparticles. Figure 4 shows the elemental mapping of both the particles.

Figure 4. STEM images of $LiCoPO_4$ nanoparticles synthesized using $Co(NO_3)_2\ 6H_2O$ and $C_4H_6CoO_4\ 4H_2O$ as starting materials.

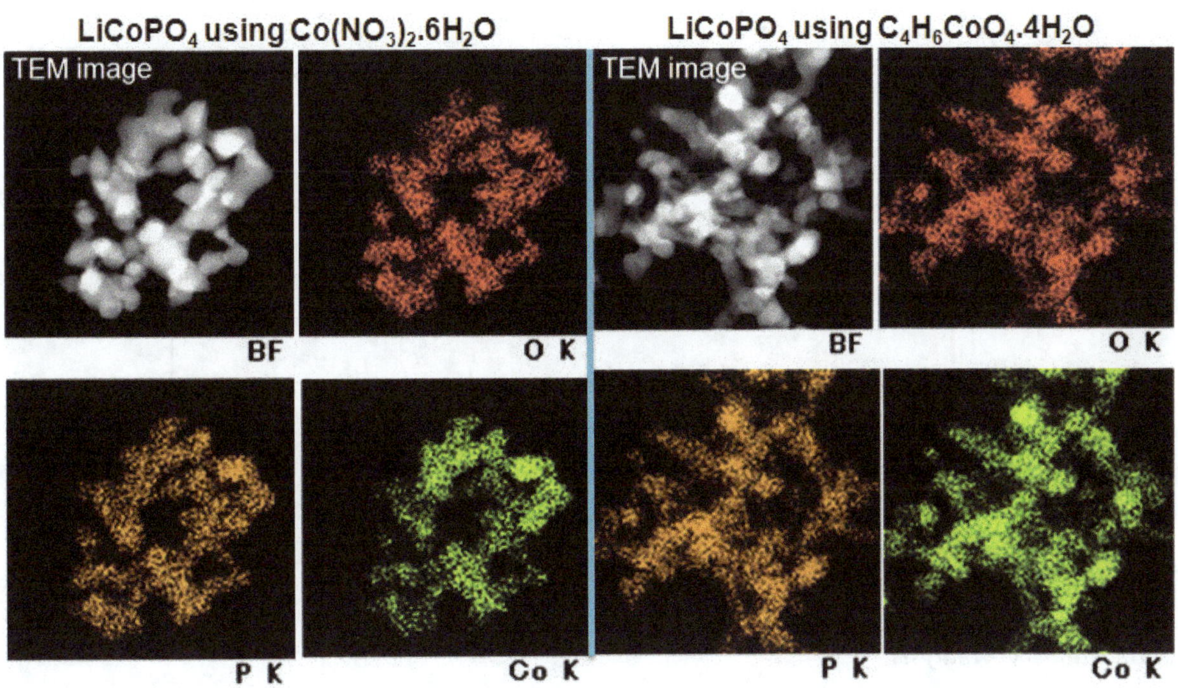

The uniform distribution of oxygen, phosphor and cobalt elements can be clearly observed in the STEM images. This result indicated that the as-synthesized particles have homogeneous elemental distribution and high purity without any impurities. The EDS spectrum of the as-synthesized using $Co(NO_3)_2\ 6H_2O$ and $C_4H_6CoO_4\ 4H_2O$ as starting materials are shown in Figure 5. The particles exhibited characteristic peaks of Co, P, and O elements in $LiCoPO_4$ nanoparticles. The peak below 0.3 eV is corresponds to the carbon peak, the intensity of C peak is higher for the $LiCoPO_4$ nanoparticles synthesized using $C_4H_6CoO_4\ 4H_2O$ as starting material due to the formation of carbon after dissociation reaction of $C_4H_6CoO_4\ 4H_2O$ at supercritical fluid conditions. However, presence of carbon in these cathode materials is beneficial in order improve the conductivity of the $LiCoPO_4$ nanoparticles during electrochemical reactions [46,47].

Figure 5. EDS of $LiCoPO_4$ nanoparticles synthesized using (**a**) $Co(NO_3)_2$ $6H_2O$ and (**b**) $C_4H_6CoO_4$ $4H_2O$ as starting materials.

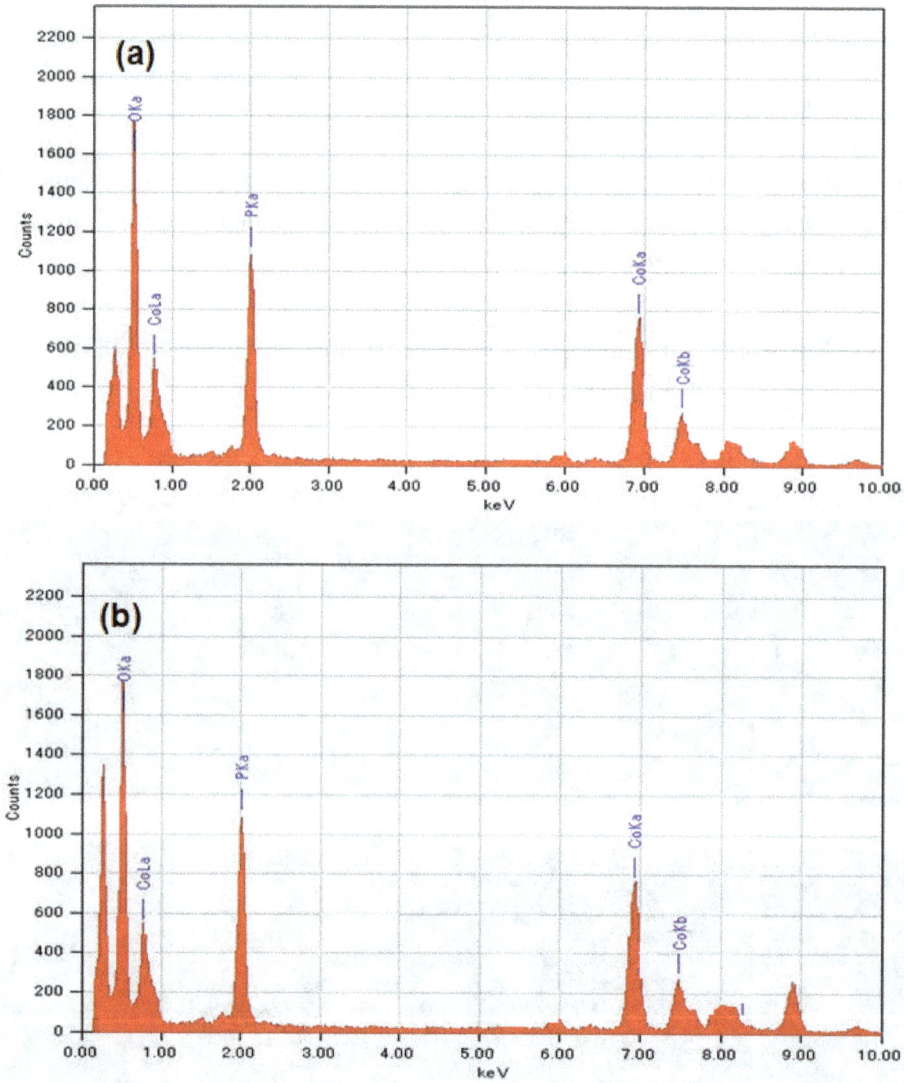

2.5. BET and TG-Analysis

The surface area of the as-synthesized particles using $Co(NO_3)_2$ $6H_2O$ and $C_4H_6CoO_4$ $4H_2O$ are analysed by BET analysis. The surface area of 14.3 m^2/g for $LiCoPO_4$ nanoparticles synthesized using $Co(NO_3)_2$ $6H_2O$ and 16.5 m^2/g for $LiCoPO_4$ particles synthesized using $C_4H_6CoO_4$ $4H_2O$ were observed. Generally, nanosize particles with higher surface area are beneficial for enhanced electrochemical reactions.

The carbon content in the as-synthesized $LiCoPO_4$ nanoparticles using $Co(NO_3)_2$ $6H_2O$ and $C_4H_6CoO_4$ $4H_2O$ are analyzed by TG analysis. The weight loss of 16 wt% for $LiCoPO_4$ nanoparticles synthesized using $Co(NO_3)_2$ $6H_2O$ and 18 wt% for $LiCoPO_4$ particles synthesized using $C_4H_6CoO_4$ $4H_2O$ were observed as show in Figure 6. The observed weight loss is the weight of carbon present in the as-synthesized particles. This carbon could be amorphous material and is not conductive, which is not useful in improving the electronic conductivity of $LiCoPO_4$ electrode materials.

Figure 6. TG analysis of $LiCoPO_4$ nanoparticles synthesized using (**a**) $Co(NO_3)_2$ $6H_2O$ and (**b**) $C_4H_6CoO_4$ $4H_2O$ as starting materials.

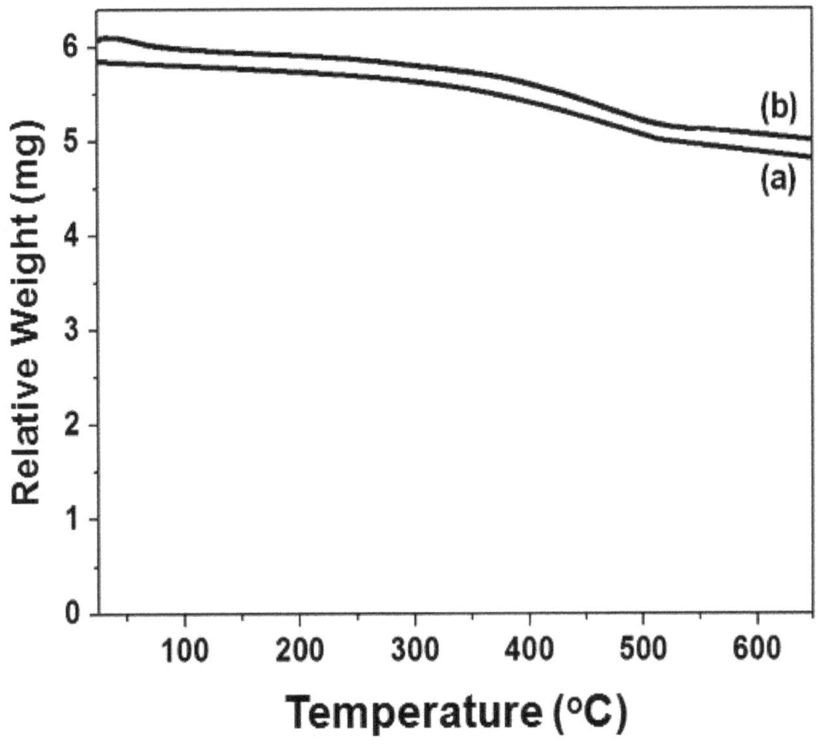

2.6. Electrochemical Performance

The electrochemical performances of as-synthesized particles using $Co(NO_3)_2$ $6H_2O$ and $C_4H_6CoO_4$ $4H_2O$ were measured by galvanostatic charge-discharge method. Figure 7 shows the discharge profiles and cyclic performance (Figure 7(c)) of $LiCoPO_4$ nanoparticles synthesized using $Co(NO_3)_2$ $6H_2O$ (Figure 7(a)) and $C_4H_6CoO_4$ $4H_2O$ (Figure 7 (b)) at 400 °C for 7 min via supercritical fluid process. The discharge measurement was carried out at 0.1C rate for $LiCoPO_4$ nanoparticles, which show a flat discharge curve around 4.75 V. The charge capacities of 150 and 165 $mAhg^{-1}$ for first and second cycle was observed for $LiCoPO_4$ nanoparticles synthesized using $Co(NO_3)_2$ $6H_2O$ as starting material. The charge capacities of 140 mA h g^{-1} and 152 mA h g^{-1} for the first and second cycle were observed for $LiCoPO_4$ nanoparticles, which were synthesized using $C_4H_6CoO_4$ $4H_2O$ as starting material. The discharge capacities of 98 mA h g^{-1} and 115 mA h g^{-1} (Figure 7 (a)) for the first cycle and 62 mA h g^{-1} and 101 mA h g^{-1} (Figure 7(b)) for the second cycles were observed. The origin of discharge capacity of 20 mA h g^{-1} around 2.5–3 V with slope like profile is might be due to the effect of particle size because of high surface area than the bulk particles, which could be responsible for secondary reactions. However, the observed discharge capacities are reasonable and even higher than that of few recently reported discharge capacities of $LiCoPO_4$ particles synthesized by different routes [20,22].

Figure 7. Discharge profiles of $LiCoPO_4$ nanoparticles synthesized using (**a**) $Co(NO_3)_2$ $6H_2O$ and (**b**) $C_4H_6CoO_4$ $4H_2O$ as starting materials; (**c**) cyclic performance.

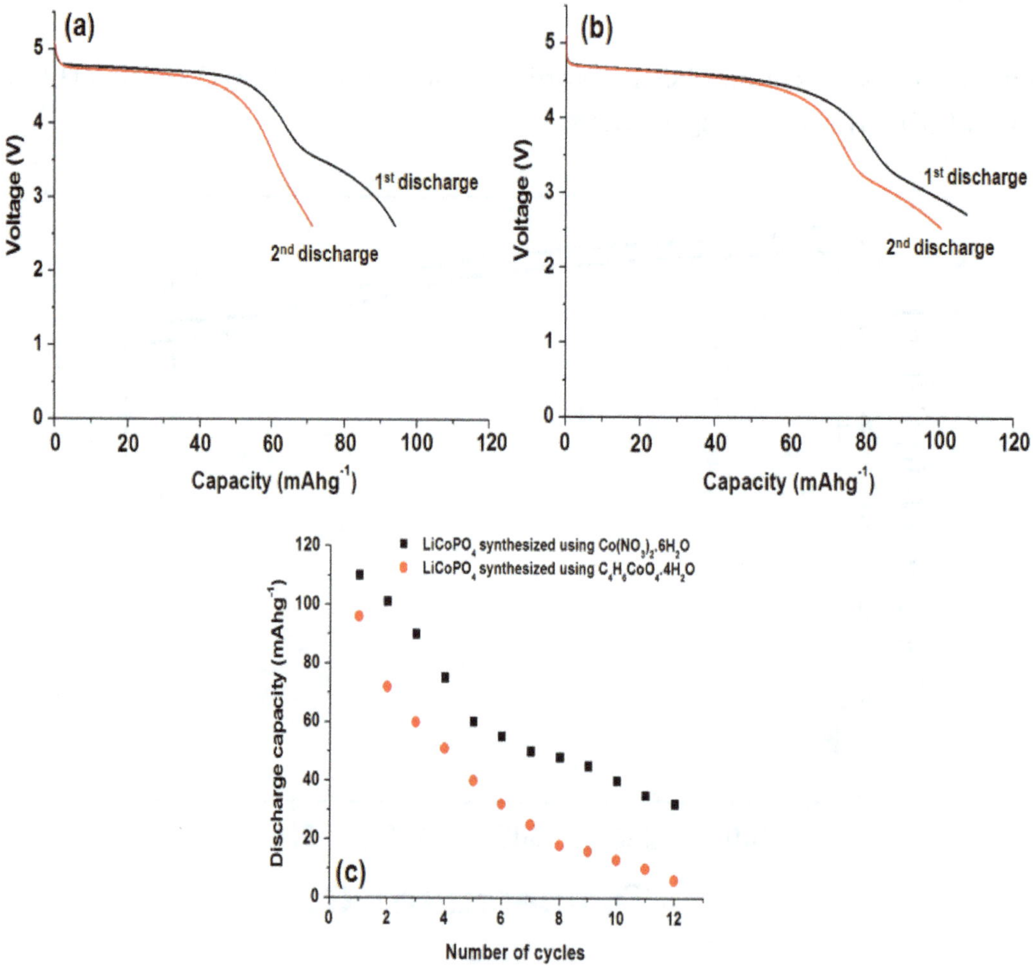

$LiCoPO_4$ nanoparticles synthesized using $C_4H_6CoO_4$ $4H_2O$ show the highest discharge capacities, which might be due to the presence of carbon residue derived from decomposition of $C_4H_6CoO_4$ $4H_2O$ at the supercritical fluid reaction. Probably, this carbon was coated onto the surface of $LiCoPO_4$ nanoparticles, which results in better electronic mobility. However, reasonable discharge capacities observed for both the $LiCoPO_4$ nanoparticles, which are due to the small particle size, good crystallinity, and due to the nature of starting materials used in the synthesis. However, poor cyclic performance was observed for $LiCoPO_4$ nanoparticles (Figure 7 (c)), which is mainly due to the instability of electrolyte at 5 V, and due to the side reactions between electrolyte and electrode and results in the formation a solid-electrolyte interface (SEI) which cause the electrolyte oxidation and lithium loss [47]. Usually, $LiCoPO_4$ cathodes suffered from poor cycling stability [49]. Another possible reason may be due to the antisite defects, where Li^+ site is occupied by Co atom upon charge process and which block the Li^+ during discharge process, so deficiency of lithium is occurred during electrochemical reaction [48]. Further, characterization is necessary to investigate on antisite defects in olivine structured cathode materials, for that ADF/ABF analysis might be very useful to locate the cobalt atoms in lithium site.

3. Experimental Section

3.1. Synthesis of $LiCoPO_4$ Nanoparticles

$LiCoPO_4$ nanoparticles were synthesized using two kinds of starting materials such as cobalt (II) nitrate hexahydrate ($Co(NO_3)_2$ $6H_2O$) and cobalt (II) acetate tetra hydrate ($C_4H_6CoO_4$ $4H_2O$). In a typical synthesis, cobalt (II) nitrate hexahydrate or cobalt (II) acetate tetra hydrate (Wako, Japan), orthophosphoric acid (o-H_3PO_4: Wako, Japan) and lithium hydroxide monohydrate (LiOH H_2O: Wako, Japan) were used in 1:1:1 molar ratio. Ascorbic acid (Wako, Japan) was used as reducing agent. The above starting materials were dissolved in mixed solvents of water and ethanol (1:1 volume ratio), the solution was stirred for 15–20 min and then 0.5 g of ascorbic acid was added to the solution. Later, the solution mixture (5 mL) was transferred to batch reactors (4 reactors, each 10 mL volume) followed by heating at 400 °C for 7 min, then reactors were quenched in cold water, the products were recovered through repeated washings using ethanol and water.

3.2. Material Characterization

The crystalline phase of the samples were characterized using powder X-ray diffraction (XRD; Rigaku RINV-2200, 40 kV and 30 mA) with CuKα radiation (λ = 1.5406 Å).The XRD patterns were analyzed by the Rietveld method using the program RIETAN-2000 [53]. The particles morphology, purity and elemental distribution was observed by transmission electron microscopy (JEOL TEM-2100F), elemental mapping and energy dispersive spectroscopy (EDS, JEM-2010F at 200 KeV). Selected area electron diffraction (SAED) and lattice fringes observation done by using high-resolution transmission electron microscopy (HRTEM, JEOL JEM 2100F, 200 kV). The surface area was measured using a Brunauer–Emmet–Teller (BET, NOVA 4200e) surface area and pore size analyzer. Thermo-gravimetric (TG) analysis was performed in air using a TG-DTA 2000S system to ascertain the exact carbon content in the $LiCoPO_4$ cathode material; the temperature was limited to 650 °C in order to avoid the oxidation of $LiCoPO_4$.

3.3. Electrochemical Performance

The electrochemical performance of $LiCoPO_4$ was investigated using coin-type cells (CR2032). The working electrodes is composed of 83 wt% $LiCoPO_4$ mixed with 10 wt.% acetylene black and 7 wt.% PTFE (poly(tetrafluoroethylene)) as a binder. The electrode paste was spread uniformly and vacuum dried for 12 h at 160 °C. Later, cathode was punched into circular discs and cut into wafers (7 mm in diameter, 0.025 mm in thickness, 3–5 mg). The tested cell was assembled inside an argon-filled glove box. For electrochemical measurements, the cell is composed of lithium metal counter, reference electrodes and a $LiCoPO_4$ positive electrode. 1 M $LiPF_6$ in a mixed solvent of ethylene carbonate (EC) and dimethyl carbonate (DMC) with 1:1 in volume ratio (Tomiyama Pure Chemical Co., Ltd., Tokyo, Japan) was used as the electrolyte. The galvanostatic charge-discharge cycling was performed between 2.5 and 5.1V *versus* Li^+/Li on multi-channel battery testers (Hokuto Denko, Japan). Current densities and specific capacities were calculated on the basis of the weight of $LiCoPO_4$ cathode used in the electrode.

4. Conclusions

$LiCoPO_4$ nanoparticles were successfully synthesized using cobalt (II) nitrate hexahydrate and cobalt (II) acetate tetrahydrate at 400 °C for 7 min via supercritical fluid process. The as-synthesized particles showed 20–50 nm in diameter with mixed type morphology observed by TEM analysis. The starting materials did not affect much on particles size and morphology of $LiCoPO_4$ particles. The phase purity of as-synthesized particles was confirmed by XRD, Rietveld analysis, STEM, and EDS analysis. Electron diffraction analysis and high-resolution image shows that, the particles are single crystalline and exhibits well-aligned crystal planes without any dislocations or defects. The experimental results show that, supercritical fluid process can synthesize single crystalline $LiCoPO_4$ nanoparticles using different kind of starting materials. Further, application of as-synthesized $LiCoPO_4$ nanoparticles as cathode part to Li-ion battery is tested by galvanostatic charge-discharge methods. $LiCoPO_4$ nanoparticles synthesized using cobalt (II) nitrate hexahydrate and cobalt (II) acetate tetra hydrate show reasonable electrochemical performances. The highest discharge capacity of 115 mA h g^{-1} for the first cycle is observed for the nanoparticles synthesized using cobalt (II) acetate tetrahydrate. However, poor cyclic performance was observed for these nanoparticles, conductive carbon coating and protection of interface would require improving the cyclic performance.

Acknowledgments

One of the authors (Murukanahally Kempaiah Devaraju) would like to acknowledge the financial support by Japan Society for Promotion of Science, Japan for their fellowship to carry out the research work.

Author Contributions

M.K.D. and I.H. conceived and designed this work. The synthesis of the material and characterization was carried out by M.K.D. and the paper was written by M.K.D., H.H performed Rietveld analysis. T.T. and Q.D.T. participated in discussion of the results.

References

1. Ceder, G.; Chiang, Y.M.; Sadoway, D.R.; Aydinol, M.K.; Jang, Y.I.; Huang, B. Identification of cathode materials for lithium batteries guided by first-principles calculations. *Nature* **1998**, *392*, 694–696.
2. Andersson, A.M.; Abraham, D.P.; Haasch, R.; MacLaren, S.; Liu, J.; Amine, K. Surface characterization of electrodes from high power lithium-ion batteries. *J. Electrochem. Soc.* **2002**, *149*, A1358–A1369.
3. Chen, J.; Vacchio, M.J.; Wang, S.; Chernova, N.; Zavalij, P.Y.; Whittingham, M.S. The hydrothermal synthesis and characterization of olivines and related compounds for electrochemical applications. *Solid State Ionics* **2008**, *178*, 1676–1693.
4. Padhi, A.K.; Nanjundaswamy, K.S.; Goodenough, J.B. Phospho-olivines as positive-electrode

materials for rechargeable lithium batteries. *J. Electrochem. Soc.* **1997**, *144*, 1188.

5. Yang, H.; Wu, X.L.; Cao, M.H.; Guo, Y.G. Solvothermal synthesis of $LiFePO_4$ hierarchically dumbbell-like microstructures by nanoplate self-assembly and their application as a cathode material in lithium-ion batteries. *J. Phys. Chem. C* **2009**, *113*, 3345–3351.
6. Delacourt, C.; Poizot, P.; Morcrette, M.; Tarascon, J.-M.; Masquelier, C. One-step low-temperature route for the preparation of electrochemically active $LiMnPO_4$ powders. *Chem. Mater.* **2004**, *16*, 93–99.
7. Fisher, C.A.J.; Prieto, V.M.H.; Islam, M.S. Lithium Battery Materials $LiMPO_4$ (M = Mn, Fe, Co, Ni): Insights into Defect Association, Transport Mechanisms and Doping Behavior. *Chem. Mater.* **2008**, *20*, 5907–5915.
8. Ren, M.M.; Zhou, Z.; Gao, X.P.; Peng, W.X.; Wei, J.P. Core–shell $Li_3V_2(PO_4)_3$@C composites as cathode materials for lithium-ion batteries. *J. Phys. Chem. C* **2008**, *112*, 5689–5693.
9. Amine, A.; Yasuda, H.; Yamachi, M. Olivine $LiCoPO_4$ as 4.8 V electrode material for lithium batteries. *Electrochem. Solid-State Lett.* **2000**, *3*, 178.
10. Rabanal, M.E.; Gutierrez, M.C.; Garcia-Alvarado, F.; Gonzalo, E.C.; Arroyo-de Dompablo, M.E. Improved electrode characteristics of olivine–$LiCoPO_4$ processed by high energy milling. *J. Power Sources* **2006**, *160*,523–528.
11. Yang, J.; Xu, J.J. Synthesis and characterization of carbon-coated lithium transition metal phosphates $LiMPO_4$ (M = Fe, Mn, Co, Ni) prepared via a nonaqueous sol-gel route. *J. Electrochem. Soc.* **2006**,*153*, A716–A723.
12. Piana, M.; Arrabito, M.; Bodoardo, S.; D'Epifanio, A.; Satolli, D.; Croce, F.; Scrosati, B. Characterization of phospho-olivines as materials for Li-ion cell cathodes. *Ionics* **2002**, *8*, 17–26.
13. Wang, D.; Wang, Z.; Huang, X.; Chen, L. Cracking causing cyclic instability of $LiFePO_4$ cathode material. *J. Power Sources* **2005**, *146*, 580–583.
14. Ruffo, R.; Mari, C.M.; Morazzoni, F.; Rosciano, F.; Scotti, R. Electrical and electrochemical behaviour of several $LiFe_xCo_{1-x}PO_4$ solid solutions as cathode materials for lithium ion batteries. *Ionics* **2007**, *13*, 287–291.
15. Wolfenstine, J.; Read, J.; Allen, J.L. Effect of carbon on the electronic conductivity and discharge capacity $LiCoPO_4$. *J. Power Sources* **2007**,*163*, 1070–1073.
16. Jang, I.C.; Son, C.G.; Yang, A.J.W.; Lee, S.M.G.; Cho, A.R.; Aravindan, V.; Park, G.J.; Kang, K.S.; Kim, W.S.; Cho, W.I.; *et al.* $LiFePO_4$ modified $Li_{1.02}(Co_{0.9}Fe_{0.1})_{0.98}PO_4$ cathodes with improved lithium storage properties. *J. Mater. Chem.* **2011**, *21*, 6510–6514.
17. Ju, H.; Wu, J.; Xu, Y. Lithium ion intercalation mechanism for $LiCoPO_4$ electrode. *Int. J. Energy Env. Eng.* **2013**, *4*, 22.
18. Yujuan, Z.; Suijun, W.; Chunsong, Z.; Dingguo, X. Synthesis and electrochemical performance of $LiCoPO_4$ micron-rods by dispersant-aided hydrothermal method for lithium ion batteries. *Rare Met.* **2009**, *28*, 117–121.
19. Huang, X.; Ma, J.; Wu, P.; Hu, Y.; Dai, J.; Zhu, Z.; Chen, H.; Wang, H. Hydrothermal synthesis of $LiCoPO_4$ cathode materials for rechargeable lithium ion batteries. *Mater. Lett.* **2005**, *59*, 578–582.
20. Kotobuki, M. Hydrothermal synthesis of carbon-coated $LiCoPO_4$ cathode material from various Co sources. *Int. J. Energy Env. Eng.* **2013**, *4*, 25.

21. Gangulibabu, D.; Bhuvaneswari, N.; Kalaiselvi, N.; Jayaprakash, P.; Periasamy, J. CAM sol–gel synthesized $LiMPO_4$ (M=Co, Ni) cathodes for rechargeable lithium batteries. *Sol–Gel Sci. Technol.* **2009**, *49*, 137–144.
22. Rajalakshmi, A.; Nithya, V.D.; Karthikeyan, K.; Sanjeeviraja, K.C.; Lee, Y.S.; Kalai Selvan, R. Physicochemical properties of V^{5+} doped $LiCoPO_4$ as cathode materials for Li-ion batteries. *J. Sol-Gel Sci. Technol.* **2013**, *65*, 399–410.
23. Sahnmukaraj, D.; Murugan, R. Synthesis and characterization of $LiNi_yCo_{1-y}PO4$ (y = 0–1) cathode materials for lithium secondary batteries. *Ionics* **2004**, *10*, 88–92.
24. Saint-Martin, R.; Franger, S. Growth of $LiCoPO_4$ single crystals using an optical floating-zone technique. *J. Cryst. Growth* **2008**, *310*, 861–864.
25. Xie, J.; Imanishi, N.; Zhang, T.; Hirano, A.; Takeda, Y.; Yamamoto, O. Li-ion diffusion kinetics in $LiCoPO_4$ thin films deposited on NASICON-type glass ceramic electrolytes by magnetron sputtering. *J. Power Sources* **2009**,*192*, 689–692.
26. Shui, J.L.; Yu, Y.; Yang, X.F.; Chen, C.H. $LiCoPO_4$-based ternary composite thin-film electrode for lithium secondary battery. *Electrochem. Commun.* **2006**, *8*, 1087–1091.
27. Li, H.H.; Jin, J.; Wei, J.P.; Zhou, Z.; Yan, J.J. Fast synthesis of core-shell $LiCoPO_4$/C nanocomposite via microwave heating and its electrochemical Li intercalation performances. *Electrochem. Commun.* **2009**, *11*, 95–98.
28. Doan, T.N.L.; Taniguchi, I. Cathode performance of $LiMnPO_4$/C nanocomposites prepared by a combination of spray pyrolysis and wet ball-milling followed by heat treatment. *J. Power Sources* **2011**,*196*, 1399–1408.
29. Devaraju, M.K.; Honma, I. Hydrothermal and solvothermal process towards development of $LiMPO_4$ (M = Fe, Mn) nanomaterials for lithium-ion batteries. *Adv. Energy Mater.* **2012**, *2*, 284–297.
30. Devaraju, M.K.; Sathish, M.; Honma, I. In *Handbook of Sustainable Engineering*; Kauffman, J., Lee, K.-M., Eds.; Springer: Dordrecht, The Netherlands, 2013; pp. 1149–1173.
31. Devaraju, M.K.; Rangappa, D.; Honma, I. Controlled synthesis of plate-like $LiCoPO_4$ nanoparticles via supercritical method and their electrode property. *Electrochem. Acta* **2012**, *85*, 548.
32. Hong, S.; Kim, S.J.; Chung, K.Y.; Chung, M.S.; Lee, B.G.; Kim, J. Continuous synthesis of lithium iron phosphate ($LiFePO_4$) nanoparticles in supercritical water: Effect of mixing tee. *J. Supercrit. Fluids* **2013**, *73*, 70–79.
33. Adschiri, T.; Lee, Y.W.; Goto, M.; Takamib, S. Green materials synthesis with supercritical water. *Green Chem.* **2011**, *13*, 1380–1390.
34. Hong, S.; Kim, S.J.; Chung, K.Y.; Lee, Y.W.; Kim, J.; Sang, B. Continuous synthesis of lithium iron phosphate nanoparticles in supercritical water: Effect of process parameters. *Chem. Eng. J.* **2013**, *229*, 313–323.
35. Nugroho, A.; Kim, S.J.; Kyung, W.C.; Chung, Y.; Kim, J. Facile synthesis of hierarchical mesoporous $Li_4Ti_5O_{12}$ microspheres in supercritical methanol. *J. Power Sources* **2013**, *244*, 164–169.
36. Nugroho, A.; Kim, S.J.; Chung, K.Y.; Kim, J. Synthesis of $Li_4Ti_5O_{12}$ in Supercritical water for Li-Ion batteries: Reaction mechanism and high-rate performance. *Electrochim. Acta* **2012**, *78*, 623–632.

37. Nugroho, A.; Chang, W.; Kim, S.J.; Chung, K.Y.; Kim, J. Superior high rate performance of core–shell $Li_4Ti_5O_{12}$/Carbon Nanocomposite synthesized by a supercritical alcohol approach. *RSC Adv.* **2012**, *2*, 10805–10808.
38. Jensen, K.; Christensen, M.; Tyrsted, C.; Brummerstedt Iversen, B. Real-time synchrotron powder X-ray diffraction study of the antisite defect formation during sub- and supercritical synthesis of $LiFePO_4$(4) and $LiFe_{1-x}Mn_x$ PO_4 nanoparticles. *J. Appl. Crystallogr.* **2011**, *44*, 287–294.
39. Rangappa, D.; Sone, K.; Kudo, T.; Honma, I. Directed growth of nanoarchitectured $LiFePO_4$ electrode by solvothermal synthesis and their cathode properties. *J. Power Sources* **2010**, *195*, 6167–6171.
40. Rangappa, D.; Sone, K.; Ichihara, M.; Kudo, T.; Honma, I. Rapid one-pot synthesis of $LiMPO_4$ (M = Fe, Mn) colloidal nanocrystals by supercritical ethanol process. *Chem. Commun.* **2010**, *46*, 7548.
41. Devaraju, M.K.; Dinesh, R.; Honma, I. Controlled synthesis of nanocrystalline Li_2MnSiO_4 particles for high capacity cathode application in lithium-ion batteries. *Chem. Commun.* **2012**, *48*, 2698–2700.
42. Dinesh, R.; Devaraju, M.K.; Tomai, T.; Unemoto, A.; Honma, I. Ultrathin nanosheets of Li_2MSiO_4 (M = Fe, Mn) as high-capacity Li-ion battery electrode. *Nano Lett.* **2012**, *12*, 1146.
43. Devaraju, M.K.; Tomai, T.; Unemoto, A.; Honma, I. Novel processing of lithium manganese silicate nanomaterials for Li-ion battery applications. *RSC Adv.* **2013**, *3*, 608–615.
44. Devaraju, M.K.; Tomai, T.; Honma, I. Supercritical hydrothermal synthesis of rod like Li_2FeSiO_4 particles for cathode application in lithium ion batteries. *Electrochem. Acta* **2013**, *109*, 75–78.
45. Devaraju, M.K.; Truong, Q.D.; Honma, I. Synthesis of Li_2CoSiO_4 nanoparticles and structure observation by annular bright and dark field electron microscopy. *RSC Adv.* **2013**, *43*, 20633–20638.
46. Devaraju, M.K.; Honma, I. One-pot synthesis of Li_2FePO_4F nanoparticles via a supercritical fluid process and characterization for application in lithium-ion batteries. *RSC Adv.* **2013**, *43*, 19849–19852.
47. Liu, J.; Conry, T.E.; Song, X.Y.; Yang, L.; Doeff, M.M.; Richardson, T.J. Spherical nanoporous $LiCoPO_4$/C composites as high performance cathode materials for rechargeable lithium-ion batteries. *J. Mater. Chem.* **2011**, *21*, 9984–9987.
48. Doan, T.N.L.; Taniguchi, I. Preparation of $LiCoPO_4$/C nanocomposite cathode of lithium batteries with high rate performance. *J. Power Sources* **2011**, *196*, 5679–5684.
49. Truong, Q.D.; Devaraju, M.K.; Tomai, T.; Honma, I. Direct observation of antisite defects in $LiCoPO_4$ cathode materials by annular dark and bright field electron microscopy. *ACS Appl. Mater. Interfaces* **2013**, *5*, 9926–9932.
50. Han, D.W.; Kang, Y.M.; Yin, R.Z.; Song, M.S.; Kwon, H.S. Effects of Fe doping onthe electrochemical performance of $LiCoPO_4$/C composites for high power-density cathode materials. *Eletrochem. Commun.* **2009**, *11*, 137–140.
51. Ehrenberg, H.; Bramnik, N.N.; Senyshyn, A.; Fuess, H. Crystal and magnetic structures of electrochemically delithiated $Li_{1-x}CoPO_4$ phases. *Solid State Sci.* **2009**, *11*, 18–23

52. Bramnik, N.N.; Nikolowski, K.; Baehtz, C.; Bramnik, K.G.; Ehrenberg, H. Phase transitions occurring upon lithium insertion−extraction of $LiCoPO_4$. *Chem. Mater.* **2007**, *19*, 908–915.
53. Izumi, F.; Ikeda, T. A Rietvled-analysis program REITAN-98 and its applications to zeolites. *Mater. Sci. Forum* **2000**, *321–324*, 198–203.

10

Syntheses of Macromolecular Ruthenium Compounds: A New Approach for the Search of Anticancer Drugs

Andreia Valente * and M. Helena Garcia

Centro de Ciências Moleculares e Materiais, Departamento de Química e Bioquímica, Faculdade de Ciências da Universidade de Lisboa, Campo Grande, 1749-016 Lisboa, Portugal;
E-Mail: lena.garcia@fc.ul.pt

* Author to whom correspondence should be addressed; E-Mail: amvalente@fc.ul.pt

---

Abstract: The continuous rising of the cancer patient death rate undoubtedly shows the pressure to find more potent and efficient drugs than those in clinical use. These agents only treat a narrow range of cancer conditions with limited success and are associated with serious side effects caused by the lack of selectivity. In this frame, innovative syntheses approaches can decisively contribute to the success of "smart compounds" that might be only selective and/or active towards the cancer cells, sparing the healthy ones. In this scope, ruthenium chemistry is a rising field for the search of proficient metallodrugs by the use of macromolecular ruthenium complexes (dendrimers and dendronized polymers, coordination-cage and protein conjugates, nanoparticles and polymer-"ruthenium-cyclopentadienyl" conjugates) that can take advantage of the singularities of tumor cells (*vs.* healthy cells).

Keywords: ruthenium; arene-derivatives; conjugates; multinuclear; organometallics; cytotoxicity; chemotherapy; anticancer; EPR effect

---

1. Introduction

There has been a growing awareness that nanotechnology applied to medicine has considerable potential to improve the treatment of several diseases. Specifically, in cancer therapy, the polymer-metal complex of oxaliplatin has been approved for the treatment of malignant tumors, including colorectal cancer, in 2003 [1].

The literature concerning macromolecules for drug delivery applications is mainly dedicated to platinum drugs [2–4], with some reports in initial study phases using copper [5], palladium [6], gold [7], tungsten [8] and ruthenium (which will be the focus of this review). Most of the approaches to the development of macromolecular drugs are based on the EPR (enhanced permeation and retention) effect, which was first identified by Maeda *et al.* in 1986 [9], and states that macromolecules selectively accumulate in tumors relative to healthy tissues, due to their defective vessel vascular structure and decreased lymphatic drainage. This passive targeting results, thus, in the passive accumulation of macromolecules in solid tumors, increasing the therapeutic index, while preventing the undesirable side effects generated by free drugs [10].This finding was a landmark in the anticancer nanomedicine field (the drug concentration in tumor can be 10 to 100 times higher than that in the blood) [11,12].

However, the Food and Drug Administration (FDA) has only approved 11 nano-therapeutics for cancer therapy so far [2–4]. One of the reasons for this situation is certainly related to the problems encountered in the development of new covalently bound macromolecule-drug conjugates, such as multi-step preparation, complicated and poor reproducibility synthesis, which often cause an inevitable loss of drug activity. It is thus of upmost importance to develop newer and simpler strategies for conjugate drugs with carriers without using such long processes. This problem can be partially overcome by using a one-step coordination strategy, as with some of the examples fully exposed in this review.

We will mostly focus on the syntheses of macromolecular ruthenium complexes (dendrimers and dendronized polymers, coordination-cage and protein conjugates, nanoparticles and polymer-"ruthenium-cyclopentadienyl" conjugates) to be used as chemotherapeutic agents in cancer treatment. Nowadays, ruthenium complexes are established alternatives to Pt-based drugs in cancer therapy, showing different mechanisms of action and spectrums of activity and possessing the potential to overcome platinum-resistance, as well as lower toxicity [13–18]. There are not yet any commercially available ruthenium drugs, even though there are two important examples that have completed Phase I clinical trials, namely KP1019 [19] ([HInd][trans-$Ru^{III}Cl_4(Ind)_2$]; Ind = indazole) and NAMI-A [20] ([HIm][trans-$Ru^{III}Cl_4$(DMSO)Im], Im = imidazole, DMSO = dimethylsulfoxide). Ruthenium is now a clear candidate for the search for new chemotherapeutics, since complexes bearing this metal core present several properties that make them attractive within this area, such as multiple oxidation states (II, III and IV) accessible under physiological conditions, favorable ligand-exchange kinetics with low toxicity, antitumor activity either *in vitro* as *in vivo*, as well as antimetastatic and intrinsic angiostatic activity. In this review, we will discuss the rationale behind the syntheses of these macromolecular ruthenium-based drugs and the coordination to metal strategies. We will finally discuss the best synthesis routes in order to shorten the gap between the huge number of papers published annually and the few compounds proceeding to clinical trials.

2. Multinuclear Approaches

The idea behind the multinuclearity in metal-conjugates is the increase of the cytotoxicity of a drug by increasing the number of metal centers. In this frame, dendrimers, coordination-cages conjugates, coordinate polymers or the coordination of a drug to a biomolecule are emerging fields in metal-based drugs, due to their multimeric scaffolds.

2.1. Ruthenium-Based Dendrimers

Dendrimers are synthetic, highly-branched macromolecules that arise from a central core and present a well-defined architecture, which can be easily tunable to present different molecular weights and sizes and can be straightforwardly functionalizable with the molecules of interest.

Scheme 1. (**a**) Tetra- and octanuclear arene ruthenium dendritic systems [21]; (**b**) Tetra- and octa-nuclear chelating neutral (*N*,*O*) and cationic (*N*,*N*) ruthenium(II) metallodendrimers [22].

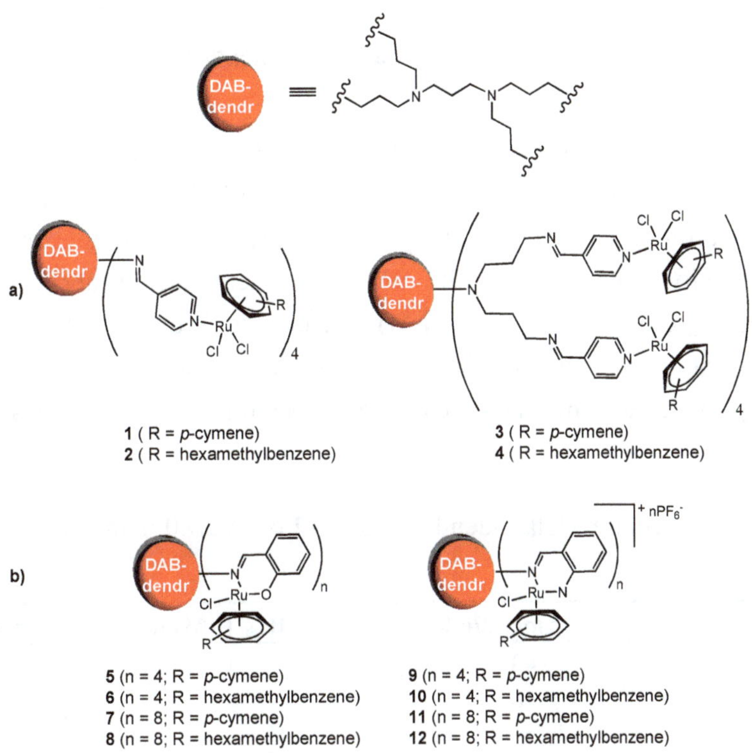

A series of first- and second-generation monodentate (*N*-donor) ruthenium(II)-arene (arene = *p*-cymene or hexamethylbenzene) metallodendrimers based on poly(propyleneimine) dendritic scaffolds was synthesized in order to exploit the EPR effect [21]. Dinuclear arene ruthenium complexes, $[Ru(arene)Cl_2]_2$, react with the dendritic scaffolds by stirring at room temperature in CH_2Cl_2 to yield the neutral tetranuclear and octanuclear ruthenium metallodendrimers (Scheme 1a) [21]. The yellow-orange products are isolated as air-stable solids in high yields (79%–98%) [21]. The complexes are soluble in most organic solvents [21]. The ^1H NMR spectra of all the compounds show broadened peaks upon complexation of the multinuclear ruthenium moieties [21].

Evidence of the coordination of the aromatic nitrogen atom to the ruthenium metal was observed through a deshielding in the doublet assigned to aromatic protons on the carbon adjacent to the pyridyl nitrogen atom [21]. This deshielding is attributed to the electron-withdrawing effects of the coordinating metal [21]. The ruthenium functionalized dendrimers were precipitated with the inclusion of solvent molecules, trapped between the dendritic arms (confirmed by elemental analysis) [21].

A second series of metallodendrimers, containing tetranuclear and octanuclear chelating neutral (*N*,*O*) and cationic (*N*,*N*) first- and second-generation ruthenium(II) arene metallodendrimers based on poly(propyleneimine) dendritic scaffolds, was also synthesized from dinuclear arene ruthenium precursors, $[Ru(arene)_2Cl_2]_2$ (arene = *p*-cymene, hexamethylbenzene), by reactions with salicylaldimine and iminopyridyl dendritic ligands in ethanol at room temperature (Scheme 1b) [22]. The *N*,*N* cationic complexes are isolated as hexafluorophosphate salts. These compounds are air-stable, the neutral complexes being soluble in most polar organic solvents and the cationic salts soluble in dimethylsulfoxide, acetone and acetonitrile [22]. The ^1H NMR spectra of all the complexes is in agreement with the proposed structures, and the infrared spectra show shifts in the $(C{=}N)_{imine}$ absorption band (~1650 cm^{-1}) to lower wavenumbers (~1620 cm^{-1}), supporting the coordination of the imine nitrogen to the ruthenium [22]. MALDI-TOF (Matrix Assisted Laser Desorption Ionization-Time of Flight) studies confirmed that all of the dendrimer end-groups were functionalized with ruthenium(II) arene moieties [22].

The cytotoxicity of Metallodendrimers **1–12** was evaluated against A2780 human ovarian cancer cells after an incubation period of 72 h using the MTT (3-(4,5-dimethylthiazol-2-yl)-2,5-diphenyltetrazolium bromide) assay (Table 1) [21,22]. The complexes showed moderate anti-proliferative activity (between 20–50 μM per metallodendrimer), with the exception of **9**, which was not cytotoxic (IC_{50} > 200 μM). As expected, there is a correlation between the nuclearity of the dendritic compound and its cytotoxicity, *i.e.*, monoruthenium compounds have only modest cytotoxicity, whereas the tetranuclear and octanuclear compounds present increasing cytotoxicities (Table 1).

Table 1. IC_{50} of ruthenium Metallodendrimers **1–12** on A2780 human ovarian cancer cells after 72 h of exposure.

Compound	IC_{50} (μM) per metallodendrimer	IC_{50} (μM) mononuclear Ru-derivative
1 [a]	43	≈100
2 [a]	40	
3 [a]	21	
4 [a]	20	
5 [b]	50	20–50
6 [b]	27	
7 [b]	22	
8 [b]	10	
9 [b]	>200	>200
10 [b]	32	
11 [b]	23	
12 [b]	4	

IC_{50} cisplatin in the same experimental conditions: [a] 1.6 μM [21]; [b] 1.5 μM [22].

Replacing *p*-cymene by hexamethylbenzene enhances the cytotoxicity of the metallodendrimers by a factor of two in the case of neutral Compounds **5–8** and by a factor of six for cationic Compounds **9–12**. In the case of Metallodendrimers **1–4**, this change does not affect the cytotoxicity.

More recently, a new class of air-stable cationic zero generation ruthenium-based metallodendrimers prepared using nitrile-functionalized poly(alkylidenamine) has been synthesized under the basis of the recognized activity of ruthenium compounds as anticancer drugs and the known stability of these dendrimers at physiological temperature [23]. Metallodendrimers **13–16** (Scheme 2) were synthesized by peripherally functionalization of the corresponding dendrimers with the ruthenium moieties, $[Ru(\eta^5\text{-}C_5H_5)(PPh_3)_2]^+$ or $[RuCl(dppe)_2]^+$ [23]. For the synthesis of **13** and **15**, solutions of the corresponding core with $[Ru(\eta^5\text{-}C_5H_5)(PPh_3)_2Cl]$ (4.5 molar ratio) and $TlPF_6$ (4.5 molar ratio) in methanol were stirred at room temperature for approximately one day [23]. The resulting mixture was filtered, the precipitate rapidly extracted with CH_2Cl_2 and the solvent evaporated, affording the tetrakis-ruthenium dendrimers, **13** or **15**, in the form of yellow products [23]. In the synthesis of **14** and **16**, the five coordinate *cis*-$[RuCl(dppe)_2][PF_6]$ complex (4.5 molar ratio) was stirred in 1,2-dichloroethane at 90 °C, for about two days [23]. After work-up, the compounds were dissolved in the minimum amount of CH_2Cl_2 and purified by re-precipitation with Et_2O, giving pure yellowish Complexes **14** or **16** [23]. All the metallodendrimers were isolated in reasonably good yields (57%–74%). Spectroscopic (UV-Vis, IR, NMR) and mass spectrometry techniques confirmed the total functionalization of the ligand cores with the respective metal complex moieties [23]. Time degradation studies of the new metallodendrimers by NMR spectroscopy in DMSO-d_6 at 37 °C showed different behaviors along time between the metallodendrimers functionalized with $[Ru(\eta^5\text{-}C_5H_5)(PPh_3)_2]^+$ or $[RuCl(dppe)_2]^+$ [23]. While Metallodendrimers **13** and **15**, functionalized with $[Ru(\eta^5\text{-}C_5H_5)(PPh_3)_2]^+$, are unstable at physiological temperature, Metallodendrimers **14** and **16**, functionalized with $[RuCl(dppe)_2]^+$, present a higher stability in DMSO. Compound **16** does not show signals of degradation with time [23]. Currently, the cytotoxic properties of Metallodendrimers **13–16** are being studied and attempts to improve their solubility and stability in aqueous media are being made [23].

Scheme 2. Metallodendrimers **13–16** [23].

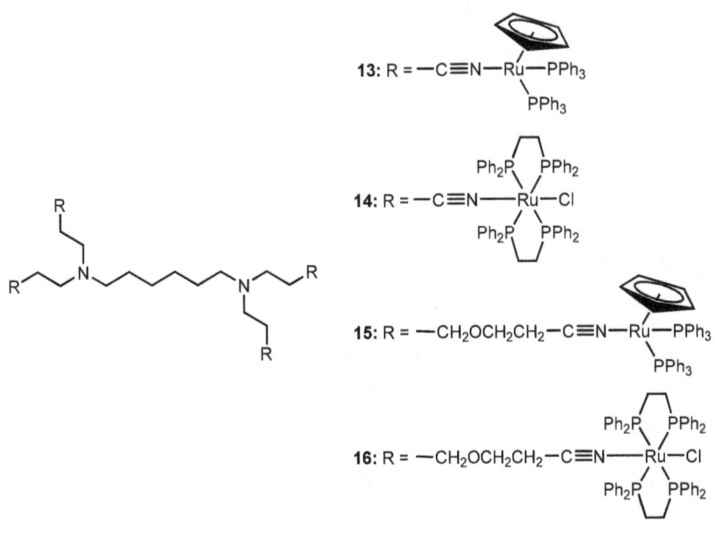

2.2. Ruthenium-Based Coordination-Cage Conjugates

Some research groups have been developing several ruthenium coordination-cages (metallaprisms, metallarectangles, metallacycles) for application in cancer chemotherapy [24–38].

The reaction of dinuclear arene ruthenium complexes [Ru_2(arene)$_2$(OO⌒OO)$_2Cl_2$] (arene = *p*-cymene, hexamethylbenzene; OO⌒OO = 2,5-dihydroxy-1,4-benzoquinonato; 2,5-dichloro-1,4-benzoquinonato) with pyrazine or bipyridine linkers (N⌒N = 4,4'-bipyridine; 1,2-bis(4-pyridyl)ethylene) in methanol, at room temperature, using $AgCF_3SO_3$ as a halide scavenger, afford the synthesis of the water soluble tetranuclear metallacyclic cations of general formula [Ru_4(arene)$_4$(N⌒N)$_2$(OO⌒OO)$_2$]$^{4+}$ (Scheme 3, Complexes **17**–**26**) [24,25,39]. The larger rectangles, incorporating the 1,2-bis(4-pyridyl)-ethylene linker, are *ca.* five times more cytotoxic ($IC_{50} \leq 6$ μM) than the 4,4'-bipyridine-containg cations ($IC_{50} \geq 30$ μM) for the A2780 human ovarian cancer cells (Table 2) [24]. The authors suggested that these variations could result from the different sized cavities, different flexibilities and different packing arrangements (observed from the X-ray diffraction of [Ru_4(hexamethylbenzene)$_4$(4,4'-bipyridine)$_2$(2,5-dihydroxy-1,4-benzo-quinonato)$_2$]$^{4+}$ **19** and [Ru_4(hexamethylbenzene)$_4$(1,2-bis(4-pyridyl)ethylene)$_2$(2,5-di-hydroxy-1,4-benzoquinonato)$_2$]$^{4+}$ **23**) [24]. In each case, the hexamethylbenzene complexes exhibit lower IC_{50} than their *p*-cymene analogues, probably due to the greater lipophilicity of the second [24].

Cationic tetra- and hexa-nuclear opened metalla-assemblies incorporating 5,15-bis(4-pyridyl)-10,20-diphenylporphyrin (Scheme 3, Complexes **25**–**26**) or 5,10,15-tris(4-pyridyl)-20-phenylporphyrin (Scheme 4, Complexes **30**–**31**) panels and dinuclear arene ruthenium clips [Ru_2(*p*-cymene)$_2$(OO⌒OO)$_2$]$^{2+}$ (OO⌒OO) = oxalate, 2,5-dioxydo-1,4-benzoquinonato, dobq) have been synthesized in the presence of $AgCF_3SO_3$ (the synthesis details are ambiguous) [25]. The compounds are sparingly soluble in water and stable in deuterated water at 60 °C for 48 h (NMR studies) [25]. All the complexes are cytotoxic against A2780 human ovarian cancer cells, the complexes with the dobq ligand (**26** and **31**) being more cytotoxic than the oxalate derivatives (**25** and **30**); this feature shows the importance of the spacer in the cytotoxic activity [25].

A solution the *N,N'*-di(4-pyridyl)-1,4,5,8-naphthalenetetracarboxydiimide donor ligand in CH_3NO_2 was added dropwise to a CH_3OH solution containing equimolar amounts of [$Ru_2(\mu^4\text{-}C_2O_4)(MeOH)_2(\eta^6$-*p*-*i*$PrC_6H_4Me)_2][O_3SCF_3]_2$, [$Ru_2(dobq)(MeOH)_2(\eta^6$-*p*-*i*$PrC_6H_4Me)_2][O_3SCF_3]_2$ or [$Ru_2(donq)(H_2O)_2(\eta^6$-*p*-*i*$PrC_6H_4Me)_2][O_3SCF_3]_2$. The mixture was stirred for 48 h at 60 °C, filtered and the solvent evaporated to dryness. Pure compounds were isolated after washing the products with diethyl ether (yield ≈ 90%) [33].UV-Vis absorption spectra presented the expected $\pi \rightarrow \pi^*$ transition bands, corresponding to the extended aromatic systems of the dipyridyl ligands [33]. The X-ray analysis of the compound bearing the [$Ru_2(dobq)(MeOH)_2(\eta^6$-*p*-*i*$PrC_6H_4Me)_2][O_3SCF_3]_2$ moiety proved the rectangle nature of this family of compounds [33]. All these compounds were tested against gastric (AGS) and colon (HCT-15) human cancer cells [33]. While compounds bearing the μ^4-C_2O_4 and dobq spacers were found to be poorly active against these cancer cell lines, Compound **27** (Scheme 3), bearing the donq linker, proved to be better than cisplatin for the AGS cells and with comparable IC_{50} values for the HCT-15 cell line [33]. These results emphasize the importance of the spacer.

Scheme 3. Metallarectangles **17–29** [24,25,33–35].

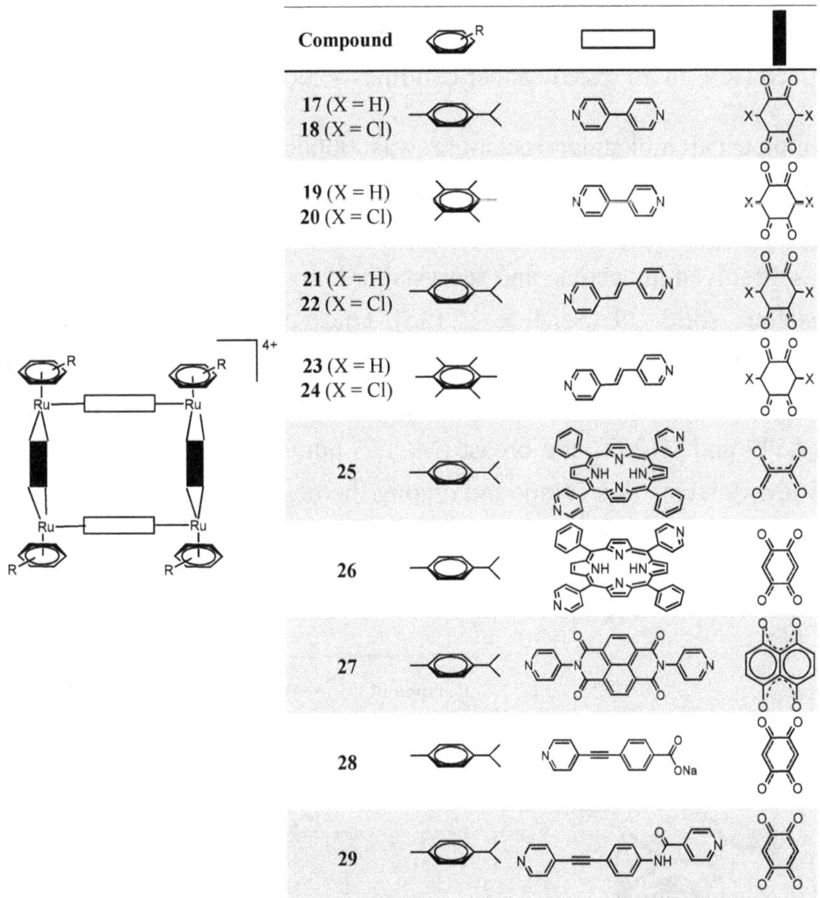

The synthesis of asymmetrical metallarectangles has also been tested [34,35]. In the first case, a solution of ambidentate donor sodium 4-(pyridin-4-ylethynyl)benzoate in MeOH was added dropwise to a solution of ruthenium acceptor $[Ru_2(donq)(H_2O)_2(\eta^6\text{-}p\text{-}i\text{PrC}_6H_4Me)_2][O_3SCF_3]_2$ in MeOH in a 1:1 molar ratio [34]. The final product was treated with diethyl ether, affording the sea-green Compound **28** (Scheme 3) [34]. Due to the asymmetry of the ambidentate donor, the formation of two isomers is possible (Scheme 5). This feature was followed by NMR spectroscopy; the four protons of the donq ligand are distinct if we have one or another isomer: for the head-to-tail isomer (**A**), two protons are oriented towards the Ru–N centers and the other two protons are oriented towards the Ru–O centers (on the same clip), thus making them chemically different; in the case of the head-to-head isomer (**B**), all the donq protons of a given clip have the same neighborhood, thus being equivalent [34]. In this frame, the head-to-head isomer presents two singlets in the ^{1}H NMR, while the head-to-tail isomer presents two sets of doublets [34]. In the particular case of Compound **28**, the ^{1}H-NMR showed the predominance of the head-to-head isomer [34]. This can be justified in terms of ring strain, as it was observed by the solid-state structure of Compound **28** [34]. The authors describe that the Ru–Ru–N angle in the head-to-tail isomer is 78.37°, while the Ru–Ru–O angle is 96.85° [34]. In this frame, the presence of two pyridyl or two carboxylate groups on the same clip would give unfitting angularities and eventually lead to the terminal ligand-ends being too close or too far apart, thus making the coordination with the second clip unfavorable [34].The *in vitro* anticancer activity of this compound was tested against lung (A549), gastric (AGS), colon (HCT-15) and liver (SK-hep-1) human cancer cell lines. The compound is active for all these cancer cell lines, in particular for the AGS human gastric cancer cell line. When the donq linker is replaced by dobq, a non-cytotoxic compound is obtained (IC_{50} in all tested cancer cell lines >200 μM) [34].

A second asymmetrical molecular rectangle was obtained by suspension of *N*-(4-(pyridine-4-ylenthynyl)phenyl)-isonicotinamide and $[Ru_2(donq)(H_2O)_2(\eta^6\text{-}p\text{-}i PrC_6H_4Me)_2][O_3SCF_3]_2$ in CH_2Cl_2/CH_3OH (1:1) for 6 h at room temperature [35]. The crude product (obtained after evaporation of the solvent) was dissolved in acetone and recrystallized by slow diffusion of diethyl ether, resulting in the green crystalline solid, **29** (Scheme 3) [35]. Once again, due to the linker asymmetry, the formation of two isomers might occur. In this case, however, it seems that there is not a preferred one (data from ^{1}H NMR show the signals of both isomers) [35]. The IC_{50} values determined for colorectal (Colo320), lung (A549 and H1299) and breast (MCF7) human cancer cell lines reveal very low IC_{50} values (0.1–10.18 μM), placing this compound among the best ruthenium-macromolecular compounds tested for *in vitro* anticancer activity [35].

Scheme 4. Structure of the Metalla-assemblies **30–34** [25,26,37].

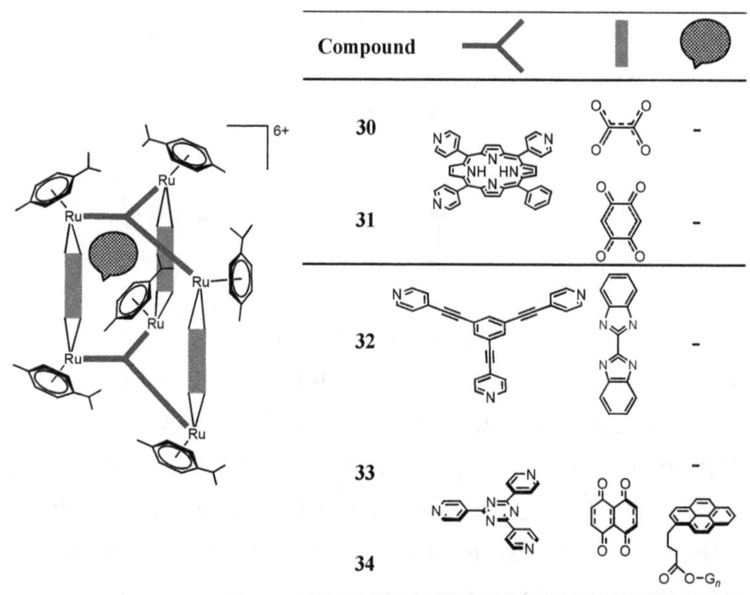

Scheme 5. Possible isomers of Compound **28**.

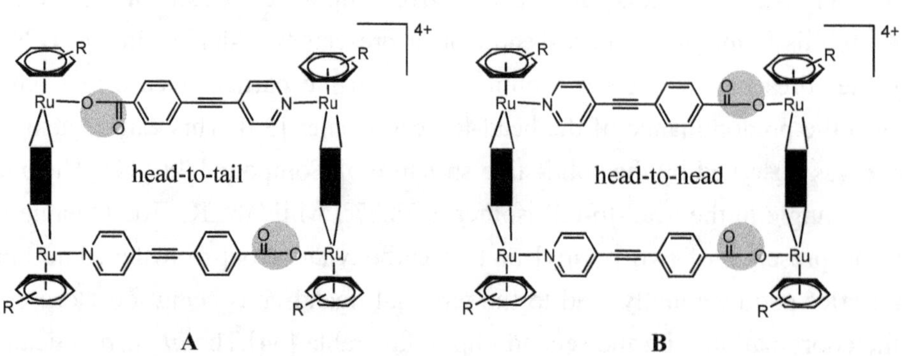

Self-assembly of the 5,10,15,20-tetra(4-pyridyl)porphyrin (tpp-H_2) tetradentate panel with the dinuclear *p*-cymene ruthenium clip, [Ru_2(*p*-cymene)$_2$(OO∩OO)Cl_2] (OO∩OO = oxalato; dobq), affords the cationic organometallic cube, [Ru_8(p-cymene)$_8$(tpp-H_2)$_2$(OO∩OO)$_4$]$^{8+}$ [26]. In addition, the reaction of the dinuclear arene ruthenium dobq clips, [Ru_2(indane)$_2$(dobq)Cl_2] and [Ru_2(nonylbenzene)$_2$(dobq)Cl_2], in MeOH for 48 h at reflux temperature, with tpp-H_2 in the presence of $AgCF_3SO_3$, affords the corresponding cationic cubes, [Ru_8(indane)$_8$(tpp-H_2)$_2$(dobq)$_4$]$^{8+}$ and [Ru_8(nonylbenzene)$_8$(tpp-H_2)$_2$(dobq)$_4$]$^{8+}$, respectively [26]. However, these octanuclear ruthenium compounds are poorly soluble in H_2O and show decreased cytotoxic activity compared with their hexanuclear homologues, showing, in this case, that there is not a direct correlation between the number of ruthenium centers *vs.* cytotoxicity.

The reaction of the [Ru_2(bis-benzimidazole)(MeOH)$_2$(η^6-*p*-*i*PrC_6H_4Me)$_2$][O_3SCF_3]$_2$ clip with 1,3,5-tris-(4-pyridylethynyl)-benzene in a 3:2 molar ratio results in self-assembled Metalla-prism **32** (Scheme 4) [37]. This compound was found to inhibit the proliferation of colon (Colo320), lung (A549 and H1299) and breast (MCF7) human cancer cell lines at low concentrations [37].

Table 2. IC_{50} of the ruthenium coordination-cage conjugates, **17–26**, **30–31** and **33–34**, on A2780 human ovarian cancer cells after 72 h of exposure.

Compound	IC_{50} (μM) per coordination-cage
17 [a]	66
18 [a]	43
19 [a]	27
20 [a]	33
21 [a]	6
22 [a]	29
23 [a]	4
24 [a]	23
25 [b]	11
26 [b]	5.6
30 [b]	3.1
31 [b]	2.1
33	3.1
34	2.4

IC_{50} cisplatin in the same experimental conditions: [a] 2 μM [24]; [b] 1.6 μM [25].

Arene-ruthenium metallacages were used to encapsulate lipophilic pyrenyl functionalized poly(benzylether) dendrimers (Scheme 4, **33** *vs.* **34**) [27,28,31]. The host-guest systems, **33**, were prepared using a two-step strategy. Firstly, the dinuclear complex, [Ru_2(*p*-cymene)$_2$(donq)Cl_2] (donq = 5,8-dioxydo-1,4-naphthoquinonato), was reacted with $AgCF_3SO_3$ in MeOH at room temperature affording the dinuclear intermediate. Then, 0.66 equivalents of 2,4,6-tris(4-pyridyl)-1,3,5-triazine (tpt) and 0.33 equivalents of the guest molecule were added, and the solution was stirred at 60 °C for 24 h to obtain the corresponding inclusion compounds. The resulting hexacationic host-guest systems are obtained in a good yield (80%) as triflate salts, [**34**][CF_3SO_3]$_6$. Both Metallacage **33** and the host-guest system, **34**, exhibit a similar cytotoxicity on the A2780 cell line (Table 2). However, these results did not clarify whether the guest is released or not after cellular internalization into the cancer cells. The cytotoxicity seemed to be inversely related with the size of the encapsulated dendrimer in the majority of cases, *i.e.*, smaller dendrimers lead to lower cytotoxicities. Replacement of donq by doaq (5,8-dioxido-1,4-anthraquinonato) or dotq (6,11-dioxido-5,12-naphthacenedionato) led to less cytotoxic compounds [30].

2.3. Ruthenium(II)-Coordinate Polymers

RAPTA-C, [RuCl$_2$(*p*-cymene)(PTA)] (PTA = 1,3,5-triaza-7-phosphaadamantane), was bounded to poly(2-chloroethyl methacrylate)(PCEMA) and poly(2-chloroethyl methacrylate-*co*-*N*-(2-hydroxypropyl) methacrylamide) (P(HPMA-CEMA)) to generate water-soluble macromolecular drugs [40]. Two strategies for the synthesis of the RAPTA-C-polymer conjugate were employed using the nitrogen groups of PTA as a site for alkylation: (a) the synthesis of the complex and direct conjugation to the polymer; or (b) attachment of PTA to the polymer and subsequent complexation with the dimer, to give the polymer-RAPTA-C conjugate [40]. The high temperature needed in the direct reaction (a) of RAPTA-C with the polymer led to the loss of the *p*-cymene ligand, invalidating this procedure [40]. In the two-step reaction (b), despite the fact that only 50% of the iodated copolymer had reacted (see Scheme 6), this method was chosen as the preferred pathway for the subsequent synthesis of the water-soluble polymer, P(HPMA$_{172}$-IEMA$_{44}$-(RAPTA-C-EMA)$_{44}$), **35** (Scheme 6) [40]. One should note that these macromolecular ruthenium complexes were only synthesized in an NMR experiment using DMSO-d$_6$ for seven days and were then recovered by dialysis [40]. The cytotoxicity of the RAPTA-C-copolymer conjugate was measured on the ovarian cancer cell line, OVCAR-3, revealing high IC$_{50}$ values (>300 μM) [40].

Scheme 6. Synthesis of complex copolymer-RAPTA-C ([RuCl$_2$(*p*-cymene)(PTA)] (PTA = 1,3,5-triaza-7-phosphaadamantane))conjugate (**35**) in DMSO-d$_6$ [40].

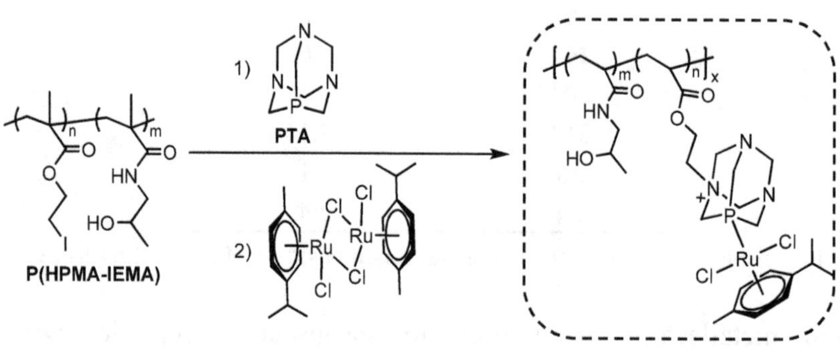

35

2.4. Ruthenium(II)-HSA Conjugates

Organoruthenium complexes of the general formula [Ru(η^6-arene)Cl(L)]Cl, where arene is 4-formylphenoxyacetyl-η^6-benzylamide and L is a cyclin-dependent kinase (Cdk) inhibitor, [3-(1H-benzimidazol-2-yl)-1H-pyrazolo[3,4-*b*]pyridines or indolo[3,2-*d*]benzazepines, were conjugated to recombinant human serum albumin (rHSA) to exploit the EPR effect (Scheme 7) [41]. The conjugation of the ruthenium moiety to modified rHSA was carried out via hydrazine bond formation according to a previous reported procedure [42]. Briefly, purified rHSA is shaken with 10 equivalents a solution of succinyl HCl terephthalic hydrazine in dimethylformamide (DMF) for 16 h at room temperature (the DMF volume did not exceed 5% (*v*/*v*)) [41]. The reaction mixture is then ultrafiltered against the conjugation buffer (100 mM MES, 2-(*N*-morpholino)ethanesulfonic acid, 0.9% NaCl,pH 6.0), and the concentration determined using the Bradford assay [41]. The modified protein solution is added to solutions of several complexes in order to achieve a 3:1 metal/protein ratio and shaken for 6 h at room temperature [41]. Afterwards, thc protein mixture solution is desalted and restored in PBS (phosphate buffered saline) [41]. MALDI-TOF-MS analysis showed that the obtained samples correspond most likely to the presence of about two bound ruthenium moieties per protein [41].

Scheme 7. Ruthenium-recombinant human serum albumin (rHSA) conjugates **36–40** [41]. Coordinating nitrogen in bold.

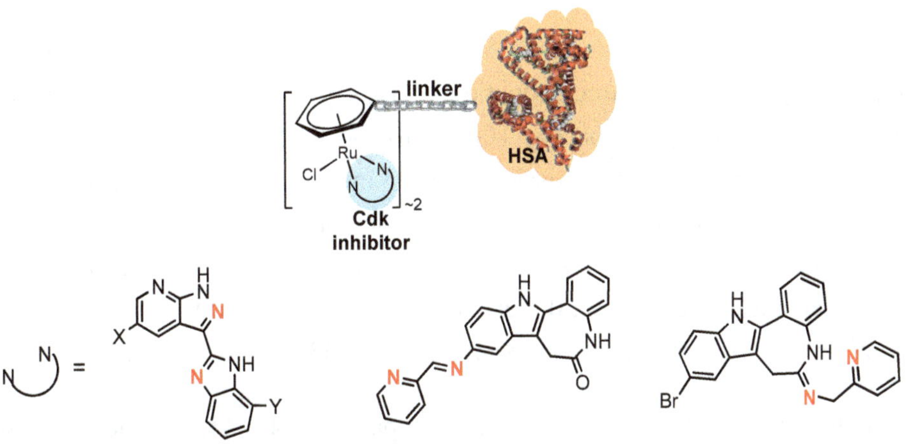

36 (X = H, Y = H)
37 (X = Br, Y = H)
38 (X = Br, Y = CH_2OCH_3)

39

40

Table 3. IC_{50} of the different ruthenium-rHSA Conjugates **36–40** on A2780 human ovarian cancer cells after 72 h exposure.

Compound	IC_{50} (μM) per compound
36	>200
36-rHSA	45
37	>200
37-rHSA	43
38	>200
38-rHSA	46
39	>100
39-rHSA	49
40	85
40-rHSA	26

The high molecular weight compounds (**36-rHSA** to **40-rHSA**), together with their non-protein complexes (**36–40**) were evaluated *in vitro* in an ovarian carcinoma cell line (A2780). From Table 3, one can observe that the complexes alone are not cytotoxic. When coordinated with rHSA, there is an increase in cytotoxicity. One should not neglect that the cyclin-dependent kinase (Cdk) inhibitor ligands are much more cytotoxic for CH1 (human ovarian carcinoma), SW480 (human colon carcinoma) and A549 (human non-small cell lung carcinoma) cell lines than their corresponding ruthenium Complexes **36–40**. Unfortunately, data for the A2780 cancer cell line is not provided.

3. Mononuclear Approaches

3.1. Ruthenium Nanoparticles

Nanoparticles find increasing applications in medicinal chemistry as drug delivery agents, medicinal imaging tools or as diagnostic agents. In addition, nanoparticles can also benefit from the enhanced permeability and retention effect and can be tunable to present specific properties.

Ruthenium(0) nanoparticles stabilized by a long-chain *N*-ligand derived from isonicotinic acid (**L**) have been prepared by the solvent-free reduction of $[Ru(\eta^6\text{-}C_6H_6)(L)Cl_2]$ in a magnetically stirred stainless-steel autoclave with H_2 (50 bar) at 100 °C for 64 h (**41**) [43]. The mean particle size was found to be 8.5 nm (established by transmission electron microscopy, TEM), which is relatively large. Smaller ruthenium nanoparticles stabilized by the isonicotinic ester ligand **L** were obtained by reducing $[Ru(\eta^6\text{-arene})(H_2O)_3]SO_4$ in ethanol in the presence of one equivalent of **L** in a magnetically stirred stainless-steel autoclave under 50 bar pressure of H_2 at 100 °C for 14 h (Table 4; **42**: arene = C_6H_6; **43**: arene = *p*-$MeC_6H_4Pr^i$; **44** arene = C_6Me_6) [43].

The *in vitro* cytotoxicity of the **L**-stabilized Ru nanoparticles, **41–44**, and their corresponding small molecules (*i.e.*, complexes of the general formula $[Ru(\eta^6\text{-arene})(L)Cl_2]$) havebeen studied in the A2780 ovarian cancer cell line using the MTT assay. While the small molecules exhibit a good to moderate cytotoxicity, the nanoparticles exhibit only moderate cytotoxicity in the studied ovarian cancer cell line, with the exception of the*p*-cymene derived system, **43**, which was unusually inactive (Table 4). For **41**, **42** and **44**, neither the nanoparticles size nor the nature of the ligands in the precursor complex appear to have an effect on cytotoxicity, since all three compounds exhibit similar IC_{50} values (29–39 μM). It is plausible to think that the *in vitro* activity of the complexes and nanoparticles is mainly due to the isonicotinic ester ligand **L**, since it presents, itself, a high cytotoxicity (IC_{50} of **L** in A2780 after 72 h of exposure = 5 μM).

Table 4. IC_{50} of the ruthenium (0) nanoparticles, **41–44**, on A2780 human ovarian cancer cells after 72 h of exposure.

Compound	Ru nanoparticles	Mean size (nm)	IC_{50} (μM)
41	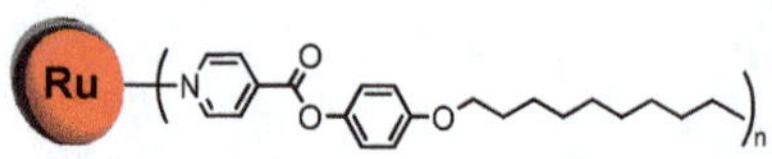	8.5	29
42		2.8	34
43		2.3	>200
44		2.2	39

3.2. Polymer-"Ruthenium-Cyclopentadienyl" Conjugate

$Ru^{II}Cp$ (Cp = cyclopentadienyl) low molecular weight drugs [44–52] are currently promising candidates in the search fornew chemotherapeutics, due to the excellent cytotoxic results against human colon adenocarcinoma (LoVo and HT29), pancreatic (Mia PaCa), promyelocytic leukemia (HL-60), breast (MDAMB231 and MCF7), prostate (PC3) and ovarian (A2780 and A2780cisR) cancer cell lines (IC_{50} values in the nano- to micro-molar range) [44–52]. Importantly, the $[CpRu(P)(bpy)]^+$ family (P = phosphane coligand, bpy = bipyridine) presents very good stability in an aqueous environment. This feature of these complexes prompted the search fornew polymer-metal complexes using the same organometallic core as potential anticancer agents.

Conceptually, polymer conjugates share several features with other macromolecular approaches (liposomes, dendrimers, nanotubes and nanoparticles), but they have the added benefit of the synthetic chemical versatility that allows the tailoring of the molecular weight and also the adding of biomimetic features [53]. In this frame, the unprecedented synthesis of **45** (Scheme 8), and its preliminary *in vitro* results have been recently published [54]. $[RuCp(P)(bpyPLA)]^+$ (RuPMC; Cp = η^5-C_5H_5, P = triphenylphosphane and bpyPLA = 2,2'-bipyridine-4,4'-D-glucose end-capped polylactide) has been synthesized in a good yield by halide abstraction from $[Ru(\eta^5\text{-}C_5H_5)(PPh_3)_2Cl]$ with silver $AgCF_3SO_3$, under reflux for 3 h in CH_2Cl_2 in the presence of the bpyPLA macroligand. The molecular weight of the bpyPLA macroligand can be easily tuned by playing with the monomer/initiator ratio.

A degradation study, by UV-visible spectroscopy, of the RuPMC performed in order to infer the polymer hydrolysis at physiological and at tumor cell pH (pH = 7.4 and 5, respectively) showed that RuPMC is stable over a period of at least 72 h in an aqueous environment at physiologic pH, while at acidic pH, some degradation of the PLA is observed. Such behavior suggests a pH-dependent degradation, which is important considering drug delivery, since the measured pH of most solid tumors range from pH 5.7 to pH 7.2, while in blood it remains well-buffered and constant at pH 7.4 [55]. Accordingly, this feature of the polymer degradation discards the need for a biodegradable linker and provides the opportunity for site-specific drug delivery, mainly within endosomal/lysosomal compartments, where the pH approaches 4.5–6.0 [56].

This polymer-"ruthenium-cyclopentadienyl" conjugate **45** is cytotoxic against human MCF7 and MDAMB231 breast and A2780 ovarian adenocarcinoma, revealing IC_{50} values in the micromolar range (IC_{50} = 3.9, 3.8 and 1.6 μM, respectively). ICP-MS (inductively coupled plasma mass spectrometry) studies showed that the Ru-polymer conjugate enters the MCF7 estrogen receptor positive cancer cells and is retained *ca.* 50% in the nucleus, foreseeing its application as a therapeutic agent in, for example, hormone-responsive cancers. On the contrary, its Ru-precursor (**TM34**, $[Ru^{II}(\eta^5\text{-}C_5H_5)(bipy)(PPh_3)]^+$, in Scheme 8) is mainly found in the membrane (*ca.* 80%), forecasting different mechanisms of cellular uptake and of cell death for these two compounds bearing the same cytotoxic fragment.

Direct comparison of the IC_{50} values between RuPMC and its low molecular weight parent drug, **TM34**, reveals a decrease on the cytotoxicity of RuPMC (3.9 *vs.* 0.29 μM for MCF7). However, one should not neglect the potential effect that the prolonged plasma half-life of the RuPMC could have on the improvement of the chemotherapeutic efficacy, allowing a positive final outcome, as has been described for many platinum-related compounds [57–60]. This new RuPMC seems to be a viable candidate for the intended drug-delivery application, yet further studies are needed to prove its higher *in vivo* accumulation in cancer cells.

Scheme 8. D-Glucose end-capped polylactide ruthenium-cyclopentadienyl (RuPMC, **45**) and **TM34**.

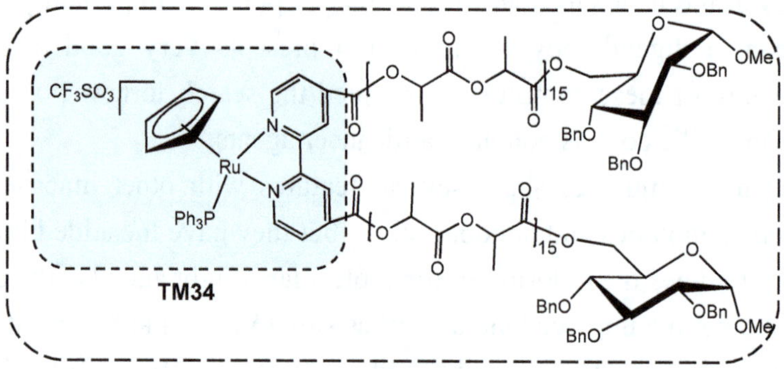

RuPMC
45

4. Conclusions

Regardless of the advances in the area of macromolecular compounds, there is still the need to develop high molecular weight and biodegradable carriers that can better exploit EPR-mediated tumor targeting. There is an urgency to move away from heterogeneous carriers towards better defined structures. In this frame, several strategies are being developed, which can be seen as a step forward to this end. An approach that has attracted much attention lately is a synergic effect between the EPR effect and the introduction of increasing nuclearity, which is expected to strengthen the cytotoxicity, while also raising the selectivity towards the cancer cells (sparing the healthy ones). Metallodendrimers have thus appeared as a promising option, since they combine the features of monodisperse nanoscale geometry with high end-group density at their surface. Furthermore, other supramolecular assemblies, like ruthenium-based coordination-cage conjugates and ruthenium-rHSA conjugates, showed cytotoxicity against several cancer cell lines. However, it seems that there is not always a direct correlation between the nuclearity and the cytotoxicity of the compounds, possibly due to solubility issues or to over-positively charged complexes that might originate retention at the cell membrane. Indeed, the only reported case of a conjugate bearing approximately forty four ruthenium centers per molecule (P(HPMA$_{172}$-IEMA$_{44}$-(RAPTA-C-EMA)$_{44}$) showed no benefit in terms of cytotoxicity towards the ovarian cancer cell line, OVACAR-3 (IC_{50} > 300 μM), compared to RAPTA-C ($IC_{50} \approx 200$ μM).

Establishing structure-activity relationships is of primordial importance, since small changes in the chemical structure might dictate significant cytotoxicity differences. This is the case of coordination-cages, where both linkers and arene ligands have a strong influence on the cytotoxicity, probably due to the different size cavities, flexibilities and packing, as well as different lipophilicities.

Some of the problems encountered in the development of new covalently bound metal-conjugates lie on the loss of drug activity. This is the case of the reported ruthenium nanoparticles, where the macromolecular drugs lead to a marked decrease in the cytotoxic properties of the low molecular weight compounds. It is thus imperative to develop simpler strategies for the coordination of drugs to the carriers. One good example was observed on the one-step coordination strategy in ruthenium cyclopentadienyl derivatives, where the cytotoxicity of the final polymer-ruthenium conjugate was maintained in the low micromolar range.

Ruthenium-conjugates seem to be a promising alternative, although many studies must still be done. Most of the cytotoxic studies were performed mainly over one cancer cell line (namely, human ovarian A2780), and there is still the need to present *in vivo* studies in order to have a proof of concept, *i.e.*, if these new macromolecular compounds are indeed better than their low molecular weight parental drugs by the so-called EPR effect. Furthermore, studies revealing the stability and speciation of these metal-conjugates in an aqueous environment and blood are mandatory.

In the chemical point of view, creating carriers that degrade under acidic conditions to trigger the drug release, by the slightly acidic tumor environment, is seen as a good strategy, already tested with good results in platinum drugs. Furthermore, this effect can also be achieved after the internalization by cancer cells, resulting in the accumulation of the polymer in the acidic endosomes and lysosomes. Finally, it is also expected that receptor-targeting ligands will lead to improved tumor targeting through the EPR effect. In this frame, innovative chemical reactions leading to "smart drugs" are powerful tools for the search of new chemotherapeutics presenting chemical diversity and original architectures.

Acknowledgments

The authors thank the Portuguese Foundation for Science and Technology (FCT) within the scope of projects PEst-OE/QUI/UI536/2011, PTDC-QUI-QUI-118077-2010 and FCT Investigator Programme.

References

1. Chau, I.; Cunningham, D. Oxaliplatin for colorectal cancer in the united states: better late than never. *J. Clin. Oncol.* **2003**, *21*, 2049–2051.
2. Vicent, M.J.; Duncan, R. Polymer conjugates: Nanosized medicines for treating cancer. *Trends Biotechnol.* **2006**, *24*, 39–47.
3. Vinditto, V.J.; Szoka, F.C., Jr. Cancer nanomedicines: So many papers and so few drugs! *Adv. Drug Deliv. Rev.* **2013**, *65*, 80–88.
4. Wang, R.; Billone, P.S.; Mullett, W.M.J. Nanomedicine in action: An overview of cancer nanomedicine on the market and in clinical trials. *Nanomaterials* **2013**, *2013*, 629681.
5. Zhao, X.; Loo, A.C.J.; Lee, P.P.-F.; Tan, T.T.Y.; Chu, C.K. Synthesis and cytotoxic activities of chloropyridylimineplatinum(II) and chloropyridyliminecopper(II) surface-functionalized poly(amido-amine) dendrimers. *J. Inorg. Biochem.* **2010**, *104*, 105–110.

6. Ahamad, T.; Mapolie, S.F.; Alshehri, S.M. Synthesis and characterization of polyamide metallodendrimers and their anti-bacterial and anti-tumor activities. *Med. Chem. Res.* **2012**, *21*, 2023–2031.
7. Robilotto, T.J.; Alt, D.S.; von Recum, H.A.; Gray, T.G. Cytotoxic gold(I)-bearing dendrimers from alkyne precursors. *Dalton Trans.* **2011**, *40*, 8083–8085.
8. Hurley, A.L.; Mohler, D.L. Organometallic photonucleases: Synthesis and DNA-cleavage studies of cyclopentadienyl metal-substituted dendrimers designed to increase double-strand scission. *Org. Lett.* **2000**, *2*, 2745–2748.
9. Maeda, H.; Matsumura, Y. A new concept for macromoecular therapeutics in cancer chemotherapy: mechanisms of tumoritropic accumulation of proteins and the antitumor agent smancs. *Crit. Rev. Ther. Drug. Carrier Syst.* **1986**, *46*, 6387–6392.
10. Kopecek, J., Kopeckova, P., Minko, T.; Lu Z.R. HPMA copolymer-anticancer drug conjugates: Design, activity, and mechanism of action. *Eur. J. Pharm. Biopharm.* **2000**, *50*, 61–81.
11. Maeda, H.; Bharate, G.Y.; Daruwalla, J. Polymeric drugs for efficient tumor-targeted drug delivery based on EPR-effect. *Eur. J. Pharm. Biopharm.* **2009**, *71*, 409–419.
12. Duncan, R.; Dimitrijevic, S.; Evagorou, E.G. The role of polymer conjugates in the diagnosis and treatment of cancer. *STP Pharma Sci.* **1996**, *6*, 237–263.
13. Ang, W.H.; Dyson, P.J. Classical and non-classical ruthenium-based anticancer drugs: Towards targeted chemotherapy. *Eur. J. Inorg. Chem.* **2006**, *20*, 4003–4018.
14. Levina, A.; Mitra, A.; Lay, P.A. Recent developments in ruthenium anticancer drugs. *Metallomics* **2009**, *1*, 458–470.
15. Bruijnincx, P.C.A.; Sadler, P.J. New trends for metal complexes with anticancer activity. *Curr. Opin. Chem. Biol.* **2008**, *12*, 197–206.
16. Süss-Fink, G. Arene ruthenium complexes as anticancer agents. *Dalton Trans.* **2010**, *39*, 1673–1688.
17. Antonarakis, E.S.; Emadi, A. Ruthenium-based chemotherapeutics: Are they ready for prime time? *Cancer Chemother. Pharmacol.* **2010**, *66*, 1–9.
18. Bergamo, A.; Gaiddon, C.; Schellens, J.H.M.; Beijnen, J.H.; Sava, G. Approaching tumour therapy beyond platinum drugs: Status of the art and perspectives of ruthenium drug candidates. *J. Inorg. Biochem.* **2012**, *106*, 90–99.
19. Hartinger, C.G.; Jakupec, M.A.; Zorbas-Seifrieda, S.; Groessl, M.; Egger, A.; Berger, W.; Zorbas, H.; Dyson, P.J.; Keppler, B.K. KP1019, a new redox-active anticancer agent—Preclinical development and results of a clinical phase I study in tumor patients. *Chem. Biodivers.* **2008**, *5*, 2140–2155.
20. Alessio, E.; Mestroni, G.; Bergamo, A.; Sava, G.; Alessio, E.; Mestroni, G.; Bergamo, A.; Sava, G. Ruthenium antimetastatic agents. *Curr. Topics Med. Chem.* **2004**, *4*, 1525–1535.
21. Govender, P.; Antonels, N.C.; Mattsson, J.; Renfrew, A.K.; Dyson, P.J.; Moss, J.R.; Therrien, B.; Smith, G.S. Anticancer activity of multinuclear arene ruthenium complexes coordinated to dendritic polypyridyl scaffolds. *J. Organomet. Chem.* **2009**, *694*, 3470–3476.
22. Govender, P.; Renfrew, A.K.; Clavel, C.M.; Dyson, P.J.; Therrien, B.; Smith, G.S. Antiproliferative activity of chelating *N,O*- and *N,N*-ruthenium(II) arene functionalised poly(propyleneimine) dendrimer scaffolds. *Dalton Trans.* **2011**, *40*, 1158–1167.

23. Rodrigues, J.; Jardim, M.G.; Figueira, J.; Gouveia, M.; Tomás, H.; Rissanen, K. Poly(alkylidenamines) dendrimers as scaffolds for the preparation of low-generation ruthenium based metallodendrimers. *New J. Chem.* **2011**, *35*, 1938–1943.
24. Mattsson, J.; Govindaswamy, P.; Renfrew, A.K.; Dyson, P.J.; Štěpnička, P.; Süss-Fink, G.; Therrien, B. Synthesis, molecular structure, and anticancer activity of cationic arene ruthenium metallarectangles. *Organometallics* **2009**, *28*, 4350–4357.
25. Barry, N.P.E.; Zava, O.; Furrer, J.; Dyson, P.J.; Therrien, B. Anticancer activity of opened arene ruthenium metalla-assemblies. *Dalton Trans.* **2010**, *39*, 5272–5277.
26. Barry, N.P.E.; Zava, O.; Dyson, P.J.; Therrien, B. Synthesis, characterization and anticancer activity of porphyrin-containing organometallic cubes. *Aust. J. Chem.* **2010**, *63*, 1529–1537.
27. Pitto-Barry, A.; Barry, N.P.E.; Zava, O.; Deschenaux, R.; Therrien, B. Encapsulation of pyrene-functionalized poly(benzyl ether) dendrons into a water-soluble organometallic cage. *Chem. Asian J.* **2011**, *6*, 1595–1603.
28. Pitto-Barry, A.; Barry, N.P.E.; Zava, O.; Deschenaux, R.; Dyson, P.J.; Therrien, B. Double targeting of tumours with pyrenyl-modified dendrimers encapsulated in an arene-ruthenium metallaprism. *Chem. Eur. J.* **2011**, *17*, 1966–1971.
29. Barry, N.P.E.; Edafe, F.; Therrien, B. Anticancer activity of tetracationic arene ruthenium metalla-cycles. *Dalton Trans.* **2011**, *40*, 7172–7180.
30. Barry, N.P.E.; Zava, O.; Dyson, P.J.; Therrien, B. Excellent correlation between drug release and portal size in metalla-cage drug-delivery systems. *Chem. Eur. J.* **2011**, *17*, 9669–9667.
31. Pitto-Barry, A.; Zava, O.; Dyson, P.J.; Deschenaux, R.; Therrien, B. Enhancement of cytotoxicity by combining pyrenyl-dendrimers and arene ruthenium metallacages. *Inorg. Chem.* **2012**, *51*, 7119–7124.
32. Furrer, M.A.; Garci, A.; Denoyelle-Di-Muro, E.; Trouillas, P.; Giannini, F.; Furrer, J.; Clavel, C.M.; Dyson, P.J.; Süss-Fink, G.; Therrien, B. Synthesis, characterisation and *in vitro* anticancer activity of hexanuclear thiolato-bridged arene ruthenium metalla-prisms. *Chem. Eur. J.* **2013**, *19*, 3198–3203.
33. Dubey, A.; Min, J.W.; Koo, H.J.; Kim, H.; Cook, T.R.; Kang, S.C.; Stang, P.J.; Chi, K.-W. Anticancer potency and multi-resistant studies of self-assembled arene-ruthenium metallarectangles. *Chem. Eur. J.* **2013**, *19*, 11622–11628.
34. Jung, H.; Dubey, A.; Koo, H.J.; Vajpayee, V.; Cook, T.R.; Kim, H.; Kang, S.C.; Stang, P.J.; Chi, K.-W. Self-assembly of ambidentate pyridyl-carboxylate ligands with octahedral ruthenium metal centers: Self-selection for a single-linkage isomer and anticancer-potency studies. *Chem. Eur. J.* **2013**, *19*, 6709–6717.
35. Mishra, A.; Jung, H.; Park, J.W.; Kim, H.K.; Kim, H.; Stang, P.J.; Chi, K.-W. Anticancer activity of self-assembled molecular rectangles via arene-ruthenium acceptors and a new unsymmetrical amide ligand. *Organometallics* **2012**, *31*, 3519–3526.
36. Vajpayee, V.; Lee S.; Kim, S.-H.; Kang, S.C.; Cook, T.R.; Kim, H.; Kim, D.W.; Verma, S.; Lah, M.S.; Kim, I.S.; *et al.* Self-assembled metala-rectangles bearing azodipyridyl ligands: Synthesis, characterization and antitumor activity. *Dalton Trans.* **2013**, *42*, 466–475.
37. Vajpayee, V.; mi Lee, S.; Park, J.W.; Dubey, A.; Kim, H.; Cook, T.R.; Stang, P.J.; Chi, K.-W. Growth inhibitory activity of a bis-benzimidazole-bridged arene ruthenium

metalla-rectangle and-prism. *Organometallics* **2013**, *32*, 1563–1566.

38. Cook, T.R.; Vajpayee, V.; Lee, M.H.; Stang, P.J.; Chi, K.-W. Biomedical and biochemical applications of self-assembled metallacycles and metallacages. *Acc. Chem. Res.* **2013**, *46*, 2464–2474.
39. Han, Y.-F.; Jia, W.-G.; Lin, Y.-J.; Jin, G.-X. Stepwise formation of molecular rectangles of half-sandwich rhodium and ruthenium complexes containing bridging chloranilate ligands. *Organometallics* **2008**, *27*, 5002–5008.
40. Blunden, B.M.; Thomas, D.S.; Stenzel, M.H. Macromolecular ruthenium complexes as anti-cancer agents. *Polym. Chem.* **2012**, *3*, 2964.
41. Stepanenko, I.N.; Casini, A.; Edafe, F.; Novak, M.S.; Arion, V.B.; Dyson, P.J.; Jakupec, M.A.; Keppler, B.K. Conjugation of organoruthenium(II) 3-(1H-benzimidazol-2-yl)pyrazolo[3,4-b]pyridines and indolo[3,2-d]benzazepines to recombinant human serum albumin: A strategy to enhance cytotoxicity in cancer cells. *Inorg. Chem.* **2011**, *50*, 12669–12679.
42. Warnecke, A.; Fichtner, I.; Garmann, D.; Jaehde, U.; Kratz, F. Synthesis and biological activity of water-soluble maleimide derivatives of the anticancer drug carboplatin designed as albumin-binding prodrugs. *Bioconjugate Chem.* **2004**, *15*, 1349–1359.
43. Süss-Fink, G.; Khan, F.-A.; Juillerat-Jeanneret, L.; Dyson, P.J.; Renfrew, A.K. Synthesis and anticancer activity of long-chain isonicotinic ester ligand-containing arene ruthenium complexes and nanoparticles. *J. Clus. Sci.* **2010**, *21*, 313–324.
44. Garcia, M.H.; Morais, T.S.; Florindo, P.; Piedade, M.F.M.; Moreno, V.; Ciudad, C.; Noe, V. Inhibition of cancer cell growth by ruthenium(II) cycplopentadienyl derivative complexes with heteroaromatic ligands, journal of inorganic biochemistry. *J. Inorg. Biochem.* **2009**, *103*, 3, 354–361.
45. Garcia, M.H.; Valente, A.; Florindo, P.; Morais, T.S.; Piedade, M.F.M.; Duarte, M.T.; Moreno, V. New ruthenium(II) mixed metallocene derived complexes: Synthesis, characterization by X-ray diffraction studies and evaluation on DNA interaction by atomic force microcopy. *Inorg. Chim. Acta* **2010**, *363*, 3765–3775.
46. Moreno, V.; Lorenzo, J.; Avilés, F.X.; Garcia, M.H.; Ribeiro, J.; Florindo, P.; Robalo, M.P. Studies of the antiproliferative activity of ruthenium(II) cyclopentadienyl derived complexes with nitrogen coordinated ligands. *Bioinorg. Chem. Appl.* **2010**, doi:10.1155/2010/936834.
47. Moreno, V.; Font-Bardia, M.; Calvet, T.; Lorenzo, J.; Avilés, F.X.; Garcia, M.H.; Morais, T.S.; Valente, A.; Robalo, M.P. DNA interaction and cytotocixity studies of new ruthenium(II) cyclopentadienyl derivative complexes containing heteroaromatic ligands. *J. Inorg. Biochem.* **2011**, *105*, 2, 241–249.
48. Tomaz, A.I.; Jakusch, T.; Morais, T.S.; Marques, F.; Almeida, R.F.M.; de Mendes, F.; Enyedy, E.A.; Santos, I.; Pessoa, J.C.; Kiss, T.; *et al.* $[Ru^{II}(\eta^5\text{-}C_5H_5)(bipy)(PPh_3)]^+$ a promising large spectrum antitumor agent: cytotoxic activity and interaction with human serum albumin. *J. Inorg. Biochem.* **2012**, *117*, 261–269.
49. Morais, T.S.; Silva, T.J.L.; Marques, F.; Robalo, M.P.; Avecilla, F.; Madeira, P.J.A.; Mendes, P.J.G.; Santos, I.; Garcia, M.H. Synthesis of organometallic ruthenium(II) complexes with strong activity against several human cancer cell lines. *J. Inorg. Biochem.* **2012**, *114*, 65–74.
50. Morais, T.S.; Santos, F.; Côrte-Real, L.; Marques, F.; Robalo, M.P.; Madeira, P.J.A.;

Garcia, M.H. Biological activity and cellular uptake of [Ru(η^5-C_5H_5)(PPh_3)(Me_2bpy)][CF_3SO_3]. *J. Inorg. Biochem.* **2013**, *122*, 8–17.

51. Morais, T.S.; Santos, F.C.; Jorge, T.F.; Côrte-Real, L.; Madeira, P.J.A.; Marques, F.; Robalo, M.P.; Matos, A.; Santos, I.; Garcia, M.H. New water-soluble ruthenium(II) cytotoxic complex: Biological activity and cellular distribution. *J. Inorg. Biochem.* **2013**, *130*, 1–14.
52. Côrte-Real, L.; Matos, A.P.; Alho, I.; Morais, T.S.; Tomaz, A.I.; Garcia, M.H.; Santos, I.; Bicho, M.P.; Marques, F. Cellular uptake mechanisms of an antitumor ruthenium compound: The endossomal/lysosomal system as a target for anticancer metal-based drugs. *Microsc. Microanal.* **2013**, *19*,1122–1130.
53. Duncan, R. Designing polymer conjugates as lysosomotropic nanomedicines. *Biochem. Soc. Trans.* **2007**, *35*, 56–60.
54. Valente, A.; Garcia, M.H.; Marques, F.; Miao, Y.; Rousseau, C.; Zinck, P. First polymer "ruthenium-cyclopentadienyl" complex as potential anticancer agent. *J. Inorg. Biochem.* **2013**, *127*, 79–81.
55. Tannock, I.F.; Rotin, D. Acid pH in tumors and its potential for therapeutic exploitation. *Cancer Res.* **1989**, *49*, 4373–4384.
56. Duncan, R. The dawning era of polymer therapeutics. *Nat. Rev. Drug Discov.* **2003**, *2*, 347–360.
57. Cabral, H.; Nishiyama, N.; Okazaki, S.; Koyama, H.; Kataoka, K. Preparation and biological properties of dichloro(1,2-diaminocyclohexane)platinum(II) (DACHPt)-loaded polymeric micelles. *J. Control. Rel.* **2005**, *101*, 223–232.
58. Uchino, H.; Matsumura, Y.; Negishi, T.; Hayashi, F.; Honda, T.; Nishiyama, N.; Kataoka, K.; Naito, S.; Kakizoe, T. Cisplatin-incorporating polymeric micelles (NC-6004) can reduce nephrotoxicity and neurotoxicity of cisplatin in rats. *Br. J. Cancer* **2005**, *93*, 678–687.
59. Nishiyama, N.; Kataoka, K. Preparation and characterization of size-controlled polymeric micelle containing cis-dichlorodiammineplatinum(II) in the core. *J. Control. Rel.* **2001**, *74*, 83–94.
60. Kim, J.-H.; Kim, Y.-S.; Park, K.; Lee, S.; Nam, H.Y.; Min, K.H.; Lo, H.G.; Park, J.H.; Choi, K.; Jeong, S.Y.; *et al.* Antitumor efficacy of cisplatin-loaded glycol chitosan nanoparticles in tumor-bearing mice. *J. Control. Rel.* **2008**, *127*, 41–49.

Permissions

All chapters in this book were first published in MDPI; hereby published with permission under the Creative Commons Attribution License or equivalent. Every chapter published in this book has been scrutinized by our experts. Their significance has been extensively debated. The topics covered herein carry significant findings which will fuel the growth of the discipline. They may even be implemented as practical applications or may be referred to as a beginning point for another development.

The contributors of this book come from diverse backgrounds, making this book a truly international effort. This book will bring forth new frontiers with its revolutionizing research information and detailed analysis of the nascent developments around the world.

We would like to thank all the contributing authors for lending their expertise to make the book truly unique. They have played a crucial role in the development of this book. Without their invaluable contributions this book wouldn't have been possible. They have made vital efforts to compile up to date information on the varied aspects of this subject to make this book a valuable addition to the collection of many professionals and students.

This book was conceptualized with the vision of imparting up-to-date information and advanced data in this field. To ensure the same, a matchless editorial board was set up. Every individual on the board went through rigorous rounds of assessment to prove their worth. After which they invested a large part of their time researching and compiling the most relevant data for our readers.

The editorial board has been involved in producing this book since its inception. They have spent rigorous hours researching and exploring the diverse topics which have resulted in the successful publishing of this book. They have passed on their knowledge of decades through this book. To expedite this challenging task, the publisher supported the team at every step. A small team of assistant editors was also appointed to further simplify the editing procedure and attain best results for the readers.

Apart from the editorial board, the designing team has also invested a significant amount of their time in understanding the subject and creating the most relevant covers. They scrutinized every image to scout for the most suitable representation of the subject and create an appropriate cover for the book.

The publishing team has been an ardent support to the editorial, designing and production team. Their endless efforts to recruit the best for this project, has resulted in the accomplishment of this book. They are a veteran in the field of academics and their pool of knowledge is as vast as their experience in printing. Their expertise and guidance has proved useful at every step. Their uncompromising quality standards have made this book an exceptional effort. Their encouragement from time to time has been an inspiration for everyone.

The publisher and the editorial board hope that this book will prove to be a valuable piece of knowledge for researchers, students, practitioners and scholars across the globe.

List of Contributors

Demetra S. Achilleos and Maria Vamvakaki
Institute of Electronic Structure and Laser, Foundation for Research and Technology–Hellas, 711 10 Heraklion, Crete, Greece
Department of Materials Science and Technology, University of Crete, 710 03 Heraklion, Crete, Greece

Theresia M. M. Richter and Rainer Niewa
Institut für Anorganische Chemie, Universität Stuttgart, Pfaffenwaldring 55, Stuttgart 70569, Germany

Roberto Rosa, Chiara Ponzoni and Cristina Leonelli
Dipartimento di Ingegneria "Enzo Ferrari", Università degli Studi di Modena e Reggio Emilia, via Vignolese 905/A, 41125 Modena, Italy

Sabine N. Neal, Samuel A. Orefuwa, Atiya T. Overton and Ahmed A. Mohamed
Department of Chemistry, Delaware State University, 1200 N. DuPont Highway, Dover, Delaware 19901, USA

Richard J. Staples
Department of Chemistry, Center for Crystallographic Research, Michigan State University, East Lansing, Michigan 48824, USA

Annika Betke and Guido Kickelbick
Inorganic Chemistry, Saarland University, Am Markt Zeile 3, Saarbrücken 66125, Germany

Dai Oyama, Ayumi Fukuda and Takashi Yamanaka
Department of Industrial Systems Engineering, Cluster of Science and Engineering, Fukushima University, 1 Kanayagawa, Fukushima 960-1296, Japan

Tsugiko Takase
Center for Practical and Project-Based Learning, Cluster of Science and Engineering, Fukushima University, 1 Kanayagawa, Fukushima 960-1296, Japan

George Marshall, Ashley J. Wooles, David P. Mills, William Lewis, Alexander J. Blake and Stephen T. Liddle
School of Chemistry, University of Nottingham, University Park, Nottingham, NG7 2RD, UK

Murukanahally Kempaiah Devaraju, Quang Duc Truong, Hiroshi Hyodo, Takaaki Tomai and Itaru Honma
Institute of Multidisciplinary Research for Advanced Materials, Tohoku University, 2-1-1, Katahira, Aoba-ku, Sendai 980-8577, Japan

Andreia Valente and M. Helena Garcia
Centro de Ciências Moleculares e Materiais, Departamento de Química e Bioquímica, Faculdade de Ciências da Universidade de Lisboa, Campo Grande, 1749-016 Lisboa, Portugal

Index

www.ingramcontent.com/pod-product-compliance
Lightning Source LLC
LaVergne TN
LVHW060544210926
848485LV00005B/1257